VOLUME ONE HUNDRED AND TWENTY SEVEN

PROGRESS IN MOLECULAR BIOLOGY AND TRANSLATIONAL SCIENCE

The Mitochondrion in Aging and Disease

VOLUME ONE HUNDRED AND TWENTY SEVEN

PROGRESS IN MOLECULAR BIOLOGY AND TRANSLATIONAL SCIENCE

The Mitochondrion in Aging and Disease

Edited by

HEINZ D. OSIEWACZ

Institute of Molecular Biosciences,
Goethe University,
Frankfurt, Germany

ELSEVIER

AMSTERDAM • BOSTON • HEIDELBERG • LONDON
NEW YORK • OXFORD • PARIS • SAN DIEGO
SAN FRANCISCO • SINGAPORE • SYDNEY • TOKYO
Academic Press is an imprint of Elsevier

Academic Press is an imprint of Elsevier
225 Wyman Street, Waltham, MA 02451, USA
525 B Street, Suite 1800, San Diego, CA 92101-4495, USA
32 Jamestown Road, London NW1 7BY, UK
The Boulevard, Langford Lane, Kidlington, Oxford OX5 1GB, UK

First edition 2014

ISBN: 978-0-12-394625-6
ISSN: 1877-1173

For information on all Academic Press publications
visit our website at store.elsevier.com

CONTENTS

CONTRIBUTORS

Annayya R. Aroor
Division of Endocrinology, Diabetes, and Metabolism, Diabetes Cardiovascular Center, and Harry S. Truman Memorial Veterans Hospital, Columbia, Missouri, USA

Georg Auburger
Experimental Neurology, Goethe University Medical School, Frankfurt am Main, Germany

Gustavo Barja
Department of Animal Physiology II, Faculty of Biological Sciences, Complutense University, Madrid Spain

Jürgen Bereiter-Hahn
Institute for Cell Biology and Neurosciences, Goethe University Frankfurt am Main, Frankfurt am Main, Germany

Isabel Denzer
Molecular and Clinical Pharmacy, and Henriette Schmidt-Burkhardt Chair of Food Chemistry, Department of Chemistry and Pharmacy, Friedrich-Alexander-Universität Erlangen-Nürnberg, Erlangen, Germany

Lan-Feng Dong
School of Medical Science, Griffith University, Southport, Queensland, Australia

Gunter P. Eckert
Department of Pharmacology, Biocenter, University of Frankfurt, Frankfurt, Germany

Kristina Friedland-Leuner
Molecular and Clinical Pharmacy, Department of Chemistry and Pharmacy, Friedrich-Alexander-Universität Erlangen-Nürnberg, Erlangen, Germany

Suzana Gispert
Experimental Neurology, Goethe University Medical School, Frankfurt am Main, Germany

S. Michal Jazwinski
Tulane Center for Aging and Department of Medicine, Tulane University Health Sciences Center, New Orleans, Louisiana, USA

Marina Jendrach
Experimental Neurology, Goethe University Medical School, Frankfurt am Main, Germany

Guanghong Jia
Division of Endocrinology, Diabetes, and Metabolism, Diabetes Cardiovascular Center, and Harry S. Truman Memorial Veterans Hospital, Columbia, Missouri, USA

Konstantin Khrapko
Harvard Medical School, Beth Israel Deaconess Medical Center, Boston, Massachusetts, USA

Edda Klipp
Theoretical Biophysics, Institute for Biology, Humboldt-Universität zu Berlin, Berlin, Germany

Axel Kowald
Theoretical Biophysics, Institute for Biology, Humboldt-Universität zu Berlin, Berlin, Germany

Alexander N. Lukashev
Institute of Mitoengineering, Lomonosov Moscow State University, and Chumakov Institute of Poliomyelitis and Viral Encephalitides, Moscow, Russia

Walter E. Müller
Department of Pharmacology, Biocenter, University of Frankfurt, Frankfurt, Germany

Jiri Neuzil
School of Medical Science, Griffith University, Southport, Queensland, Australia, and Institute of Biotechnology, Academy of Sciences of the Czech Republic, Prague, Czech Republic

Victoria Ostapenko
Institute of Mitoengineering, Lomonosov Moscow State University, Moscow, Russia

V.V. Pavshintsev
Institute of Mitoengineering, Lomonosov Moscow State University, and Faculty of Biology, Lomonosov Moscow State University, Moscow, Russia

Alla Yu. Savchenko
I.M. Sechenov First Moscow State Medical University, Moscow, Russia

Maxim V. Skulachev
Institute of Mitoengineering, Lomonosov Moscow State University, and Faculty of Biology, Lomonosov Moscow State University, Moscow, Russia

Vladimir P. Skulachev
A.N. Belozersky Institute of Physico-Chemical Biology, Lomonosov Moscow State University, Moscow, Russia

James R. Sowers
Division of Endocrinology, Diabetes, and Metabolism, Diabetes Cardiovascular Center; Harry S. Truman Memorial Veterans Hospital, and Department of Medical Pharmacology and Physiology, Columbia, Missouri, USA

Carola Stockburger
Department of Pharmacology, Biocenter, University of Frankfurt, Frankfurt, Germany

Doug Turnbull
LLHW Centre for Ageing and Vitality, Newcastle University, Newcastle, United Kingdom

PREFACE

Because of their role in energy transduction and conservation, mitochondria are best known as the "power plant" of the eukaryotic cell. In fact, in oxygenic eukaryotes including humans, by far most of the universal cellular "energy currency" adenosine triphosphate (ATP), that is essential to carry out all energy-consuming processes of a living being, is generated in mitochondria. Apart from this most appreciated function, mitochondria are involved in various other essential processes, like iron/sulfur cluster synthesis, calcium storage and signaling, copper homeostasis, and control of programmed cell death.

Mitochondria are semiautonomous organelles: their function depends on the genetic information in the mitochondrion, the mitochondrial DNA (mtDNA), and in the chromosomes of the nucleus, the nuclear DNA (nDNA). More than 99% of the proteins, about 1200–1500 in humans, are encoded by the nDNA, synthesized in the cytoplasm, transported to mitochondria, and correctly delivered to the different mitochondrial subcompartment where they function. Collectively, these processes are essential to keep mitochondrial functional over the whole lifetime of an individual. During development, mitochondria are distributed to the growing number of cells. This is possible because mitochondria are highly dynamic organelles: they are "growing" via biosynthesis of components that become inserted into existing units which are constantly dividing and fusing. Mitochondria thus do not represent static units but a genetically balanced population of organelles that meets the different physiological situations of a cell. Accordingly, mitochondrial morphology can change from filamentous to punctate or to highly complex networks. Also, the ultrastructure of mitochondria can strongly differ. In particular, the structure of the inner mitochondrial membrane can form different kinds of invaginations (cristae) of tubular or lamellar structure. Even more, as revealed by recent data, the inner membrane can form vesicles prior to the induction of the mitochondrial transition pore and programmed cell death in aged conditions.

Given the different essential functions of mitochondria, it is not surprising that various molecular pathways are effective in keeping mitochondria functional over time as long as possible. Such pathways are active in controlling the abundance of reactive oxygen species (ROS) which are generated in mitochondria themselves as by-products of respiration and are essential for molecular signaling, but, at higher concentrations, dangerous because they can cause molecular damage. In addition, other pathways are involved in repair of damaged molecules, and yet other ones control the degradation of damaged molecules, whole mitochondria, or even whole cells. All of these pathways are limited in their capacity but, if one pathway is overwhelmed, other pathways may become activated. While still not elucidated in detail, a view of an effective network of interacting quality control pathways is emerging that keeps mitochondria "healthy" over time. However, if the whole network becomes overwhelmed for any reason, functionally impaired mitochondria accumulate and after passing certain threshold give rise to degeneration of the biological system as it occurs during aging and the development of diseases.

This volume of *Progress in Molecular Biology and Translational Sciences* presents a current view of selected aspects about the role of mitochondria in aging and disease. The first part of the book addresses several general aspects linked to aging, while the rest of the chapters deal with specific age-related diseases and mitochondrial therapy. In Chapter 1, the "mitochondrial free radical theory of aging," which strongly influenced aging research over decades since it has been postulated in the 1970s, is critically reviewed and conclusions for interventions into aging via dietary restriction based on recent experimental data are provided. Also, Chapter 2 deals with a long-standing topic in aging research: the role of somatic mutations of the mtDNA. Specific emphasis lies on the experimentally demonstrated accumulation of mutations in different organs and a discussion of how such mutations, which first occur in a single mtDNA molecule, are "taking over" by the mechanism of clonal expansion. After a general introduction of mathematical and computational modeling approaches, the same topic is also addressed in Chapter 3 in which mitochondria as a population of dynamic units are mathematically modeled. Chapter 4 reviews aspects of mitochondrial dynamics including mitochondrial trafficking and localization as well as with fission and fusion. Next, the ability of biological systems to sense and respond to functional impairments of mitochondria, a signaling pathway known as the "retrograde response," is introduced and discussed as a compensatory mechanism (Chapter 5). The next four chapters deal with the impact of mitochondria on specific age-related diseases like Parkinson's (Chapter 6) and Alzheimer's disease (Chapter 7), cancer (Chapter 8) and the cardiorenal metabolic syndrome (Chapter 9). In addition to general considerations about the underlying mechanisms and the role of mitochondria in the development of these diseases, strategies towards the development of therapeutic interventions are part of these chapters. The progress in developing rechargeable mitochondrial antioxidants for therapeutic use is finally reviewed in Chapter 10.

I would like to thank all authors, all experts in their field of research, for taking their valuable time to critically review important new developments in the field of mitochondrial biology. My special thanks to the editor-in-chief of *Progress in Molecular Biology and Translational Science*, Dr. P. Michael Conn, for initiating this enterprise and the editorial team of Elsevier, in particular Mary A. Zimmerman and Helene Kabes, for their help in the realization of this project.

Heinz D. Osiewacz
Frankfurt/Main, Germany

CHAPTER ONE

The Mitochondrial Free Radical Theory of Aging

Gustavo Barja
Department of Animal Physiology II, Faculty of Biological Sciences, Complutense University, Madrid Spain

Contents

Abstract

The mitochondrial free radical theory of aging is reviewed. Only two parameters currently correlate with species longevity in the right sense: the mitochondrial rate of reactive oxygen species (mitROS) production and the degree of fatty acid unsaturation of tissue membranes. Both are low in long-lived animals. In addition, the best-known manipulation that extends longevity, dietary restriction, also decreases the rate of mitROS production and oxidative damage to mtDNA. The same occurs during protein restriction as well as during methionine restriction. These two manipulations also increase maximum longevity in rodents. The decrease in mitROS generation and oxidative stress that takes place in caloric restriction seems to be due to restriction of a single dietary substance: methionine. The information available supports a mitochondrial free radical theory of aging focused on low generation of endogenous damage and low sensitivity of membranes to oxidation in long-lived animals.

Progress in Molecular Biology and Translational Science, Volume 127
ISSN 1877-1173
http://dx.doi.org/10.1016/B978-0-12-394625-6.00001-5

1. INTRODUCTION

Many different theories of aging have been proposed, but the mitochondrial free radical theory of aging (MFRTA)[1] can still afford the best explanation for aging and longevity in mammals, birds, and multicellular animals in general. Any aging theory must explain why maximum longevity (referred here as "longevity") varies so widely in animals: 30-fold from mice to men, 200-fold from shrews to the longest-living whales, or more than 5000-fold from perhaps a few days in some invertebrates to *Arctica islandica* mussels (longevity around 400 years). Such huge differences indicate that longevity is markedly regulated and flexible during species evolution. Copying only a small fraction of this natural capacity would make possible in the future to obtain negligible senescence in humans. It is known that mean lifespan or the life expectancy at birth of the individuals of a population depends more on the environment than on the genes. On the contrary, longevity and its inverse—the species aging rate—depend more than 90% on the genotype, like in the case of any other species-specific trait. Longevity and aging rate are the main parameters that matter concerning the endogenous process of aging, which is situated at the main root of all the degenerative killer diseases. Presently, only two known factors correlate in the right sense with animal longevity in vertebrates including mammals and birds: (a) the rate of mitochondrial reactive oxygen species production (mtROSp)[1–4] and (b) the degree of fatty acid unsaturation of tissue cellular membranes including the mitochondrial ones.[5,6] The longer the longevity of a species, the smaller these two parameters are. The decrease in mtROSp in long-lived animal species lowers their generation of endogenous (free radical) damage at mitochondria. The decrease in the fatty acid double bond index (DBI) and peroxidizability index (PI) lowers the sensitivity of the cellular and mitochondrial membranes to free radical attack. No other theory of aging has parameters like these correlating in the right sense with longevity across species and offering plausible mechanistic explanations for the accumulation of damage from endogenous origin. The two known parameters appropriately correlating with animal longevity appertain to the MFRTA, not to any alternative theory. This is important since any theory trying to explain aging must explain why longevity varies so widely among different animal species. Species closely related by phylogeny can have very different longevities, indicating that evolution of longevity is a relatively flexible and fast process, and thus can be subjected to experimental manipulation.

2. ANTIOXIDANTS AND LONGEVITY

Studies about MFRTA first focused in antioxidants, mainly because they could be measured with rather simple laboratory assays. In 1993, it was found that both enzymatic and nonenzymatic endogenous tissue antioxidants, including catalase, GSH-peroxidases, GSH-reductases, GSH, or ascorbate, correlated with longevity across vertebrates. However, and rather surprisingly, such correlation was negative[7] instead of positive as it was then widely believed. That review on the relationship between endogenous antioxidants and vertebrate longevity[7] also included all the then available published data on the subject obtained in mammals by other different laboratories. All those data from different sources consistently agreed: the longer the longevity, the lower were the levels of endogenous tissue antioxidants. Posterior reappraisals of the subject[8] have confirmed the early findings on the existence of a generally negative correlation between tissue antioxidants and longevity in all kinds of animals. It was most interesting that long-lived animals have lower instead of higher antioxidant levels. Among 27 studied correlations, 21 negatively correlated with longevity, 6 did not show significant differences, and not a single positive correlation with longevity was found.[7] Superoxide dismutase was among the antioxidants tending to show no association with longevity. Previous believe that this enzymatic activity was positively associated with longevity was due to referring the SOD (total SOD, CuZn plus Mn) activity values to the oxygen consumption (VO_2) of the whole animal (to the aerobic metabolic rate). Since metabolic rate strongly decreases as body size increases, the larger SOD/VO_2 of humans compared to rats was due to the lower value of the denominator in the humans instead of to a higher value of the numerator. In fact, tissue SOD (total SOD, CuZn plus Mn) activities were not correlated to longevity in mammals in the original publication,[9] although in the brain and lung of vertebrate species—but not in liver—the correlation between SOD (total SOD, CuZn plus Mn) and longevity was again negative like for other antioxidants. Recent studies in different mammals including long-lived naked mole-rats, as well as ants, honey bees, and marine bivalves also found a negative correlation with longevity for this antioxidant enzyme—SOD.[8] In this more recent review of the subject, among a total of 78 correlations between endogenous tissue antioxidants and longevity, 72 were negative, 6 did not show significant differences, and only a single one was positive,[8] corroborating studies performed almost two decades ago.[7] Therefore, high endogenous antioxidant levels are clearly not the cause of the high longevity of long-lived animal species.

3. MITOCHONDRIAL ROS PRODUCTION AND OXIDATIVE DAMAGE IN mtDNA

What is the reason why long-lived animals need less antioxidant levels in their vital organs? It was proposed[10] that the rate of mtROSp could be negatively correlated with longevity and that this would be the critical factor for aging. Long-lived animals would not need to maintain high antioxidant enzyme levels, which is energetically expensive, because they would produce mtROS at a low pace (and they could transitorily induce them if needed). This was indeed experimentally corroborated both when comparing different mammalian species[3] and when comparing short-lived rodents (rats and mice) with 8-fold longer-lived birds (pigeons, parakeets, and canaries) of similar size and weight-specific metabolic rate.[11,12] A posterior more complete investigation studying up to 12 different mammalian species confirmed these findings even after correcting for body size differences.[4]

The investigations in birds are especially important because the studies performed in mammals used species following the Pearl rate of living law of aging: "the lower the whole body weight-specific metabolic rate the longer the longevity." Thus, the species with longer longevity entered in those comparisons could show low rates of mtROSp simply because their rates of oxygen consumption were also lower than those of the short-lived ones. In fact, mtROSp was positively correlated with mitochondrial O_2 consumption and with global metabolic rate in those studies.[3] It was then important to study the problem in some of the many species that deviate from the Pearl rate of living law. Three groups of warm-blooded vertebrates have much higher longevity than expected for their body size or metabolic rate compared to most mammals: birds, bats, and primates. Birds have both a high rate of global oxygen consumption and a high longevity. This makes them ideal to solve the problem mentioned earlier. The lower mtROSp of pigeons, canaries, and parakeets, when compared to rats in the first case and with mice in the second and third, strongly reinforces the MFRTA since it indicates that the low mtROSp of long-lived animals occurs both in comparisons between animals following Pearl's law and in those not following it. A high longevity is not a simple consequence of a slow rate of living. It can be obtained—as the birds case shows—together with high rates of oxygen consumption and activity by lowering the rate of mtROSp both in absolute terms and as percentage of mitochondrial oxygen consumption (the percent free radical leak, %FRL).

During a long time, it has been widely thought that complex III of the respiratory chain was the respiratory complex responsible for ROS production in the mitochondrial electron transport chain (mtETC).[13] Later it was found, working with freshly isolated and well-coupled functional mitochondria, that complex I also produces ROS in heart or brain mitochondria isolated from rats, mice, pigeons, canaries, and parakeets,[12,14] which was soon confirmed in rats by other laboratories[15,16] and soon became established knowledge (Fig. 1.1). A key experiment to detect complex I ROS production was to measure mtROSp with succinate alone as well as with succinate + rotenone. In the second situation, the rate of mtROSp acutely decreases because rotenone does not allow the electrons to flow back to complex I from succinate-complex II through reverse electron flow.[17] But the common procedure of adding succinate alone, followed or not by antimycin A, and rarely using complex I-linked substrates, led to the general believe during a long time that ROS came mainly from complex III-semiquinone.

It was also found that the higher mtROS generation rate observed in mammals compared to birds of similar body size and metabolic rate occurred only at complex I,[12,14,17] not at complex III. This was especially interesting taking into account the finding of analogous results in dietary restriction

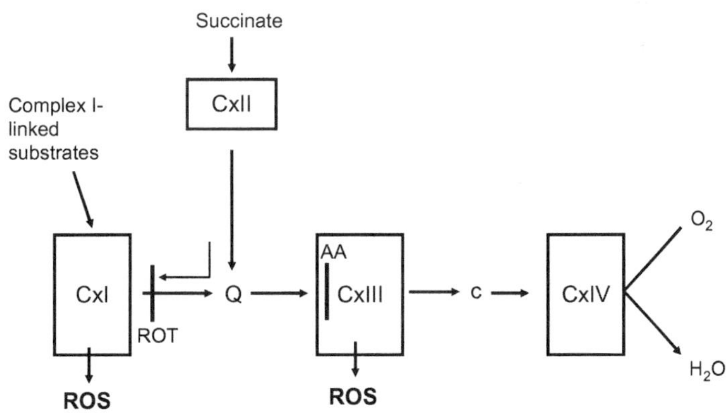

Figure 1.1 Sites of ROS production at the mitochondrial electron transport chain. The figure shows the four complexes of the respiratory chain (CxI–IV). ROS are mainly produced at CxI and CxIII. Rotenone (ROT) blocks electron transport from ubiquinone (Q) to complex I, thus avoiding complex I ROS production when succinate is used as substrate. AA, antimycin A.

(DR) models. Concerning the precise site within complex I where ROS are produced, three generators have been suggested: the flavin at the beginning of the electron path within the complex, the FeS clusters of the hydrophilic matrix arm, and the ubiquinone located in the membrane arm. Various investigators have supported the role of the flavin based on experiments with the inhibitor diphenyliodonium, which strongly decreases mtROSp. However, the site of action of diphenyliodonium, at the beginning of the electron path, avoids electrons to reach the other two possible generators, the various FeS clusters and the ubiquinone, which therefore cannot be discarded. Electron leak to oxygen could occur between the ferricyanide reduction site and the rotenone binding site of complex I both in intact mitochondria[12,14,17] and in submitochondrial particles.[18] Iron–sulfur clusters with a higher midpoint potential than FeSN1a, which could be situated in the electron path after the ferricyanide reduction site, or the unstable semiquinone known to be present in the membrane domain of complex I and possibly functioning in $H+$ pumping coupled to electron transport,[19] could be the complex I oxygen radical generators. If this last possibility were true, ubisemiquinones could be responsible for oxygen radical generation at both complex I and complex III, although the ROS source at complex I would be the important one for aging. But many other complex I FeS clusters can also be implicated because, under physiological conditions, (a) their reduced and oxidized states will not be present in equal concentrations; (b) interactions with many different factors and surrounding macromolecules can modify the final redox potential of the carriers *in vivo*; and (c) the exact position of many of FeS clusters in the complex I electron path is still unknown. Thus, the important aging-related question whether flavin, FeS clusters, or ubisemiquinone, or a combination of these, are responsible for complex I ROS generation is still unanswered.

The location where mitochondrial DNA (mtDNA) is situated is very close to the site of mtROS generation, the inner mitochondrial membrane. ROS production also occurs at other cellular sites like microsomes, peroxisomes, or membrane-bound NADPH-oxidases, and the rate of ROS generation at those sites can exceed in various situations that coming from mitochondria. However, the ROS produced at mitochondria can be still the most important ones for longevity due to the presence of mtDNA within the mitochondria but not at those other organelles or parts of the cell. Since long-lived animal species have low rates of mtROS generation, it is logical to expect that this should have an effect on the amount of oxidative damage in their mtDNA. Therefore, the level of

8-oxodG (8-oxo-7,8-dihydro-2′-deoxyguanosine) was measured in the heart and brain mitochondrial and nuclear DNA of eight different mammalian species differing by up to 13-fold in longevity. The results showed that the level of 8-oxodG in the mtDNA of both organs is negatively correlated with longevity.[20] The longer the longevity of a species, the smaller is its mtDNA oxidative damage degree. In contrast, the 8-oxodG level in nuclear DNA (nDNA) did not correlate with longevity in any organ even though mitochondrial and nuclear DNA were measured in the same samples taken from each individual animal.[20] Therefore, the different mtROSp rates from the different species seem to have a direct impact on mtDNA, not on nDNA, concerning oxidative damage. This makes sense since the site of ROS generation at mitochondria is very close to mtDNA, whereas nDNA is situated far away from it.

The rate of mtROSp is measured in isolated mitochondria *in vitro* due to the lack of available common methods for direct *in vivo* mitochondrial H_2O_2 production determination. However, the fact that the variations in levels of 8-oxodG in mtDNA closely reflect the variations in mtROSp both in comparative and in DR studies suggests that the mtROSp *in vitro* measurements closely reflect the situation *in vivo*. On the other hand, the level of 8-oxodG in mtDNA was generally lower in the heart and brain of three long-lived birds when compared to two short-lived mammals of similar body size and metabolic rate, in agreement with the superior longevity of the birds, whereas again this was not the case for nDNA.[21] These investigations also showed that the intensity of oxidative damage is severalfold higher in mtDNA than in nDNA in the heart and brain of all the 11 different species of mammals and birds studied,[20,21] which is again consistent with the close proximity between mtDNA and the sites of mtROS generation.

Initial studies about MFRTA were mainly focused on antioxidants because they were easier to measure and because sensitive techniques to assay mtROSp in different species with enough margin over the detection limits were generally not available at that time mainly due to a frequent use of spectrophotometry over fluorometry. Most studies on the effect of adding dietary antioxidants to the diet were performed during the 1970s and 1980s. The general result was that antioxidants did not increase (maximum) longevity. In some experiments they increased only mean longevity. Interestingly, this tended to occur when the (maximum) longevity of the control rodents was short, usually less than 3 years. This suggests that antioxidants, when the husbandry conditions were less than optimum, could protect from causes of early death, and thus they were capable of making more rectangular the survival curve, similar to what happened in humans during the twentieth century in many developing western countries when mean life expectancy increased from 30–40 to 80 years without decreases in aging rate. Antioxidants, in such cases, were bringing back toward optimum the diminished survival of the controls reared under suboptimum environmental conditions, which is interesting but not the goal of gerontology. Ironically, the poorer the survival curve of the controls, the largest is the chance of obtaining a positive result in terms of mean longevity. Like in the comparative interspecific studies described earlier, antioxidants clearly lacked the capacity to decrease the aging rate and to increase (maximum) longevity. When the techniques to obtain knockout or transgenic mice with increased or lack of expression of antioxidant enzymes, like SODs, catalase, or GSH-peroxidases, were applied to this problem, the results were similarly disappointing.[22,23] The increased antioxidant enzymes, like the nonenzymatic dietary antioxidants, lacked the capacity to slow down aging. Independent of the way in which the antioxidants were manipulated, dietary or genetic, the result was the same: a lack of effect of antioxidants on mammalian longevity. This has been misinterpreted by some authors as the "death" of the MFRTA,[23] but this conclusion did not take into account that what correlates with longevity in the right sense is not the level of the antioxidants, but the mtROSp rate and the fatty acid unsaturation degree of the cellular membranes. Studies in simpler organisms like the fungus *Podospora anserina* have also provided evidence for a role of mitochondrial ROS in senescence including DR effects.[24]

Thus, mtROS production seems to be involved in modifying longevity, while antioxidants are not. This is counterintuitive only if we think of the cell as a homogeneous system without considering compartmentation. But cells are not like that. The global level of cellular oxidative stress should depend on both ROS production and ROS elimination. Both contribute to determine cell survival or death according to the general "balance" between them. However, the ROS concentration in particular compartments like mitochondria, especially very near to the places of ROS generation like complex I, should be much more dependent on mtROSp than on antioxidants as the free radical generation source is approached at microlevel. At such places, it is mtROSp what would mainly determine the local ROS concentration present. This is especially important because the main target for aging, mtDNA, is located very close in the vicinity, perhaps even in contact with the free radical generation source. This could help to explain why lowering the rate of mtROSp instead of increasing antioxidants was selected for during the evolution of longevity in mammals, birds, and other species.

4. LONGEVITY AND MEMBRANE FATTY ACID UNSATURATION

In addition to mtROSp, there is a second known parameter that also correlates with longevity in the right way, the fatty acid unsaturation degree of cellular (including mitochondrial) membranes. This is also firmly established since it has been studied many times and concordant results were always obtained. The degree of fatty acid unsaturation can be summarized as the DBI or alternatively as the PI. The longer the longevity of the species, the smaller the total number of fatty acid double bonds (the smaller the DBI), which strongly decreases the sensitivity of the cellular and mitochondrial membranes to lipid peroxidation, a toughly destructive process that, in addition, produces mutagenic and toxic metabolites. This was first described in 1996 in rat compared to pigeon and human mitochondria[25] followed by many studies in mammals and birds.[5] A total of 23 studies extended the first seminal observation to many different mammals, various bird species, and some invertebrates, without finding a single exception.[6] Since the low degree of unsaturation occurs both in mitochondrial and in total cellular membranes in long-lived animals, it can diminish lipoxidation-derived damage in various cellular compartments including the mitochondrial one where there is strong abundance of membranes.

Among the different fatty acids composing the different cellular membranes (plasma, mitochondrial, and other internal membranes), many are responsible for this strong decrease in DBI (and PI) as longevity increases among species. But the most important ones, both due to their content in double bonds (high or low) and for their larger quantitative presence and variation among species, are 18:2*n*-6, 18:3*n*-3, and 22:6*n*-3, and sometimes 18:1*n*-9 (in some birds) and 20:4*n*-6. When they vary among species, 18:1*n*-9, 18:2*n*-6, and 18:3*n*-3 increase and 20:4*n*-6 and 22:6*n*-3 decrease as longevity increases. Among them, the decrease in 22:6*n*-3 in long-lived animals is the most relevant to explain their low DBI and PI values. The final result is that the total amount of unsaturated and saturated fatty acids does not change among species. Instead, it is the unsaturation degree of the polyunsaturated fatty acids present what decreases from short- to long-lived animals. With this kind of redistribution, long-lived animals obtain a strong decrease in the sensitivity of their cellular membranes to the dangerous process of lipid peroxidation, while maintaining essentially unaltered the fluidity of their membranes, the so-called homeoviscous–

longevity adaptation.[5] The low DBI of long-lived animals likely protects not only the lipids but also other kinds of cellular macromolecules. Since lipid peroxidation is a relatively massive process compared to oxidative damage to other kinds of macromolecules, long-lived animals, thanks to their low DBI, will produce far less amounts of highly toxic and mutagenic lipid peroxidation products like hydroxynonenal or malondialdehyde (MDA) among many others. These in turn, having carbonyl groups, can modify free amino groups in proteins and DNA. At least the first of these two kinds of modifications seems to be involved in aging, since comparisons among different mammalian species have found that the amount of MDA–lysine adducts in heart proteins negatively correlates with longevity.[26]

What is the mechanism causing the negative correlation between the fatty acid unsaturation degree and species longevity? A role for acylation/deacylation of the constitutive membrane fatty acids cannot be discarded. However, since the more unsaturated 20:4*n*-6 and 22:6*n*-3 are essential fatty acids synthesized from their dietary precursors 18:2*n*-6 and 18:3*n*-3, the enzymatic processes that control the corresponding biosynthetic pathways seem to be involved. In this respect, in various comparative studies relating the degree of fatty acid unsaturation to longevity, the results suggest that desaturase and elongase enzymatic activities in the $n-3$ and $n-6$ series (which are rate limiting for those biosynthetic pathways) are low in long-lived animals. In some cases, decreases in peroxisomal beta-oxidation could also be involved. It is now considered that this last process is responsible for the last steps in the synthesis of the highly unsaturated 22:6*n*-3 in the *n*-3 pathway. The low delta-5 and delta-6 activities (which are rate limiting enzymes in the $n-3$ and $n-6$ fatty acid synthesis pathways) of long-lived animals will decrease the conversion of the less unsaturated 18:2*n*-6 and 18:3*n*-3 precursors to the highly unsaturated 20:4*n*-6 and 22:6*n*-3 products. Thus, 18:2*n*-6 and 18:3*n*-3 would accumulate and 20:4*n*-6 and 22:6*n*-3 will diminish, which is just the general kind of result that is found in long-lived animals. In summary, the membrane fatty acid unsaturation degree is low in tissues from long-lived animals. This is the only other known factor, in addition to mtROSp, which correlates with longevity in the right sense and has mechanistic capacity to contribute to the widely different aging rates of the different animal species. This is true concerning the MFRTA as well as any other theory of aging. And what is the situation concerning experimental aging-related studies in single species?

5. DR, mtROS PRODUCTION, AND OXIDATIVE DAMAGE IN mtDNA

It is well known that standard (40%) DR increases not only mean but also maximum longevity (by around 40%) and decreases and delays the incidence of degenerative diseases in laboratory rodents, rotifers, flies, spiders, worms, fish, and other mammalian species.[27] In rhesus monkeys, it was observed that 30%DR strongly decreases age-related mortality (from 37% to 13%), age-related diseases, and age-associated brain atrophy.[28] Many mechanisms of action of DR on longevity have been proposed including modifications in GH and insulin/IGF-1-like signaling, changes in gene expression profiles, sirtuins, apoptosis, and many different signaling molecules like mTOR, FOXO, S6K, AKT, PKA, nutrition, and amino acid-sensing pathways. Many of these changes are interconnected and related to mitochondrial oxidative stress.[29]

In the previous sections, it was described that long-lived animals have lower rates of mtROSp and lower mtDNA oxidative damage than short-lived ones. But what occurs in DR concerning these parameters? If the MFRTA is correct, these two parameters should also decrease during DR. Initial studies, like in the case of the comparison between different species, focused mainly on antioxidants. They showed that DR in rodents does not lead to a generalized increase in antioxidants. Instead, increases, decreases, or lack of changes depending on the particular antioxidant have been reported even within the same study.[30] Therefore, similar to the interspecies case, the key to longevity does not seem to lie on the side of the antioxidants during DR either. A different situation concerns mitochondrial ROS generation. The effect of DR on the rate of mtROSp was repeatedly investigated in mice and especially in rats by many different laboratories. The results of these investigations consistently agreed that long-term (40%) DR significantly decreases the rate of mtROS generation in rat organs including skeletal muscle, kidney, liver, heart, and brain.[31] This agrees again with the concept that lowering mtROSp increases longevity. This decrease was detected in freshly isolated functional mitochondria under similar conditions including the substrate concentration used to feed electrons to the *ad libitum* and DR mitochondria. Thus, DR mitochondria are different from those obtained from *ad libitum*-fed animals, and this difference (due to DR) is responsible for the lowered mtROSp detected *in vitro*. In addition, some data suggest that complex I substrates like pyruvate decrease during DR in the tissue.[32] If that is correct, the matrix NADH level would decrease in DR altering the redox state of the mtETC, lowering its reducing potential including that of the complex I ROS generator, since NADH directly feeds electrons to this complex. Indeed, DR also decreases the NADH concentration,[32] a change that is known to decrease the rate of mtROSp.[16] This will lead to a further decrease in the rate of mtROSp *in vivo* which would add to that due to the lowered qualitative capacity of DR mitochondria to generate ROS detected *in vitro*. Interestingly, the decrease in mtROSp in DR rats specifically occurred at complex I in all the organs studied.[33–35] Thus, a low rate of mtROSp is a trait both of long-lived species and of DR animals. In contrast, the low DBI only occurs in long-lived species since 40%DR does not change the membrane unsaturation degree.

DR, in addition to the lowering of mtROSp, also decreases the %FRL, indicating that the efficiency of the mitochondrial respiratory chain in avoiding ROS generation increases in DR animals. Especially, long-lived animals like canaries and pigeons also show lower %FRL values than the much shorter-lived rats or mice,[12,17] suggesting that this can constitute a highly conserved mechanism of life-span extension both between and within species. On the other hand, since mtROSp is lower in DR than in the *ad libitum*-fed control animals, oxidative damage should also be lower in the mtDNA of the restricted animals. In agreement with this, it was found that the level of 8-oxodG in DNA was significantly lower in the liver, heart, and brain of the long-term DR old rats in which mitochondrial ROS production was also diminished.[36] Depending on the organ studied, such decrease in 8-oxodG occurred only in mtDNA, or in both mtDNA and nDNA.

6. PROTEIN AND METHIONINE RESTRICTION

6.1. Effect on longevity

It has been generally agreed for a long time that calorie intake *per se* would be exclusively responsible for the increase in life-span induced by DR in rodents. However, now many studies question this classical consensus. The results of many investigations are consistent with the possibility that part of the life-extending effects on DR is due to the decreased intake of particular components of the diet, such as proteins, and more specifically the amino acid methionine.[37–41] Neither life-long isocaloric carbohydrate nor lipid restriction seems to increase rodent life-span. Two available investigations of carbohydrate restriction or supplementation reported contradictory and minor changes in rat longevity,[42,43] whereas it was found that the longevity of Fisher 344 rats does not change after life-long lipid restriction.[44] In contrast, the large majority of the investigations on the effects of isocaloric protein restriction (PR) in rats and mice found increases in longevity. Ten of 11 PR studies in rats or mice (16 of 18 different life-long survival experiments) reported increases in longevity,[39] although the mean magnitude of this increase (around 19%) was lower than that usually found in 40%DR (around 40% increase). Thus, PR would be responsible for around half of the life-extension effect of DR.

Among the different dietary amino acids, which is the one acid responsible for the increase in longevity exerted by PR? It has been demonstrated that isocaloric 80% methionine restriction (MetR) increases longevity in F344 rats[37] and mice3[38,40] to a similar extent than PR (around 18% mean increase). This occurs even when MetR is started as late as at 12 months of age in C6BF1 mice.[40] Studies performed in *Drosophila melanogaster* have also shown that casein restriction[45] and MetR[46] extend longevity independent of the caloric intake. Moreover, other recent studies link essential amino acids, and again especially methionine, with the positive effect of DR on longevity in yeast[47] and *D. melanogaster*.[48] Interestingly, PR performed in rats, results in profound changes in methionine and serine metabolism (including lowering cystathionine β-synthase and cystathionine γ-lyase activities), and increases in the oxidation of fatty acids.[49]

In addition to extending life-span, 80%MetR also decreases disease-associated markers and the incidence of age-related degenerative diseases.[50,51] The beneficial effects of this intervention in rodents include decreases in serum glucose, insulin, IGF1, cholesterol, triglycerides, and leptin. Besides, MetR protects against age-related changes in immunity, slows cataract development,[38] improves colon tight junction barrier function,[52] and improves metabolic flexibility and increases respiratory uncoupling.[53] MetR may also be an important strategy to inhibit tumor growth particularly in many cancers that exhibit the known phenomenon of "methionine dependence." These include bladder, breast, colon, glioma, kidney, melanoma, prostate, and other cancers in which tumor cells have a much greater reliance on methionine than normal cells do.[54] They need this amino acid for survival and proliferation and their growth seems seriously limited or inhibited in the absence of methionine.[55,56]

MetR (80%) also decreases total adipose tissue mass and lowers visceral fat by 70% (by more than 40% after correcting for the decrease in body mass) in association with an improvement in insulin sensitivity.[57] In addition, MetR decreases leptin and increases adiponectin in rodents in agreement with the decrease in visceral adiposity and the size of white adipose tissue depots. These beneficial effects seem to be mediated by tissue-specific responses that favor increased mitochondrial function and biogenesis, fatty acid oxidation, and total energy expenditure possibly mediated by β-adrenergic receptor signaling and changes in lipid homeostasis.[58] In this line, a recent metabolomic and genomic MetR study found changes in the expression of a large number of genes and proteins that led the authors to conclude that MetR increases lipid metabolism in adipose tissue and muscle, whereas it decreases lipid synthesis in the liver.[59] Therefore, these changes in lipid metabolism seem to be involved in the strong decrease in adiposity and increased insulin sensitivity observed in isocaloric restriction of dietary methionine.

Altered levels of sulfur-containing amino acids have also been described in MetR: serum levels of methionine, cysteine, cystathionine, and taurine decrease in MetR rats, whereas homocysteine levels[60] and glutathione[37] increase. Interestingly, adding cysteine to the MetR diet reverses most of the studied beneficial changes on adiposity and insulin resistance[60] and increases the transcription of various genes associated with inflammation and carcinogenesis.[59] Therefore, the beneficial changes of MetR diet have been attributed to the decrease of cysteine in serum[60] or liver[59] observed in animals subjected to MetR. On the other hand, excessive intake of dietary methionine is toxic. This toxicity far exceeds that produced by any other amino acid,[61] leading to damage in some vital organs and increases in tissue oxidative stress[62,63] with similar negative effects to those observed in rats fed diets with a high protein content. Chronic and excessive methionine supplementation increases plasma hydroperoxides and low-density lipoprotein cholesterol,[64] induces vascular[65] and kidney damage with tubular hypertrophy,[66] raises iron accumulation and lipid peroxidation, and leads to liver dysfunction,[67] besides other alterations in other organs. In addition, methionine supplementation strongly increases methionine and its two more nearly derived methionine cycle metabolites, S-adenosylmethyonine (SAM) and S-adenosylhomocysteine (SAH), in rat liver and kidney.[63] Some of the harmful effects have been attributed to these methionine-related metabolites like SAM, SAH, or homocysteine, rather than to methionine itself, although in other cases a direct methionine toxic effect has been suggested.[61,65] This last case fits well with the observation that directs addition of methionine to isolated mitochondria *in vitro* increases their rate of mtROSp in liver and kidney mitochondria.[68]

Oxidation of methionine residues in proteins generates methionine sulfoxide depriving them of their function as methyl donors and may lead to loss of their biological activity.[69] This modification can be repaired by methionine sulfoxide reductase in a thioredoxin-dependent reaction. In this context, it is interesting that overexpression of methionine sulfoxide reductase increases longevity in *D. melanogaster*[70] and the opposite manipulation, knocking out the same enzyme, increases protein carbonyls and decreases longevity.[71] There is evidence that this enzyme plays an important role in protection against oxidative, cold, and heat stress and in the regulation of aging in *D. melanogaster.*[72] Also in agreement with a methionine role in aging, it has been reported that long-lived Ames dwarf mice have an altered methionine metabolism showing a marked increase in the trans-ulfuration pathway compared to their wild-type siblings.[73] All the above results point to methionine as the single dietary factor responsible for part of the longevity extension effect of DR.

6.2. Role of mtROS generation and oxidative damage

DR decreases oxidative stress in mitochondria. But, what is the specific dietary component responsible for the decreases in mtROS production and oxidative damage to mtDNA during DR? In agreement with their lack of effect on longevity,[42–44] neither isocaloric 40% lipid restriction[74] nor isocaloric 40% carbohydrate restriction[75] changes mtROSp or 8-oxodG in mtDNA. However, isocaloric 40%PR does decrease mtROSp and oxidative damage to mtDNA in rat liver[76] in a strikingly similar way, quantitatively and qualitatively to 40%DR. The effect of PR was studied in rat liver without changing the amount eaten per day of the other dietary components and it was found, like in 40%DR, that 40%PR decreases liver mtROSp specifically at complex I, lowers %FRL and 8-oxodG in mtDNA,[76] and decreases the five specific markers of protein purely oxidative, glycoxidative, and lipoxidative modification studied, as well as the complex I protein content in rat liver mitochondria and tissue.[77] Strikingly, the direction of change, the magnitude, mechanisms, and site of action exerted by PR on mtROSp and 8-oxodG in mtDNA are very similar to those found in 40%DR.[39] Taken together, those studies suggest that proteins are the dietary components responsible for the decreases in mtROSp and oxidative damage to mitochondrial macromolecules that takes place in DR, and for part of the increase in longevity induced by this dietary intervention.

It was logical to suspect that dietary methionine could be involved in those PR and DR effects since it was already known that MetR, independent of energy restriction, increases rat (maximum) longevity[37] while such effect had not been described for any of the other dietary amino acids. This is why in our laboratory we decided to study the effects of MetR on mtROSp and oxidative stress (Table 1.1). Isocaloric MetR (80% and 40%), applied to young rats during 7 weeks, lowers mtROSp (mainly at complex I), the % FRL, the complex I content, 8-oxodG in mtDNA, and specific markers of protein oxidative, glycoxidative, and lipoxidative modification in rat heart (40% and 80%MetR)[78,81] or liver (40%MetR)[79,82] mitochondria, similar to what occurs after 7 weeks of 40%MetR in rat kidney and brain mitochondria.[80,84] In order to obtain these decreases, it was enough to restrict methionine by 40%. 80%MetR leaded to a similar decreases in 8-oxodG than 40%MetR, while the decrease in mtROSp in 80%MetR was only somewhat more intense than in 40%MetR, being the decrease in mtROSp from control to 40%MetR more pronounced than that occurring between 40%MetR and 80%MetR.

Consistently with the role of methionine, in another experiment, when all the dietary amino acids except methionine were restricted (by 40%) during 7 weeks, neither the rate of mtROSp nor the level of 8-oxodG in mtDNA was modified.[83] In addition, we have recently found that 40% MetR also decreases mtROSp, %FRL, and 8-oxodG in mtDNA and reverses aging-related increases in protein modification when implemented at old age (during 7 weeks in 24-month-old rats).[82] All those results, taken together, indicate that the lowered ingestion of methionine during MetR (and PR and DR) is responsible for the decreases in mitochondrial ROSp and oxidative stress observed in MetR (and PR and DR), and possibly for all (during PR and MetR) or part (during DR) of the life-extension effect observed during these dietary manipulations. Moreover, the extraordinary capacity of a "single dietary molecule" to induce the decrease in mtROSp is still present when the animals reach old age.

Various mechanisms can be responsible for the decrease in mtROS production during MetR. A simple one is based on a decrease in the content of the complex I protein in MetR that would directly lead to a decreased rate of mtROSp. This has been reported under 40%MetR in the majority of tissues studied (Table 1.1), also during DR and PR, as well as in long-lived birds (pigeons, canaries, and parakeets) compared to the much short-lived mammals (rats and mice) of similar body size.[85,86] But this cannot be the whole explanation. MetR also induces qualitative changes in mitochondria since

Table 1.1 Effect of methionine restriction (MetR) on changes in mitochondrial oxidative stress-related parameters in rats

Dietary manipulation	Organ	mtROSp change and site	Variation in amount of respiratory complexes	FRL (%)	8-oxodG in mtDNA	Protein oxidative damage	References
80%MetR	Liver	↓at CxI ↓at CxIII	↓CxI/IV	↓at CxI	↓	↓	78
	Heart	↓at CxI	↓CxI/IV	↓at CxI	↓	↓	
80%MetR	Liver	↓at CxI ↓at CxIII	↓CxI/II/III/IV	↓at CxI	↓	↓	79
40%MetR	Liver	↓at CxI ↓at CxIII	↓CxI/II/III/IV	↓at CxI	↓	↓	79
40%MetR	Brain	↓at CxI	↓CxI/II/III/IV	↓at CxI	=	↓	80
	Kidney	↓at CxI	↓CxIV	↓CxI	↓	↓	
40%MetR	Heart	↓at CxI	=	n.d.	↓	↓	81
40%MetR at old age	Liver	↓at CxI	↑CxIV	↓	↓	↓	82
40% RESTAAS	Liver	=	=	=	=	↓	83

=, no change; ↑, increase; ↓, decrease; n.d., not determined. Cx, respiratory complex; mtROSp, rate of mitochondrial ROS production; %FRL, percent free radical leak at the respiratory chain; 8-oxodG, 8-oxo-7,8-dihydro-2′-deoxyguanosine. Diets were given during 6–7 weeks starting at 6–7 weeks of age except in Ref. 82 (started at 24 months of age). The RESTAAS diet contained 40% less of all the dietary amino acids, except for methionine that was present at same concentration in the two diets.

they not only generate less ROS, but they also have a lower %FRL and a lower electronic reduction state of the complex I ROS generator (the decrease in mtROSp is observed, like in DR, with partial complex I reduction but not with full reduction). Thus, MetR mitochondria (from both young and old animals) are more efficient in avoiding mtROS generation. They leak less radicals per unit of electron flow in the respiratory chain, similar to what has been found in long-lived compared to short-lived animals as well as in DR and PR rats compared to *ad libitum*-fed ones.[36] These quantitative and qualitative changes can be due to (i) direct interaction of methionine, or more likely, of a more chemically reactive methionine metabolite with the mitochondria or some critical complex I polypeptide/s; (ii) changes in cellular signaling molecules and the ensuing modification of specific gene expression; and (iii) decreases in the matrix NADH (which feeds electrons to complex I), thus lowering the state of electronic reduction of the complex I generator, due to decreases in the amounts of mitochondrial substrates.

Concerning possible direct interaction of methionine, recent studies have shown that the addition of methionine to isolated functional mitochondria freshly obtained from rats increases their rates of mtROSp.[68] Therefore, a rather direct and rapid effect of methionine (or a closely derived reactive metabolite) on complex I *in vivo* seems to occur. However, this action could be due to a chemically reactive methionine metabolite. This possibility is most relevant because in the methionine molecule, differing from homocysteine or cysteine, the potentially reactive sulfur is located inside the molecule and it is not available for direct covalent chemical reaction with protein thiols. Interestingly, it has been recently observed that the reaction of methionine with hydroxyl radicals generates methionine radical carbon-, nitrogen-, and sulfur-centered radicals as intermediates in the formation of the methanetiol product, as detected by EPR spin trap techniques and GC-FID and GC-MS techniques.[87] These radicals or methanetiol (CH_3SH) itself could react with complex I or some of its subunits leading to increases in mtROS generation. Since it is known that GSSG thiolization of isolated complex I increases its rate of ROS production,[88] a similar reaction of methanethiol or cysteine (which also has a free thiol group available for direct reaction) with complex I thiol groups could be involved in the decrease in mtROSp in MetR. This dietary manipulation decreases hepatic methionine and cysteine[59] and likely methanetiol levels, which can decrease thiolization of complex I subunits and then their rates of mtROSp. Alternatively, cysteine could also interact with the protein cysteines of some of the FeS clusters of the hydrophilic arm of complex I, leading to iron release or availability for reaction and then ROS generation. Lower cysteine levels in MetR could also decrease mtROSp through this kind of mechanism.

Changes in gene expression can be also involved in the MetR effects (mechanism "ii") since a recent genomic MetR study found changes in the expression of a large number of genes and proteins involved in lipid metabolism.[59] In addition, modifications of DNA methylation could be also involved.[89,90] Methionine is an essential amino acid with many key roles in mammalian metabolism including protein synthesis and function, as well as protein and DNA methylation.[91] Since aging seems to be associated with site-specific changes in DNA methylation,[92–96] MetR diets could extend longevity in rodents through various changes including modulation of DNA methylation patterns, specific changes in gene expression, and changes in translation rates, whose effects could include decreases in mtROS generation and oxidative damage. In agreement with that, we have recently detected that MetR induces a small but statistically significant decrease in global genomic DNA methylation in rat heart of young immature rats,[81] whereas when this manipulation was performed in old rats, the decrease in this parameter was not statistically significant in the liver.[82] Concerning mechanism "iii": decreased NADH, it is more likely in DR than in MetR, due to the large number of metabolites than can potentially be decreased because of the lower caloric ingestion. In fact, there is a published study in which it was shown that pyruvate, malate, and succinate, as well as NADH and the NADH/NAD+ ratio, are decreased in the tissues of rodents subjected to DR.[32]

Summarizing the described results, DR, PR, and MetR are nutritional interventions that increase longevity in rodents, although the magnitude of the longevity extension of MetR and PR in rodents is around 50% that of DR. This lower but significant life-extension effect in MetR than in DR would agree with the widely held notion that aging and longevity have more than one cause. Restriction of methionine intake can be responsible for part of the aging-delaying effects of DR by decreasing mtROSp and oxidative damage to mtDNA and macromolecules, acting at least in this sense as a "DR-mimic." All that suggests that methionine is the single dietary substance responsible for the beneficial changes of DR on mitochondrial oxidative stress. The remaining effects of DR on aging rate could be due to decreases in other dietary components or in the calories themselves through different additional mechanisms. In any case, it is interesting that 40%MetR can decrease mitochondrial oxidative stress because this dietary manipulation, or PR, does not imply the strong behavioral and nutritional stress of caloric restriction and thus seems a much more feasible option for wide application to human populations. Most importantly, negative effects such as delays in puberty and decreases in growth rate and final body size are shared by DR and 80%MetR but do not occur in 40%MetR. MetR at the 40% level is potentially advantageous because it lowers mtROSp and 8-oxodG in mtDNA to a similar extent than 80%MetR, while totally avoiding the decreases in body and organ weight, growth rate, maturation, and likely final body size that takes place in 80%MetR and 40%DR.

Health benefits can be obtained using "prudent" diets largely based on the consumption of complex carbohydrates, emphasizing vegetables with proteins rich in essential amino acids but low in the sulfur-containing amino acids methionine and cysteine (like pulses), or almost totally lacking methionine and cysteine (like fruits and vegetables), and avoiding the presently excessive intake of animal proteins (as well as fats). The results already available about PR in humans seem to be positive for human health and of similar character than those found in DR after up to 6 years of intervention in human beings.[97] These studies suggest that DR and PR can protect from obesity, mortality, and degenerative diseases including at least cardiovascular ones, diabetes and cancer, and can increase the human healthspan.

7. CONCLUSIONS

An updated version of the MFRTA described above is schematized in Fig. 1.2. Long-lived mammals and birds have species-specific low mitochondrial ROS generation rates at complex I and low fatty acid unsaturation degrees in the cellular and mitochondrial membranes. These are the only two known traits correlating with animal longevity in the right sense concerning not only MFRTA but also all theories of aging in general. The close vicinity or even contact between the site of ROS generation and mtDNA avoids antioxidants to interfere with ROS induced final forms of irreversible damage in mtDNA and this is likely why antioxidants do not modify longevity. It is well known that DR also decreases mtROSp and 8-oxodG in mtDNA. This is exclusively due to the lower methionine intake (MetR) of the animals subjected to DR. Around 50% of the longevity extension effect of DR is due to MetR and this 50% effect seems to work through decreases in mtROSp; the other 50% effect of DR on longevity would act through other mechanisms. The constantly produced mtROS throughout life at a

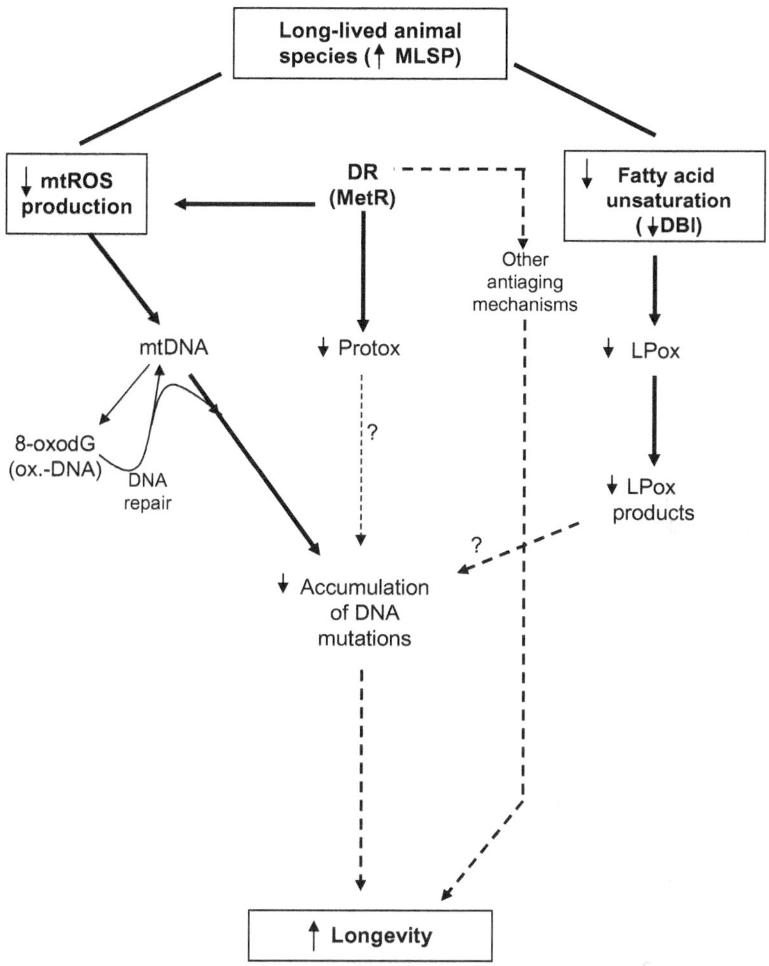

Figure 1.2 Aging and oxidative stress. The figure summarizes the main known causes and mechanisms of oxidative damage finally leading to aging that are known to be associated with animal longevity or dietary restriction (DR) in rodents. mtROS, mitochondrial ROS; DR, dietary restriction; MetR, methionine restriction; DBI, double bond index; 8-oxodG, 8-oxo-7,8-dihydro-2′-deoxyguanosine; Protox., protein oxidative modification; LPox, lipid peroxidation; LPox products, lipid peroxidation products.

different rate in each species leads to the generation of oxidative damage in mtDNA (e.g., 8-oxodG) which is repaired and can lead to point mutations in the process. In addition, mtROS can directly generate single- and double-strand breaks leading also to irreversible forms of damage (mutations) like mtDNA deletions or mtDNA fragments. Mutations can also arise due to processes unrelated to oxidative stress like mtDNA synthesis and repair.

However, it is unknown if these last mechanisms of damage generation are related to longevity or not. The low fatty acid unsaturation degree of mitochondrial membranes from long-lived animals leads to relatively low rates of endogenous lipid peroxidation *in vivo* (a strongly destructive membrane process), which in turn decreases the generation of lipid peroxidation products like MDA, hydroxynonenal, and many others. Some of these products have the potential to modify mtDNA, e.g., through direct interaction of the carbonyl group from the aldehydes with free amino groups in mtDNA, which would add secondary damage to that coming from the primary mtROS. There is a paucity of studies concerning this interesting possibility,[98] especially due to technical limitations. DR (and MetR) also lowers protein oxidation, glycoxidation, and lipoxidation, perhaps due to the induced decrease in mtROSp or due to increased protein catabolism. Protein oxidation can also potentially contribute to the accumulation of mtDNA mutations although there is scarcity of published support concerning this kind of process. If irreversibly damaged mtDNA reaches a high threshold level (approaching homoplasmy of mutated mtDNA), there is the possibility that oxidative mitochondrial ATP generation through oxidative phosphorylation is decreased to levels great enough to contribute to aging. There is no consensus if this classical concept of the MFRTA[99] can explain aging and longevity. However, there is an additional possibility concerning the accumulation of mtDNA fragments inside nuclear DNA which is known to increase with age in rat liver and brain.[100] Such fragments would alter the information coded in nDNA, thus contributing to aging.

ACKNOWLEDGMENTS

Results obtained at the author laboratory described in this review have been supported by Grant No. BFU2011-23888 from the Ministry of Science and Innovation to G. Barja.

REFERENCES

1. Harman D. The biological clock: the mitochondria? *J Am Geriatr Soc.* 1972;20:145–147.
2. Barja G. *Longevity and Evolution.* New York, NY: Nova Science Publishers, Inc.; 2011.
3. Ku HH, Brunk UT, Sohal RS. Relationship between mitochondrial superoxide and hydrogen peroxide production and longevity of mammalian species. *Free Radic Biol Med.* 1993;15:621–627.
4. Lambert A, Boysen H, Buckingham JA, et al. Low rates of hydrogen peroxide production by isolated heart mitochondria associate with long maximum lifespan in vertebrate homeotherms. *Aging Cell.* 2007;6:607–618.
5. Pamplona R, Barja G, Portero-Otín M. Membrane fatty acid unsaturation, protection against oxidative stress, and maximum life span: a homeoviscous-longevity adaptation. *Ann N Y Acad Sci.* 2002;959:475–490.

6. Naudi A, Jove M, Ayala V, Portero-Otín M, Barja G, Pamplona R. Regulation of membrane unsaturation as antioxidant adaptive mechanisms in long-lived animal species. *Free Radic Antioxid.* 2011;1:3–12.
7. Pérez-Campo R, López-Torres M, Cadenas S, Rojas C, Barja G. The rate of free radical production as a determinant of the rate of aging: evidence from the comparative approach. *J Comp Physiol B.* 1998;168:149–158.
8. Pamplona R, Constantini D. Molecular and structural antioxidant defenses against oxidative stress in animals. *Am J Physiol.* 2011;301:R843–R863.
9. Tolmasoff JM, Ono T, Cutler RG. Superoxide dismutase: correlation with life span and specific metabolic rate in primate species. *Proc Natl Acad Sci USA.* 1980;77:2777–2781.
10. Barja G, Cadenas S, Rojas C, López-Torres M, Pérez-Campo R. A decrease of free radical production near critical sites as the main cause of maximum longevity in animals. *Comp Biochem Physiol.* 1994;108B:501–512.
11. Barja G, Cadenas S, Rojas C, Pérez-Campo R, López-Torres M. Low mitochondrial free radical production per unit O_2 consumption can explain the simultaneous presence of high longevity and high metabolic rates in birds. *Free Radic Res.* 1994;21:317–328.
12. Herrero A, Barja G. H_2O_2 production of heart mitochondria and aging rate are slower in canaries and parakeets than in mice: sites of free radical generation and mechanisms involved. *Mech Ageing Dev.* 1998;103:133–146.
13. Boveris A, Cadenas E, Stoppani AOM. Role of ubiquinone in the mitochondrial generation of hydrogen peroxide. *Biochem J.* 1976;156:435–444.
14. Barja G, Herrero A. Localization at Complex I and mechanism of the higher free radical production of brain non-synaptic mitochondria in the short-lived rat than in the longevous pigeon. *J Bioenerg Biomembr.* 1998;30:235–243.
15. Genova ML, Ventura B, Giuliano G, et al. The state of production of superoxide radical in mitochondrial complex I is not a bound semiquinone but presumably iron-sulphur cluster N2. *FEBS Lett.* 2011;505:364–368.
16. Kushnareva Y, Murphy A, Andreyev A. Complex I-mediated reactive oxygen species generation: modulation by cytochrome c and NAD(P+) oxidation-reduction state. *Biochem J.* 2002;368:545–553.
17. Herrero A, Barja G. Sites and mechanisms responsible for the low rate of free radical production of heart mitochondria in the long-lived pigeon. *Mech Ageing Dev.* 1997;98:95–111.
18. Herrero A, Barja G. Localization of the site of oxygen radical generation inside the Complex I of heart and non-synaptic brain mammalian mitochondria. *J Bioenerg Biomembr.* 2000;32:609–615.
19. Lambert A, Brand M. Inhibitors of the quinine-binding site allow rapid superoxide production from mitochondrial NADH:ubiquinone oxidoreductase (complex I). *J Biol Chem.* 2004;279:39414–39420.
20. Barja G, Herrero A. Oxidative damage to mitochondrial DNA is inversely related to maximum life span in the heart and brain of mammals. *FASEB J.* 2000;14:312–318.
21. Herrero A, Barja G. 8-Oxodeoxyguanosine levels in heart and brain mitochondrial and nuclear DNA of two mammals and three birds in relation to their different rates of aging. *Aging Clin Exp Res.* 1999;11:294–300.
22. Muller FL, Lustgarten MS, Jang Y, Richardson A, Remmen HV. Trends in oxidative aging theories. *Free Radic Biol Med.* 2007;43:477–503.
23. Perez VI, Bokov A, Van Remmen H, et al. Is the oxidative stress theory of aging dead? *Biochim Biophys Acta.* 2009;1790:1005–1014.
24. Van Diepeningen AD, Maas MF, Huberts DH, et al. Calorie restriction causes healthy life span extension in the filamentous fungus *Podospora anserina*. *Mech Ageing Dev.* 2010;131:60–68.

25. Pamplona R, Prat J, Cadenas S, et al. Low fatty acid unsaturation protects against lipid peroxidation in liver mitochondria from longevous species: the pigeon and human case. *Mech Ageing Dev*. 1996;86:53–66.
26. Ruiz MC, Ayala V, Portero-Otín M, Requena JR, Barja G, Pamplona R. Protein methionine content and MDA-lysine protein adducts are inversely related to maximum life span in the heart of mammals. *Mech Ageing Dev*. 2005;126:1106–1114.
27. Fontana L, Partridge L, Longo VD. Extending healthy life span—from yeast to humans. *Science*. 2010;328:321–326.
28. Colman JR, Anderson RM, Johnson SC, et al. Caloric restriction delays disease onset and mortality in rhesus monkeys. *Science*. 2009;325:201–204.
29. Someya S, Yu W, Hallows WC, et al. Sirt3 mediates reduction of oxidative damage and prevention of age-related hearing loss under caloric restriction. *Cell*. 2010;143:802–812.
30. Sohal RS, Ku HH, Agarwal S, Forster MJ, Lal H. Oxidative damage, mitochondrial oxidant generation and antioxidant defenses during aging and in response to food restriction. *Mech Ageing Dev*. 1994;74:121–133.
31. Gredilla R, Barja G. The role of oxidative stress in relation to caloric restriction and longevity. *Endocrinology*. 2005;146:3713–3717.
32. Burch HB, Lowry OH, Bradley ME, Max Jr PF. Hepatic metabolites and cofactors in riboflavin deficiency and calorie restriction. *Am J Physiol*. 1970;219:409–415.
33. Gredilla R, Sanz A, Lopez-Torres M, Barja G. Caloric restriction decreases mitochondrial free radical generation at Complex I and lowers oxidative damage to mitochondrial DNA in the rat heart. *FASEB J*. 2001;15:1589–1591.
34. López-Torres M, Gredilla R, Sanz A, Barja G. Influence of aging and long-term caloric restriction on oxygen radical generation and oxidative DNA damage in rat liver mitochondria. *Free Radic Biol Med*. 2002;32:882–889.
35. Sanz A, Caro P, Ibáñez J, Gómez J, Gredilla R, Barja G. Dietary restriction at old age lowers mitochondrial oxygen radical production and leak at complex I and oxidative DNA damage in rat brain. *J Bioenerg Biomembr*. 2005;37:83–90.
36. Barja G. Aging in vertebrates and the effect of caloric restriction: a mitochondrial free radical production-DNA damage mechanism? *Biol Rev*. 2004;79:235–251.
37. Richie Jr JP, Leutzinger Y, Parthasarathy S, Malloy V, Orentreich N, Zimmerman JA. Methionine restriction increases blood glutathione and longevity in F344 rats. *FASEB J*. 1994;8:1302–1307.
38. Miller RA, Buehner G, Chang Y, Harper JM, Sigler R, Smith-Wheelock M. Methionine-deficient diet extends mouse lifespan, slows immune and lens aging, alters glucose, T4, IGF-I and insulin levels, and increases hepatocyte MIF levels and stress resistance. *Aging Cell*. 2005;4:119–125.
39. Lopez-Torres M, Barja G. Lowered methionine ingestion as responsible for the decrease in rodent mitochondrial oxidative stress in protein and dietary restriction. Possible implications for humans. *Biochim Biophys Acta*. 2008;1780:1337–1347.
40. Sun L, Amir A, Akha S, Millar RA, Harper J. Life-span extension in mice by preweaning food restriction and by methionine restriction in middle age. *J Gerontol A Biol Sci Med Sci*. 2009;64A:711–722.
41. Piper MDW, Partridge L, Raubenheimer D, Simpson SJ. Dietary restriction and aging: a unifying perspective. *Cell Metab*. 2011;14:154–160.
42. Ross MH. Nutrition and longevity in experimental animals. In: Winick M, ed. *Nutrition and Aging*. New York, NY: Wiley; 1976:43–57.
43. Khorakova M, Deil Z, Khausman D, Matsek K. Effect of carbohydrate-enriched diet and subsequent food restriction on life prolongation in Fischer 344 male rats. *Fiziol Zh*. 1990;36:16–21.
44. Shimokawa I, Higami Y, Yu BP, Masoro EJ, Ikeda T. Influence of dietary components on occurrence of and mortality due to neoplasms in male F344 rats. *Aging Clin Exp Res*. 1996;8:254–262.

45. Min KJ, Tatar M. Restriction of amino acids extends lifespan in Drosophila melanogaster. *Mech Ageing Dev*. 2006;127:643–646.
46. Troen AM, French EE, Roberts JF, et al. Lifespan modification by glucose and methionine in Drosophila melanogaster fed a chemically defined diet. *Age (Dordr)*. 2007;29:29–39.
47. Petti AA, Crutchfield CA, Rabinowitz JD, Botstein D. Survival of starving yeast is correlated with oxidative stress response and nonrespiratory mitochondrial function. *Proc Natl Acad Sci USA*. 2011;108:E1089–E1098.
48. Kabil H, Kabil O, Banerjee R, Harshman LG, Pletcher SD. Increased transulfuration mediates longevity and dietary restriction in Drosophila. *Proc Natl Acad Sci USA*. 2011;108:16831–16836.
49. Kalhan SC, Uppal SO, Moorman JL, et al. Metabolic and genomic response to dietary isocaloric protein restriction in the rat. *J Biol Chem*. 2011;286:5266–5277.
50. Malloy VL, Perrone CE, Mattocks DA, Ables GP, Caliendo NS, Orentreich DS. Methionine restriction prevents the progression of hepatic steatosis in leptin-deficient obese mice. *Metabolism*. 2013;62:1651–1661.
51. Orgeron ML, Stone KP, Wanders D, Cortez CC, Vant NT, Gettys TW. The impact of dietary methionine restriction on biomarkers of metabolic health. *Prog Mol Biol Transl Sci*. 2014;121:351–376.
52. Ramaligan A, Wang X, Gabello M, et al. Dietary methionine restriction improves colon tight junction barrier function and alters claudin expression pattern. *Am J Physiol Cell Physiol*. 2010;299:C1028–C1035.
53. Hasek BE, Stewart LK, Henagan TM, et al. Dietary methionine restriction enhances metabolic flexibility and increases uncoupled respiration in both fed and fasted states. *Am J Physiol Regul Integr Comp Physiol*. 2010;299:R728–R739.
54. Komninou D, Leutzinger Y, Reddy BS, Richie Jr JP. Methionine restriction inhibits colon carcinogenesis. *Nutr Cancer*. 2006;54:202–208.
55. Cavuoto P, Fenech MF. A review of methionine dependency and the role of methionine restriction in cancer growth control and life-span extension. *Cancer Treat Rev*. 2012;38:726–736.
56. Cao WX, Ou JM, Fei XF, et al. Methionine-dependence and combination chemotherapy on human gastric cancer cells in vitro. *World J Gastroenterol*. 2002;8:230–232.
57. Malloy VL, Krajcik RA, Bailey SJ, Hristopoulos G, Plummer JD, Orentreich N. Methionine restriction decreases visceral fat mass and preserves insulin action in aging male Fischer 344 rats independent of energy restriction. *Aging Cell*. 2006;5:305–314.
58. Perrone CE, Malloy VE, Orentreich DS, Orentreich N. Metabolic adaptations to methionine restriction that benefit health and lifespan in rodents. *Exp Gerontol*. 2013;48:654–660.
59. Perrone CE, Mattocks DAL, Plummer JD, et al. Genomic and metabolic responses to methionine-restricted and methionine-restricted, cysteine-supplemented diets in Fischer 344 rat inguinal adipose tissue, liver and quadriceps muscle. *J Nutrigenet Nutrigenomics*. 2012;5:132–157.
60. Elshorbagy AK, Valdivia-Garcia M, Mattocks DA, et al. Cysteine supplementation reverses methionine restriction effects on rat adiposity: significance of stearoyl-coenzyme A desaturase. *J Lipid Res*. 2011;52:104–112.
61. Harper AE, Benevenga NJ, Wohlhueter RM. Effects of ingestion of disproportionate amounts of amino acids. *Physiol Rev*. 1970;50:428–558.
62. Park CM, Cho CW, Rosenfeld ME, Song YS. Methionine supplementation accelerates oxidative stress and nuclear factor kappaB activation in livers of C57BL/6 mice. *J Med Food*. 2008;11:667–674.
63. Gomez J, Caro P, Sanchez I, et al. Effect of methionine dietary supplementation on mitochondrial oxygen radical generation and oxidative DNA damage in rat liver and heart. *J Bioenerg Biomembr*. 2009;41:309–321.

64. Hidiroglou N, Gilani GS, Long L, et al. The influence of dietary vitamin E, fat, and methionine on blood cholesterol profile, homocysteine levels, and oxidizability of low density lipoprotein in the gerbil. *J Nutr Biochem.* 2004;15:730–740.
65. Troen AM, Lutgens E, Smith DE, Rosenberg IH, Selhub J. The atherogenic effect of excess methionine intake. *Proc Natl Acad Sci USA.* 2003;100:15089–15094.
66. Kumagai H, Katoh S, Hirosawa K, Kimura M, Hishida A, Ikegaya N. Renal tubulointerstitial injury in weanling rats with hyperhomocysteinemia. *Kidney Int.* 2002;62:1219–1228.
67. Mori N, Hirayama K. Long-term consumption of a methionine-supplemented diet increases iron and lipid peroxide levels in rat liver. *J Nutr.* 2000;130:2349–2355.
68. Gomez J, Sanchez-Roman I, Gomez A, et al. Methionine and homocysteine modulate the rate of ROS generation of isolated mitochondria in vitro. *J Bioenerg Biomembr.* 2011;43:377–386.
69. Ciorba MA, Heinemann SH, Weissbach H, Brot N, Hoshi T. Modulation of potassium channel function by methionine oxidation and reduction. *Proc Natl Acad Sci USA.* 1997;94:9932–9937.
70. Chung H, Kim AK, Jung SA, Kim SW, Yu K, Lee JH. The Drosophila homolog of methionine sulfoxide reductase A extends lifespan and increases nuclear localization of FOXO. *FEBS Lett.* 2010;584:3609–3614.
71. Moskovitz J, Bar-Noy S, Williams WM, Requena J, Berlett BS, Stadtman ER. Methionine sulfoxide reductase (MsrA) is a regulator of antioxidant defense and lifespan in mammals. *Proc Natl Acad Sci USA.* 2001;98:12920–12925.
72. Lim DH, Han JY, Kim JR, Lee YS, Kim HY. Methionine sulfoxide reductase B in the endoplasmic reticulum is critical for stress resistance and aging in Drosophila. *Biochem Biophys Res Commun.* 2012;419:20–26.
73. Uthus EO, Brown-Borg HM. Methionine flux to transsulfuration is enhanced in the long living Ames dwarf mouse. *Mech Ageing Dev.* 2006;127:444–450.
74. Sanz A, Caro P, Sanchez JG, Barja G. Effect of lipid restriction on mitochondrial free radical production and oxidative DNA damage. *Ann N Y Acad Sci.* 2006; 1067:200–209.
75. Sanz A, Gomez J, Caro P, Barja G. Carbohydrate restriction does not change mitochondrial free radical generation and oxidative DNA damage. *J Bioenerg Biomembr.* 2006;38:327–333.
76. Sanz A, Caro P, Barja G. Protein restriction without strong caloric restriction decreases mitochondrial oxygen radical production and oxidative DNA damage in rat liver. *J Bioenerg Biomembr.* 2004;36:545–552.
77. Ayala V, Naudi A, Sanz A, et al. Dietary protein restriction decreases oxidative protein damage, peroxidizability index, and mitochondrial complex I content in rat liver. *J Gerontol A Biol Sci Med Sci.* 2007;62:352–360.
78. Sanz A, Caro P, Ayala V, Portero-Otin M, Pamplona R, Barja G. Methionine restriction decreases mitochondrial oxygen radical generation and leak as well as oxidative damage to mitochondrial DNA and proteins. *FASEB J.* 2006;20:1064–1073.
79. Caro P, Gómez J, López-Torres M, et al. Forty percent and eighty percent methionine restriction decrease mitochondrial ROS generation and oxidative stress in rat liver. *Biogerontology.* 2008;9:183–196.
80. Caro P, Gomez J, Sanchez I, et al. Forty percent methionine restriction decreases mitochondrial oxygen radical production and leak at complex I during forward electron flow and lowers oxidative damage to proteins and mitochondrial DNA in rat kidney and brain mitochondria. *Rejuvenation Res.* 2009;12:421–434.
81. Sanchez-Roman I, Gomez A, Gomez J, et al. Forty percent methionine restriction lowers DNA methylation, complex I ROS generation, and oxidative damage to mtDNA and mitochondrial proteins in rat heart. *J Bioenerg Biomembr.* 2011;43:699–708.

82. Sanchez-Roman I, Gómez A, Pérez I, et al. Effects of aging and methionine restriction applied at old age on ROS generation and oxidative damage in rat liver mitochondria. *Biogerontology*. 2012;13:399–411.
83. Caro P, Gomez J, Sanchez I, et al. Effect of 40% restriction of dietary amino acids—except methionine—on mitochondrial oxidative stress and biogenesis, AIF and SIRT1 in rat liver. *Biogerontology*. 2009;10:579–592.
84. Naudi A, Caro P, Jove M, et al. Methionine restriction decreases endogenous oxidative molecular damage and increases mitochondrial biogenesis and uncoupling protein 4 in rat brain. *Rejuvenation Res*. 2007;10:473–484.
85. St-Pierre J, Buckingham JA, Roebuck SJ, Brand MD. Topology of superoxide production from different sites in the mitochondrial electron transport chain. *J Biol Chem*. 2002;277:44784–44790.
86. Pamplona R, Portero-Otín M, Sanz A, Ayala V, Vasileva E, Barja G. Protein and lipid oxidative damage and complex I content are lower in the brain of budgerigards and canaries than in mice. Relation to aging rate. *Age (Dordr)*. 2005;27:267–280.
87. Spasojević I, Bogdanović Pristov J, Vujisić L, Spasić M. The reaction of methionine with hydroxyl radical: reactive intermediates and methanethiol production. *Amino Acids*. 2012;42:2439–2445.
88. Taylor ER, Hurrell F, Shannon RJ, Lin TK, Hirst J, Murphy MP. Reversible glutathionylation of complex I increases mitochondrial superoxide formation. *J Biol Chem*. 2003;278:19603–19610.
89. Robert L, Labat-Robert J, Robert AM. Genetic, epigenetic and posttranslational mechanisms of aging. *Biogerontology*. 2010;11:387–399.
90. Passarino G, Rose G, Bellizzi D. Mitochondrial function, mitochondrial DNA and ageing: a reappraisal. *Biogerontology*. 2010;11:575–588.
91. Brosnan JT, Brosnan ME. The sulfur-containing amino acids: an overview. *J Nutr*. 2006;136:1636S–1640S.
92. Christensen BC, Houseman EA, Marsit CJ, et al. Aging and environmental exposures alter tissue-specific DNA methylation dependent upon CpG island context. *PLoS Genet*. 2009;5:e1000602.
93. Wakeling LA, Ions LJ, Ford D. Could Sirt1-mediated epigenetic effects contribute to the longevity response to dietary restriction and be mimicked by other dietary interventions? *Age (Dordr)*. 2009;31:327–341.
94. Maegawa S, Hinkal G, Kim HS, et al. Widespread and tissue specific age-related DNA methylation changes in mice. *Genome Res*. 2010;20:332–340.
95. Cedar H, Bergman Y. Programming of DNA methylation patterns. *Annu Rev Biochem*. 2012;81:97–117.
96. Heyn H, Li N, Ferreira HJ, et al. Distinct DNA methylomes of newborns and centenarians. *Proc Natl Acad Sci USA*. 2012;109:10522–10527.
97. Fontana L, Weiss EP, Villareal DT, Klein S, Holloszy JO. Long-term effects of calorie or protein restriction on serum IGF-1 and IGFBP-3 concentration in humans. *Aging Cell*. 2008;7:681–687.
98. Chaudhary AK, Nokubo M, Reddy GR, et al. Detection of endogenous malondialdehyde-deoxyguanosine adducts in human liver. *Science*. 1994;265:1580–1582.
99. Fleming JE, Miquel J, Cottrell SF, Yengoyan LS, Economos AC. Is cell aging caused by respiration-dependent injury to the mitochondrial genome? *Gerontology*. 1982;28:44–53.
100. Caro P, Gómez J, Arduini A, et al. Mitochondrial DNA sequences are present inside nuclear DNA in rat tissues and increase with age. *Mitochondrion*. 2010;10:479–486.

CHAPTER TWO

Mitochondrial DNA Mutations in Aging

Konstantin Khrapko*, Doug Turnbull†
*Harvard Medical School, Beth Israel Deaconess Medical Center, Boston, Massachusetts, USA
†LLHW Centre for Ageing and Vitality, Newcastle University, Newcastle, United Kingdom

Contents

Progress in Molecular Biology and Translational Science, Volume 127
ISSN 1877-1173
http://dx.doi.org/10.1016/B978-0-12-394625-6.00002-7

Abstract

The relationship of mitochondrial DNA mutations to aging is still debated. Most mtDNA mutations are recessive: there are multiple copies per cell and mutation needs to clonally expand to cause respiratory deficiency. Overall mtDNA mutant loads are low, so effects of mutations are limited to critical areas where mutations locally reach high fractions. This includes respiratory chain deficient zones in muscle fibers, respiratory-deficient crypts in colon, and massive expansions of deleted mtDNA in substantia nigra neurons. mtDNA "mutator" mouse with increased rate of mtDNA mutations is a useful model, although rates and distribution of mutations may significantly deviate from what is observed in human aging. Comparison of species with different longevity reveals intriguing longevity-related traits in mtDNA sequence, although their significance is yet to be evaluated. The impact of somatic mtDNA mutations rapidly increases with age, so their importance is expected to grow as human life expectancy increases.

1. INTRODUCTION: THE DIFFERENT FACES OF THE MITOCHONDRIAL THEORY OF AGING

Mitochondrial involvement in aging was proposed over 30 years ago by Denham Harman, based on his original theory that aging is caused by the accumulation of damage resulting from reactive oxygen species (ROS).[1] ROS are the inevitable by-products of normal cellular processes, most notably the process of oxidative phosphorylation, which is the primary function of the mitochondrion. Harman noted that as a major source of ROS, mitochondria should also be its major target.[1] Hence, as the part of the cell most vulnerable to ROS, mitochondria could play the role of the "aging clock," the limiting aging component of the system. Later, Linnane and others[2] specifically implied accumulating mitochondrial DNA (mtDNA) *mutations* (as opposed to general damage, which usually means also chemical modifications of DNA and other macromolecules) as the main culprit in aging. The logic is that mitochondria are renewable organelles. Some of them grow, replicate their DNA, and divide. Other mitochondria are destroyed and replaced by newly divided ones. This means that all their damaged components are constantly replaced by newly synthesized ones. The only part that saves the record of past damage despite this turnover is the sequence of mtDNA. Indeed, mtDNA mutations once they arise are replicated and faithfully transmitted to the daughter mitochondria. This supports the hypothesis that accumulation of mtDNA mutations is one of the "primary" aging processes. The significance of mtDNA mutations for aging is also corroborated by a few other observations, of which most notable is the presence in old tissues, muscle, brain, and colon in particular, of cells that are deficient in mitochondrial function caused by mtDNA mutations (Fig. 2.1).

Of note, mitochondrial *mutational* hypothesis of aging should be distinguished from a few other mitochondrial hypotheses of aging, which do not explicitly refer to mtDNA mutations (*genetic* damage), but rather to *chemical* damage, such as oxidation, cross-links, and other covalent modification of macromolecules (including DNA). The various concepts are not mutually exclusive and all may very well have their share in the aging process. For example, "mitochondrial-lysosomal axis" theory of aging[3] maintains that in aged cells, accumulating oxidated cellular waste (primarily damaged mitochondria) "chocks" cellular autophagy systems. This further hampers turnover of damaged mitochondria creating a self-accelerating loop that eventually cumulates in dysfunctional cells or cell death.

Another, more traditional "mitochondrial free radical theory of aging" (Chapter 1), postulates that aging is caused by accumulating oxidative damage, which does not necessarily involve a self-accelerating component. In this view, aging can be caused by a constant rate of oxidative damage. In this theory, while mitochondria are also considered a critical component, the

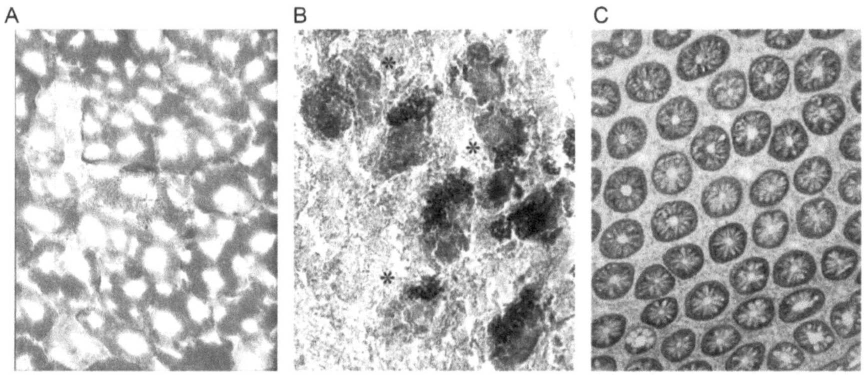

Figure 2.1 Examples of mitochondrial defects caused by mtDNA mutations in aged tissues. Cells lacking an mtDNA-encoded enzyme of electron transport chain, cytochrome-c oxidase are colored blue (blue color is visible only in the colored image) by special staining. (A) Skeletal muscle, (B) pigmented neurons of the substantia nigra (asterisks mark the deficient cells), and (C) clolon crypts. (See the color plate.)

main source and main target of oxidative damage, mtDNA mutations do not play a central role in this theory, they are just a part of damage.

Despite its attractiveness, the idea that mutations in mtDNA cause at least some aspects of aging remains controversial. Mitochondrial defects in aging tissues (Fig. 2.1) are yet to be convincingly related to age-related pathologies. To make the case of mtDNA mutations even more challenging, there are a few examples where artificially increased number of somatic mtDNA mutations in genetically engineered mice does not result in accelerated aging,[4] implying that at least some types mtDNA mutations are not involved in aging. Furthermore, controversy is fueled by the difficulties in accurately measuring of the fraction of mutant molecules. Some estimates yield rather unimpressively low overall fractions of mtDNA mutations, which seem to challenge the idea that these mutations may be of any significance. High estimates by other methods are sometimes perceived as gross overestimates.[78] To make things even more complicated, to estimate the functional impact of mtDNA mutations, one needs to know, in addition to overall mutant fraction, the detailed cell-to-cell distribution of mutations. This is because mutant mtDNA tend to affect cellular physiology only once they reach certain percentage within the mtDNA population of each individual cell, called the physiological threshold (Section 3). Here, we will review the evidence in support of the mtDNA mutational hypothesis of aging, which ranges from data on the detailed cell-by-cell distribution of mtDNA mutations to the inverse correlation of the number of repeats in mtDNA sequence and species' longevity.

2. MITOCHONDRIA, mtDNA, AND mtDNA MUTATIONS

2.1. Mitochondrial biology and mtDNA

Mitochondria are subcellular organelles responsible for generation of ATP, the cell's universal energy carrier, in a process of oxidative phosphorylation. This process is performed by a set of five multisubunit enzyme complexes (I, II, III, IV, and V) called the respiratory chain (RC), which are located on the inner membrane of the mitochondrion (for illustration see Ref. 5). Depending on the metabolic requirement of the cell, the number of mitochondria can vary markedly. In the majority of cells, mitochondria appear as a thread-like network throughout the cell and are constantly undergoing fission and fusion, a process that seems to be very important in terms of physiology and turnover of mitochondria.[6,7]

Mitochondria carry their own genome, i.e., a small (about 16,500 base pairs in animals) circular DNA molecule, which is replicated by the mtDNA polymerase gamma. Animal mtDNA encodes 13 polypeptides, which are subunits of all but one (complex II) of the RC complexes, as well as mitochondrial 22 tRNA and 2 rRNA.[8] All other subunits of the RC enzymes (~80 of them) are encoded in the nuclear genome,[5] as are the rest of several hundred mitochondrial proteins that are located in mitochondria but are not directly involved in the RC. Of note, outside the animal kingdom, mtDNA types are rather diverse.[9] MtDNA is maternally inherited in mammals since all the mitochondria from the sperm are destroyed on entry into the ovum. MtDNA is present in multiple copies in all cells. The number of copies of mtDNA varies markedly between cells, with tissues with high energy requirements, such as brain, heart, and skeletal muscle, harboring the largest numbers of mitochondrial genomes. MtDNA is believed to be organized into "nucleoids," i.e., compact nucleoprotein particles containing from one to a few mtDNA molecules. The exact composition, structure, and function of nucleoids are under intense investigation.[10]

2.2. mtDNA mutations

The mitochondrial genome principally suffers from two types of mutations: point mutations and large genome rearrangements, of which most studied are deletions, i.e., loss of large portions of the genome (from a few hundred base pairs to almost the entire mitochondrial genome).

There are several possible sources of mtDNA mutations, which can be broadly classified into spontaneous errors and damage-induced mutations. Spontaneous polymerase errors result from inherent polymerase infidelity. For example, an incorrect nucleotide may be inserted opposite to a normal nucleotide of the DNA template to generate base substitution. DNA damage is the other source of mutations. In the simplest case, DNA polymerase may be prompted to insert an incorrect nucleotide opposite to a chemically modified nucleotide. Most common example is the insertion of an A during DNA replication opposite to a deaminated cytosine (uracil), which is structurally similar to thymidine, resulting in a G to A change in the nascent strand (and a C to T change in the complementary strand, which will occur in the next replication cycle).[11] Most of mtDNA damage apparently results from endogenous processes, and exogenous mutagens play insignificant role: only unnaturally high doses of most powerful mutagens induce mutant fractions comparable to endogenous mutational background.[12]

mtDNA deletions formed via repair of a double-stranded break are another example of damage-induced mutation.[13,14] Apparently, hybridization interaction between distant complementary sequences (repeats) in mtDNA, which are exposed exonuclease removal of one of the strands during break repair stimulates formation of deletions.[15] Finally, damage to proteins, e.g., damage to DNA polymerase, its accessory factors and repair enzymes, can also potentially result in defective enzymes that are more prone to introducing errors than intact enzymes, although this field remains unexplored.

The distinction between spontaneous and damage-related mutations is particularly important in the aging field, because chemical damage to DNA and other cellular components increases with age. One therefore expects in the young tissues, mutations should be primarily spontaneous and that relative proportion of damage-induced mutations should increase with progressing age.

While there are many known mechanisms that may be potentially responsible for mtDNA mutations, it is largely unknown, which of these multiple sources of mutations are significant in a particular tissue or cell type, and at a particular age. In the above example of cytosine deamination, while deamination is the likely culprit, it is not known what primarily causes deamination—spontaneous hydrolysis, interaction with ROS (oxidative deamination),[16] or action of specific enzymes, i.e., deaminases.

It appears that the types of mutations are highly cell type and age specific. For example, deletions tend to accumulate in nondividing cell types, such as neurons and muscle fibers, and are essentially absent in actively proliferating cells, such as colon mucosa. There are significant differences on a more detailed scale, e.g., between different types of neurons. Pigmented neurons of substantia nigra in the brain from elderly subjects contain very high proportion of mtDNA deletions. Other neurons, e.g., large neurons of dentate nucleus, at the same age are essentially deletion free.[17] Deletions are essentially absent in young tissues and rapidly accumulate with age.

Importantly, actual abundances of the various types of mutations in different tissues are affected not only by the rate of the corresponding mutational events but also by the subsequent propagation of mutant genomes among other mtDNA genomes of the cell.[18] Different mutations may have selective advantages or disadvantages with respect intracellular dynamics of mitochondrial genomes (more in Section 4). An example is preferential propagation of mtDNA deletions in muscle fibers that may explain at least in part the relatively high abundance of deletions in aged muscle. In contrast to muscle, dividing haemopoetic cells tend to loose deletions because cells with higher deletion load either have proliferative disadvantage or even die, which may explain why proliferative tissues usually contain virtually no deletions.

In conclusion, mitochondria contain their own DNA that codes for components of the RC enzymes responsible for synthesis of ATP. mtDNA is prone to mutations, i.e., point mutations and large deletions. Mutations are highly cell type and age specific. Mutations are caused by spontaneous errors and chemical damage of mtDNA molecule, although precise mutational mechanisms are still under study.

3. PHYSIOLOGICAL OUTCOMES OF mtDNA MUTATIONS

3.1. RC deficiency

In contrast to nuclear DNA, mtDNA cannot be studied using standard genetic engineering approaches. It is not possible yet to introduce engineered mtDNA into a functional mitochondrion. New mtDNA variants can only be introduced into cells as a part of live mitochondria, for example, by fusion with mtDNA mutant cells. Thus our knowledge is limited to naturally occurring mtDNA mutations, those that can be obtained from living carriers, i.e., for the most part, patients with inherited mitochondrial disease caused by mtDNA mutations.

As mtDNA encodes exclusively components of the mitochondrial RC and RNAs needed for their translation, mtDNA mutations should primarily disrupt RC and therefore impede cellular respiration. Point mutations in protein coding regions should affect one particular enzyme of the RC. Point mutations in tRNA or rRNA genes should affect translation as a whole. As for mtDNA deletions, note that tRNA genes are distributed all over the mitochondrial genome, so most mtDNA deletions disrupt at least one tRNA gene. Thus, effects of all the different mtDNA deletions should be similar and constitute a general translational defect in the first place, leading to deficiency in mtDNA-encoded subunits and ultimately also to RC deficiency.

Indeed, all detrimental mtDNA mutations studied to date cause RC deficiency. In affected tissues of patients with mtDNA disease, this often manifests in a mosaic of respiratory-deficient cells, which can be detected using histochemical (i.e., performed in a thin section of tissue on a microscopic slide) reaction to the complex IV, or cytochrome-*c* oxidase (COX), an enzyme that has three catalytic subunits encoded by the mitochondrial genome. Interestingly, the appearance of cellular respiratory mosaic in mtDNA disease is similar to that observed in aged tissues (Fig. 2.1) except the frequency of deficient cells in mtDNA disease is usually higher than that in aging.

3.2. Mitochondrial DNA disease

What are the consequences, at the tissue/organism level, of the cellular respiratory deficiency caused by mtDNA mutations? Patients with mtDNA mutations present with a wide variety of symptoms due to multiple organ systems being affected. The mtDNA mutations causing disease maybe inherited or arise sporadically presumably in the oocyte and then get distributed among different organs during development. Usually, this results in "*heteroplasmy*," i.e., mixture of mutant- and wild-type mtDNA. In this case, all affected tissues of an individual contain the same mutation, though possibly at different heteroplasmy levels. Effects of these mutations depend not only on the type of mutations but also on the total mutational load and which tissues are at higher heteroplasmy levels. Often tissues heavily dependent on oxidative metabolism, such as brain and skeletal muscle are most affected. It is worth keeping in mind that effects of mtDNA mutations are not limited to effects on individual cells, but may include systemic effects. A recent mouse model, for example, proposed that RC deficiency induces a mitochondrial stress response, with local and global changes mimicking starvation.

In addition to inherited and early developmental mtDNA mutations, mtDNA disease can also occur because somatic mtDNA mutations arise at increased rate due to defects in nuclear gene involved in mtDNA replication and maintenance. In this case, there is no predominant mtDNA mutation; instead, there is a complex mixture of mutations that have arisen independently, which is similar to what is observed during normal aging. In patients with these mtDNA mutations secondary to the nuclear gene defects, the clinical features maybe similar to those seen in patients with primary mtDNA disease suggesting common mechanisms.

3.3. Phenotypic threshold

The cells from a patient tissue with mosaic respiratory deficiency can be collected individually and the fraction of mutated mtDNA in each can be quantified. Not unexpectedly, deficient cells typically contain high fractions of mutant mtDNA. More interestingly, respiration-proficient cells also frequently contain substantial (though lower) proportion of mutations. What proportion of mutant is sufficient to confer respiratory deficiency?

The quantitative effect of a detrimental mtDNA mutation on the RC activity of a cell has been studied in some detail, e.g., by measuring of respiratory activity of cultured cells with various fraction of mutant mtDNA. Usually, the effect is nonlinear: there is little decrease in RC activity with increasing proportion of mutant mtDNA, until heteroplasmy reaches a certain value, called the *phenotypic threshold*, above which the activity drops steeply with increasing mutant fraction (Fig. 2.2, solid blue curve). For deletions, phenotypic threshold is about 60%,[19] for point mutations, it typically is ~90%.[20]

The mechanism of the phenotypic threshold is not fully understood. Nonlinearity necessarily implies come kind of *exchange* between mitochondria within a cell. Indeed, if mitochondria were independently rendered respiratory inactive once their mtDNA were mutated, the total respiratory capacity of the cell would necessarily decrease linearly with the fraction of inactive mitochondria and, therefore, the fraction of mutated mtDNA (gray

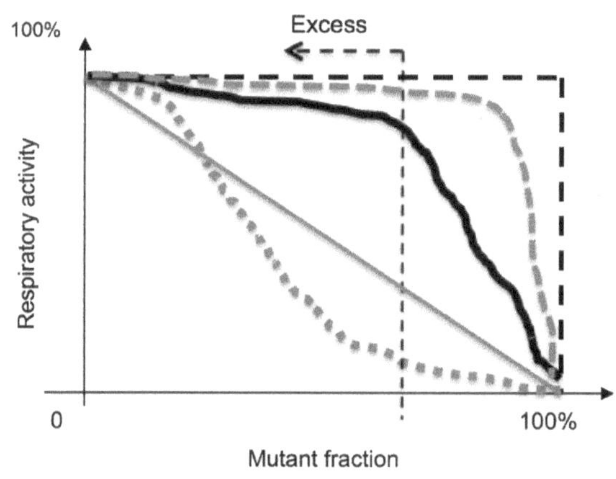

Figure 2.2 The nature of phenotypic threshold. Gray diagonal line: hypothetic situation where mitochondria are independently inactivated by mutations. Solid blue curve: mitochondria share an excess of a critical component. Once excess is exhausted at certain mutant fraction (broken vertical line), all mitochondria in the cell start loose ETC activity as availability of the component drops below the necessary minimum. Green-dashed curve: if mitochondria upregulate mtDNA in response to deficiency, full activity can be maintained at higher mutant fractions, but then suffer a steeper fall off as the capacity to upregulate is exhausted, too. The red-dotted curve illustrates the behavior of a "dominant" mutation, where even small fraction of mutation disrupts RC function of the cell. (See the color plate.)

line in Fig. 2.3). The mechanism whereby interaction between mitochondria creates the threshold effect is most likely an interplay of several factors.[21] First, mitochondrial components for respiration apparently are present in *excess*. A mitochondrion with mutant DNA can remain functional by exchange with nonmutant mitochondria, where the missing component is in excess. This exchange occurs via fusion and fission of mitochondria or through mitochondrial network[22] (see, however, point 3 in Section 4.3). As mutation burden increases, more mitochondria need to be supplemented, but all mitochondria in a cell are kept functional as long as nonmutant mitochondria produce sufficient excess of the necessary component to share. As a result, the cell appears almost fully functional even at relatively high mutant fractions. This is represented by the gently descending part of the solid blue curve in Fig. 2.2. Once excess has been exhausted, further increase of the mutant fraction results in rapid decrease of respiration (the steep portion of the solid blue curve in Fig. 2.2).

The threshold behavior can be enhanced if the cell attempts to compensate for the defect caused by the mutation. For example, muscle fibers are capable of upregulation of the amount of mtDNA in response to decrease of nonmutant component(s) and/or failing RC.[23] Upregulation of total mtDNA keeps the amount of wt mtDNA at levels sufficient to sustain respiration at even higher fractions of mutant DNA. The ability of the cell for DNA upregulation is limited, and as soon as the limit is reached, further increase of mutant fraction results in the an even steeper decrease of the wt mtDNA and respiration (dashed green curve in Fig. 2.2).

3.4. Recessive and dominant mtDNA mutations

Note that, for the sake of simplicity, we assumed that mtDNA mutations were loss-of-function mutations that merely decrease the number of normal molecules by "crowding them out." These mutations are called "recessive" by analogy with nuclear mutations, meaning that these mutations work only when present in great majority. However, not all mtDNA mutations are like that. In very rare cases, even small fraction of mutation disrupts RC function of the cell, so the mutation can be called "dominant." In other words, the threshold is very low (Fig. 2.2, dotted red curve). This happens, for example, if mutation affects anticodon and thus specificity of a tRNA.[24] Mutant tRNA delivers incorrect amino acid, and even a small fraction of such tRNA results in frequent production of defective proteins. In this case, the exchange between mitochondria becomes a curse rather than a blessing (if these mutants were confined to one or a few mitochondria, the rest would remain fully functional and the cell would be fine).

3.5. Dominant *lethal* mtDNA mutations?

Are effects of mtDNA mutations limited to respiratory deficiency? Mitochondria are involved in many other processes other than respiration. Of those, most relevant to aging are perhaps apoptosis and generation of ROS. However, absolute majority of mtDNA mutations studied so far apparently neither cause cell death (in fact, RC-deficient cells apparently survive for long periods of time), nor cause pronounced increase of ROS. This does not necessarily mean, however, that mtDNA mutations with such properties do not exist. Our knowledge about the effects of mtDNA mutations mostly comes from studies of inherited mtDNA mutations, and thus may be incomplete. Inherited mutations have to be compatible with development, i.e., be "mild" in certain sense. Somatic mutations are free from this limitation and, therefore, can potentially be more "severe" and even lethal. A demonstrative example is a mutation of the mouse mtDNA called 13885insC (an insertion of a C at position 13885). This is a "severe" mutation: it creates a translation frameshift in a complex I gene, ablates assembly of complex I,[25] and increases ROS production.[26] This is not a natural inherited mutation: it was isolated by in vitro selection in a cell line (this mutation confers resistance to a certain chemical that is toxic only when complex I is functional). When artificially introduced into the germ line, this severe mtDNA mutation is purged from progeny, as it causes oocyte death. Interestingly, oocytes die even if they carry as low as ~12% of the mutant genome. That is, the mutation is dominant, but unlike dominant mutations discussed in Section 3.4, its primary action is not merely in disabling of RC, but in killing the oocyte. It is therefore tempting to speculate that there exists a whole class of mtDNA mutations that are lethal to the cell even at low fractions. Such hypothetical mutation were termed "dominant lethal."[27] Dominant lethal mutations might act, for example, by causing an increase in ROS production that would be toxic for the cell and cause apoptosis.

Dominant lethal mutations would be difficult to detect because of their trend to self-eliminate. While these mutations may be rare, they could potentially have contributed to some important aging phenomena, in particular, progressive loss of stem cells. Dominant mitochondrial mutations were implied in a model explaining early heart pathology in mice with increased rate of somatic mtDNA mutations.[27] While dominant lethal mtDNA mutations are very intriguing, they are yet to be discovered.

In summary, the most common phenotypic effect of mtDNA mutations is the inactivation of the RC. To disturb RC, mutation needs to exceed a threshold of 50–90%, depending on mutation and cell type. Dominant mutations that are not subject to phenotypic threshold are less common. It is tempting to speculate that some dominant mutations are lethal, and these mutations, if they existed, might be very important for the aging process.

4. CLONAL EXPANSION AND AGE-DEPENDENT DYNAMICS OF mtDNA MUTATIONS

4.1. Somatic mtDNA mutations need to be clonal to be relevant to cell physiology

While discussing phenotypic threshold in Section 3.3, we assumed (for the sake of simplicity) that mutations present in a cell were all identical. This is correct in case of *inherited* mtDNA mutations causing mitochondrial disease. In aging, however, mtDNA mutations are *somatic*, i.e., they occur at random. Thus different mtDNA molecules within a cell are expected to acquire different somatic mutations. These different mutations will most likely affect different aspects of mitochondrial genome, e.g., one may deactivate a subunit of COX, another—a tRNA, still another—a rRNA, and so on. When a mixture of such mutations is present in a cell, the different mutant genomes mutually compensate their deficiencies,[22] by exchanging their contents via fusion and fission of mitochondria. This idea is illustrated in Fig. 2.3A. For example, imagine a cell where three different subsets of mtDNA molecules carry three different detrimental mutations (tRNA, rRNA, and protein coding mutation). A mitochondrion carrying DNA with a tRNA mutation can get functional tRNA from mitochondria carrying COX and rRNA mutations, while the tRNA mutant provides functional COX and rRNA to mitochondria carrying corresponding mutations. Even if every mtDNA molecule in a cell carries an inactivating mutation (of even a few), the cell may remain completely unaffected, as long as these mutations in different mtDNA molecules disrupt different aspects of the mitochondrial genome. Does this mean that somatic mutations are too inefficient to be detrimental and thus have nothing to do with mitochondrial dysfunction and therefore aging?

To efficiently disrupt mitochondrial function of the cell, mtDNA mutations on different mtDNA molecules within the cell must be identical, or at

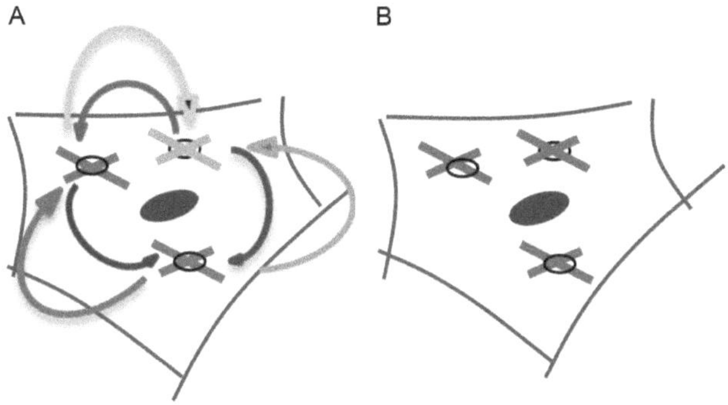

Figure 2.3 Clonal expansion is needed for phenotypic expression of somatic mutations. (A) An imaginary cell where every mtDNA molecule is carrying one of three detrimental somatic mutations inactivating different components of the mitochondrion, e.g., tRNA (green), rRNA (red), and protein coding (blue). Arched arrows show the direction of exchange of the corresponding missing components between mutant mitochondria. Exchange allows every mitochondrion to compensate the defect and remain functional. (B) Mitochondria in a cell with identical mutations in every mtDNA molecule lack the ability to exchange and compensate and thus become dysfunctional and the cell as a whole fails. (See the color plate.)

least affect the same aspect of the genome (Fig. 2.3B), which will prevent compensatory exchange. Surprisingly, analysis of single cells usually reveals that all or a substantial proportion of mtDNA molecules in a cell indeed carry one particular mtDNA mutation, but mutations are different in different cells. This puzzling observation was first made in muscle fibers,[28,29] but has since been confirmed almost for all cell types that were investigated. How possibly could the same mutation occur in almost all molecules in a cell? It is highly unlikely that the same mutation hit the same cell multiple times. Thus, the only logically remaining possibility is that identical mutant mtDNA molecules within a cell are *clonal*, i.e., descend from a single molecule that had acquired the mutation probably many mitochondrial generations ago. The process whereby a single mutant molecule multiplies and eventually overcomes the cell is thus an example of *clonal expansion*. Clonal expansion of mtDNA mutations is a puzzling phenomenon: it would be beneficial to a cell to suppress, not support expansion of detrimental mutations. We will now discuss the possible mechanisms of this counterintuitive phenomenon.

4.2. MtDNAs in a cell: A dynamic population of molecules

MtDNA molecules within a cell can be considered a "population." Indeed, new mtDNA molecules are "born" via replication of existing mtDNA molecules, which happens when mitochondria grow and divide. MtDNA molecules also "die," by being removed from the population through cell division, when mitochondria are partitioned between daughter cells.[30] For mtDNA "populations" of each of the daughter cells, mtDNA that got partitioned into the sister cell are gone forever, i.e., in effect, "dead." Importantly, not only dividing cells have a dynamic population of mtDNA. In nondividing cells, mtDNA molecules also keep being born via replication and "die," when mitochondria that carry that mtDNA molecule are destroyed in the lysosomes by a process called mitophagy. As a result, mtDNA of long-lived nonproliferating cells types is relatively short lived. For example, the half-life of mtDNA in rat cardiomyocytes is about 2 weeks.[31] There are several plausible mechanisms, most of which are likely to play at least some role *in clonal expansions*. In the next section, we will discuss these mechanisms, give examples of the tissues and/or conditions in which expansion is most likely to occur and explore the partial contributions of each mechanism to the dynamics of mtDNA mutations.

4.3. Clonal expansion via positive selection

The most straightforward potential mechanism of clonal expansion is through positive selection, i.e., when a mutation confers a selective advantage on the mtDNA molecule. The nature of such a selective advantage is a matter of debate. There are several proposed models, which are not mutually exclusive.

1. One possibility is that a base change directly affects the probability of the mutant molecule to get replicated, potentially by increasing affinity to a regulatory factor or via affecting DNA secondary structure. Examples of mutations that appear to behave in this way include certain mutations in the control region that markedly increase with age in some but not other tissues.[32,33]
2. Another possibility is specific for mtDNA *deletions*. mtDNA deletions result in genomes that are much shorter (typically by 30% or more) than normal mtDNA. Traditionally, it was assumed that deletions enjoyed selective advantage because shorter genomes need less time to replicate. The replication time, however, does not appear to be limiting, because mtDNA spends only a small fraction of time replicating.[34] Perhaps a more likely explanation is that shorter molecules have a lower chance to acquire DNA damage. The probability for a mtDNA molecule to sustain a lesion, e.g., a lesion interfering with

replication (such as DNA interstrand crosslink or DNA double-strand break) is proportional to its length. In an environment where such damage is relatively common, this can ensure a replicative advantage for shorter (deleted) molecules. Consistent with this idea, pigmented neurons of the substantia nigra, which are thought to provide a particularly hostile environment for mtDNA (Section 5.2), also carry the highest load of *clonal* mtDNA deletions.[17,35] There is no consensus, however, with respect to whether longer deletions indeed expand faster than short ones, as one would predict if based on the above hypotheses. In muscle, deletions of all sizes appear to expand at similar rate.[36] On the other hand, there were some advantages for long deletions in cell culture.[14]

3. Local feedback mechanisms. The above hypotheses are based solely on the properties of the mtDNA molecule itself. In principle, one may think of a mechanism including the endpoint effects of the mtDNA mutation (e.g., lack of a protein or tRNA, or overall RC deficiency), rather than mtDNA sequence *per se*, are responsible for positive selection of a mutant. The simplest logical possibility is that a mutation results in a shortage of a certain RC component and a compensatory feedback stimulates mtDNA replication in response. Such feedbacks appear natural and plausible, and indeed, mtDNA upregulation is observed in RC-deficient cells.[23] However, all simple feedback mechanisms of this kind had a common problem: because of mitochondrial fusion and fission, any shortage is expected to be shared among all mitochondria in the cell. For example, shortage of a component caused by a mutation in one or a few mtDNA is expected to become "global" within the cell and equally stimulate of all mtDNA molecules (mutant and wt) to replicate. Thus, any hypothesis involving an endpoint effect of a mutation needs to explain why such effect could be *local* and will feedback preferentially onto the original mutant molecule.

Recently, it has been proposed[37] that deletions may gain advantage because a certain region of mtDNA, when deleted, stimulates compensatory transcription and, as a consequence, replication of mtDNA. This model assume that, despite fusion and fission, mixing of mitochondrial components remains limited, specifically that polypeptides that were synthesized off a certain mtDNA nucleoid remain physically linked to that nucleoid.[37,38] Thus, locality of the positive feedback of this mechanism is postulated via presumable physical connection between a nucleoid and the RC complexes encoded by its mtDNA.[38]

4.4. Clonal expansions via random genetic drift

Clonal expansion can occur without selection. After a sufficient number of generations, in a constant size population of mtDNA molecules, the progeny of a single molecule will replace the whole population by pure chance. This is analogous to genetic drift in population dynamics. Indeed, with one daughter on average per mtDNA genome per generation (which is necessary to keep constant population size!) many mtDNA will not replicate while other will replicate more than once. This is because mtDNA replication is "relaxed," meaning that there is no mechanism ensuring that each mtDNA molecule replicates once per generation. When mtDNA molecule that fails to replicate before it "dies," its lineage is irreversibly terminated. As more generations pass, more mtDNA lineages in the cellular population get terminated, until only one line survives, which is equivalent to complete clonal expansion. This principle is illustrated in Fig. 2.4 on a particular case of clonal expansion in a small imaginary mtDNA population of three individuals.

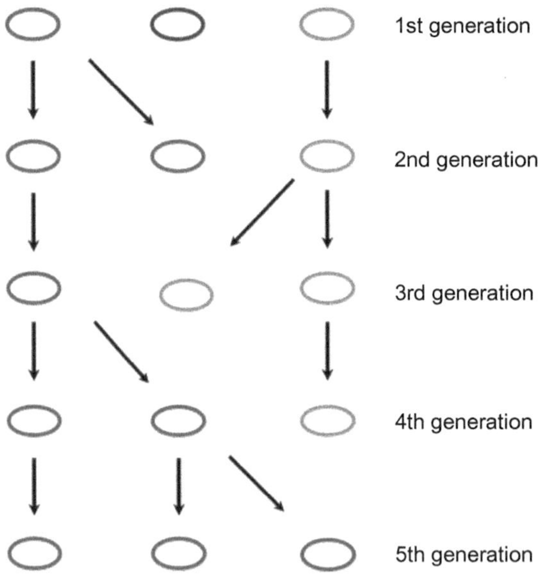

Figure 2.4 A series of plausible random events (lineage terminations) result in an irreversible clonal expansion of one of the lineages in a small imaginary population of mtDNA molecules. The original mixture of three different types (red, blue, and green) of mtDNA molecules quickly evolves to a homoplasmic red population. In a real population, clonal expansion takes more generations, approximately in proportion to the population size. (See the color plate.)

As can be seen in Fig. 2.4, the main feature of genetic drift is the termination of individual lineages. Indeed, any nascent somatic mutation most likely will disappear soon after its emergence with the lineage on which it occurred. These emergence-disappearance events will probably happen many times before a lucky mutation emerges in a mtDNA molecule whose progeny will take over the whole population of mtDNA. Note that on average, the process of random clonal expansion does not result in any increase of mutant fraction beyond what is expected from ongoing *de novo* mutagenesis. The frequency trajectory of a mutation in a cell is a random walk; it can go up or down at any moment. Mutations that have succeeded to expand, e.g., to 50% of the population, still have a 50% chance to eventually disappear. Random clonal expansion in effect is a process of redistribution of mutations, where some cells get 100% mutant and others are kept mutation free, while without the expansion phenomenon, all cells would have carry an average mutant load. Determining the expected rate of expansion in an arbitrary population is a complex mathematical problem,[39] but it is intuitively obvious (and this is indeed observed in various mathematical models) that genetic drift is faster in populations/cells with small numbers of mtDNA and/or with faster generations or mitochondrial turnover.[40]

The above conceptual framework helps to make sense of the observed dynamics of mtDNA mutations, although it should be used with caution to predict what exactly happens in particular cells. This is because the critical parameters that define the dynamics of the mtDNA molecules within cells have not been studied sufficiently. For example, it is not known what proportion of mtDNA molecules participate in mtDNA replication: by some accounts, only a small subset of mitochondria may be designated for reproduction purposes (the "stem mitochondria" hypothesis). Furthermore, mtDNA is organized into bundles called "nucleoids," which apparently replicate as a unit and most likely all carry the same genotype,[41] which implies a lower effective copy number and a higher genetic drift rate. All these features are cell type specific. There may or may not be selection with respect to which mitochondria are destroyed in the lysosomes. And, finally, the number of mtDNA generations is not known with any precision.

Several computational models have been developed that emulate dynamics of the intracellular mtDNA population. The conclusion from these *in silico* experiments is that under realistic assumptions, an entirely random process is able to expand single somatic mtDNA mutants to full cell homoplasmy within only a few dozen cell generations, while significant fraction of a mutation may be achieved much sooner.[40]

There is no doubt that random drift is indeed operating at least in some cells: it is the only mechanism yet that explains the observation of clonal expansion of phenotypically neutral mtDNA mutations in buccal[42] and colon epithelial cells because for these mutations there is no possible selective process.[43] Random drift most likely participates (perhaps together with other processes) in the dynamics of *all* mtDNA mutations, including the detrimental ones, as described in the next paragraph.

4.5. Random genetic drift with nonlocal compensatory feedback explains unidirectional expansion of detrimental mutations, without resorting to selective advantage

The hallmark of truly random genetic drift is the lack of direction: a mutation may either expand or disappear. As a result, overall mutant fraction does not change, unless new mutations are generated (see discussion below). Perhaps paradoxically, overall fraction of detrimental mutations does increase with age. This seems to imply that random drift is not sufficient to account for dynamics of these most important mutations—but this inference is incorrect. A very promising model[44] was able to explain directional expansion of detrimental mutation, based on random drift with no reference to selective advantage and/or local feedback mechanism (point 3 in Section 4.3). This model postulates a reasonable feedback mechanism that strives to keep the number of wild-type mtDNA copies per cell constant. This mechanism nonselectively increases mtDNA replication (both normal and mutant molecules) once the number of normal (only normal!) mtDNA drops below the target number. This simple rule results in an interesting dynamics of detrimental mutations. Once a detrimental mutation appears in a cell, it enters the random drift process. Most of these mutations immediately disappear, but eventually one of them succeeds to moderately expand, and in doing so replaces some of the wt molecules. Feedback mechanism senses the resulting decrease of normal mtDNA and upregulates mtDNA replication, which increases copy number of both normal *and* mutant mtDNA. Eventually mutant mtDNA may reach physiological threshold, and the cell becomes respiratory deficient. As time goes, such mutation-expansion-upregulation events keep happening. This model explains both the increase of the overall fraction of detrimental mutations and increased total mtDNA copy number in cells with high mutant load.[23] This is because while detrimental and neutral mutations are equally likely to expand (drift is random!), only cells with expanded *detrimental* mutations further upregulate their mtDNA by the feedback mechanism. Because in this model cells with expanded detrimantal mutations upregulate their mtDNA, while normal cells do not, overall fraction of detrimental mutations in tissue will progressively increase with age, in accord with observations.

5. EFFECTS OF SOMATIC mtDNA MUTATIONS IN AGING TISSUES

Perhaps the weakest point of the mitochondrial mutational hypothesis of aging is the relatively low overall fraction of mtDNA mutations in aged tissues. For example, typical fraction of mtDNA deletions in aged muscle is about 1%, which may seem incompatible with any significant role in tissue physiology and thus in the aging process. To make things worse for the hypothesis, the mouse models that contain higher fractions of mtDNA deletions,[45] or point mutations[4] do not display any of the expected aging phenotypes. This seems to imply that mtDNA mutations cannot be of any significance for the aging process. In this section, we will show how a highly nonuniform distribution of mtDNA mutations creates "weak spots," which can conceivably cause significant deterioration, despite their overall low fraction.

5.1. Muscle fibers

Studies of clonal expansions of mtDNA mutations are particularly applicable to skeletal muscle. On a histological section, the mitochondrial status of a fiber can be first determined by special staining that reveals a lack of COX activity, and then segments of individual fibers cay be collected by microdissection to measure the presence/abundance of mutations in mtDNA.[20] By studying thousands of consecutive sections, Aiken's laboratory reconstructed incredibly detailed respiratory activity/mutant fraction profiles of muscle fibers.[46]

The distribution of COX activity in muscle that they discovered was remarkable: on a transverse section there are rare distinctively blue fibers (COX deficient), while a great majority of fibers appear normal (Fig. 2.1A). Mutational analysis reveals that each individual COX-deficient fiber contains large proportion of mutant mtDNA (reaching close to 100%), typically a deletion.[23] Each COX-deficient fiber contains one type of deletion while different fibers contain different deletions, which clearly confirms their clonality. Most importantly, longitudinally, the fibers are not entirely normal or entirely COX deficient. Instead, COX deficiency appears in segments of a few microns to a couple millimeters long. At a segment's border, both COX activity increases and fraction of deleted mtDNA drops off.

What is the role of intrafiber clonal expansions in the aging process? There is no doubt that expanded mutations cause most if not all of the observed COX defects. The proportion of defective fibers, however, is very low. In young human muscle, the fraction is typically well below 0.1%. It increases dramatically with age, but even in the oldest individuals rarely exceeds ~1%. Can 1% of defective fibers have a significant impact on muscle aging? It is hard to believe that 1% defective fibers may result in any gross deficiency, such as overall muscle weakness. Much more attractive is an idea that short COX-deficient zones cause the entire muscle fiber to degenerate. Note that while only about 1% of gross fiber length is affected by clonal expansions of mtDNA deletions, because the COX-deficient zones are quite short, a large proportion of fibers carry a COX-deficient zone somewhere along its length. Careful calculations show that in a typical old muscle, about a third of all muscle fibers are affected by a clonal expansion of mtDNA mutations somewhere along their length. Aiken's group has demonstrated that at least in some instances, the development of a COX-deficient zone within a fiber eventually results in thinning followed by complete degeneration of the fiber.[23] In theory, a discontinuous fiber with a dead zone in the middle will tend to eventually completely degenerate. In other words, even if death of COX-deficient fiber is rather a slow process, a very substantial proportion of fibers may be eventually destined to wasting, and thus clonal expansion of mtDNA mutations may be responsible, at least in part, in sarcopenia, a common age-dependent loss of muscle mass.

5.2. Neurons

Perhaps one of the most intriguing observations in terms of the role of mtDNA mutations and human aging is the presence of very high levels of mtDNA deletions in the substantia nigra neurons. The substantia nigra is a key part of the basal ganglia, and this system is essential for the control of movement. Loss of cells in the substantia nigra is a sign of Parkinson's disease. By the eighth decade of life, in a high proportion of these neurons, the level of deleted mtDNA reaches 80% of total mtDNA (Fig. 2.5).[17] This is at least an order of magnitude higher than any other cell type studied so far. High levels of clonally expanded deletions result in COX-deficient neurons (Fig. 2.1B). Studies have also be done in brains from patients with

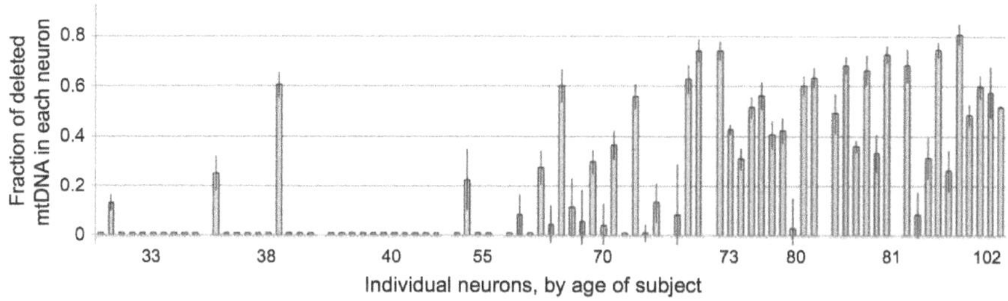

Figure 2.5 Cell-by-cell distribution of clonal expansions of mtDNA deletions in pigmented neurons of substantia nigra at various ages. Every bar represents a single-pigmented neuron isolated at random from human tissue using laser-capture technology. For example, 11 of such cells were isolated and analyzed from brain tissue of one 38-year-old individual. The height of a bar indicates the fraction of deleted mtDNA molecules in a particular cell. Thus, for example, in substantia nigra from the 38-year-old individual, there was one cell with ~25% deletions and another with ~65%, while other cells analyzed from this individual contained no deletions.

Parkinson's disease, and these have shown that there are also high levels of deleted mtDNA but not significantly higher than age-related controls.[35]

There are two obvious questions that arise from this work—why do we see much higher levels of deletions in substantia nigra neurons and are these deletions relevant to aging and/or Parkinson's disease? There have been several studies looking at mtDNA deletion levels in other neuronal populations, and the level in substantia nigra neurons is much higher. This suggests that there is something inherently different about this neuronal subtype. Substantia nigra neurons are constantly firing, and calcium entry through L-type channels in substantia nigra dopaminergic neurons occurs throughout the pacemaking cycle, in contrast to adjacent neurons. Calcium entry during pacemaking comes at a metabolic cost, as it must be extruded by ATP-dependent processes. This demand is met primarily by mitochondria through oxidative phosphorylation. Superoxide and ROS are by-products of oxidative phosphorylation, raising the possibility that calcium entry creates mitochondrial oxidant stress. Recent studies[47] in a transgenic mouse have shown that this is indeed the case with increased oxidant stress that was specific to the substantia nigra neurons. This raises the possibility that the susceptibility of these neurons to mtDNA deletions occurs due to increased damage to the mitochondrial genome. There are other potentially important factors such as the presence of dopamine but this in itself would not explain why the level of deletion is so much higher in these cells compared to other dopaminergic neurons.

The next question arises as to whether the high level of mtDNA deletions and respiratory-deficient neurons has any consequence. As highlighted earlier, we gain some interesting insights into the effect of mtDNA mutations by studying patients with mitochondrial diseases. Patients with defects of polymerase γ accumulate high levels of mtDNA deletions in several postmitotic tissues, and these are thought to be the cause of many of the symptoms in these patients. An important clinical observation in some of these patients is the presence of symptoms very similar to Parkinson's disease with slowness of movement and tremor.[48] Thus, in a disease where high levels of deletions accumulate in the substantia nigra, there are symptoms and this supports a role for the deletions in some symptoms seen in aging. However, the pathogenesis of Parkinson's disease itself is complex with defects in α-synuclein being prominent, and the role of any mitochondrial changes remains speculative.

High levels of mtDNA deletions in pigmented neurons may cause symptoms that are much milder than what is commonly observed in Parkinson's disease itself. Over 50% of individuals who are 85 years and older suffer from such symptoms, and they are familiar to almost anyone who has a very old relative: slowness of movement, tremors, altered gait, collectively called "mild parkinsonian signs."[49] Taken together, the fact that a majority of old individuals carry high levels of mtDNA deletions in their pigmented neurons and that premature increase of the levels of deletions causes a parkinsonian syndrome in patients with polG mutations and makes accumulation of deletions and attractive potential explanation of the pathogenesis of mild parkinsonian signs of the elderly.

5.3. Colonic crypts

The observations in muscle and brain show that clonal expansion and RC deficiency can occur in postmitotic tissues. Similar respiration-deficient cells are seen in mitotic tissues, with perhaps the best example being human colon. The colonic epithelium is a particularly good tissue to explore this phenomenon, since it is formed from stem cells positioned at the base of the colonic crypt. The progeny of the stem cells move up the crypt before being shed into the colonic lumen. The observation of respiration-deficient colonic crypts (Fig. 2.1C) that increase with age (Fig. 2.6) is therefore convincing evidence that mitochondrial defects occur in dividing cells as well as in postmitotic.

There are several interesting observations in the aged human colon relevant to the role of mtDNA mutations in aging. The observation that the

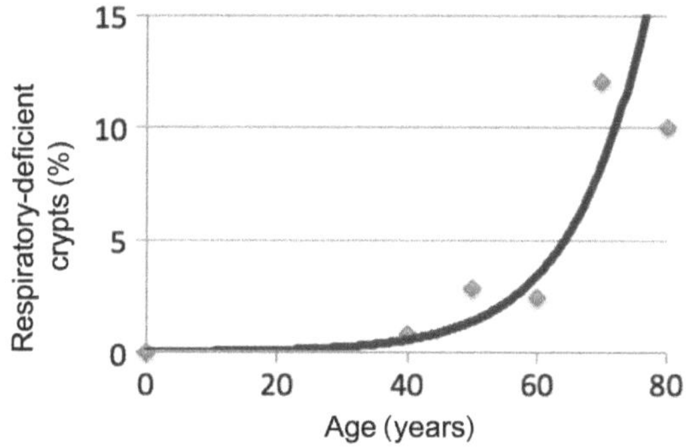

Figure 2.6 The frequency of respiratory-deficient crypts in human colon grows steeply with age. Zero data point was added arbitrary. *Data from Taylor* et al.[43]

whole crypt is often respiration deficient indicates that the mtDNA defect is present in the stem cells, since these are the only cells that can survive long enough to develop the mutations and then populate the crypt. Extensive studies on crypt cells from both COX-normal and -deficient crypts show that there are clonally expanded mtDNA mutations, but in this tissue, these are point mutations and not large-scale deletions. This further confirms that there is a significant difference between tissues with respect to the nature of the mtDNA defect causing the RC defect. However, similar to postmitotic tissues, the mtDNA mutation has clonally expanded in the individual cells to very high levels and high levels of pathogenic mtDNA point mutations correlate with the presence of a RC defect.

The studies in human colon also add information to the discussion on the mechanism of clonal expansion. There is no opportunity for selection based on size, and in this tissue, there are many examples of clonal expansion of neutral mtDNA mutations (such as synonymous changes in protein encoding genes) that have also expanded to high levels in cells with no biochemical defect. The only mechanism to explain this is the random drift hypothesis. The observation that the whole crypt is frequently involved means that the stem cell is the origin of the clonal expansion. As might have been predicted from discussion earlier, the colonic crypt stem cells are quite small, and thus the number of replicating mtDNA molecules is likely to be quite small, which favors rapid genetic drift/clonal expansion, as discussed earlier (Section 4.4).

As with the discussion of the consequence of mtDNA mutations on the aging phenotype in postmitotic cells, we should have the same discussion about the relevance of the observations in human colon and other mitotic tissues. In human colon, this is particularly relevant since on average 15% of all colonic crypts are respiratory deficient in subjects in their mid-80s (Fig. 2.6). Surprisingly, there are relatively minor effects on crypt size, and crypt function is very difficult to measure. It is also intriguing that so many respiratory-deficient crypts accumulate with age, and this of course might mean that these stem cells are pretty resistant to the effects of the mutations. COX-deficient crypts apparently are very long lived, they even have a tendency to divide and create clusters of related COX-deficient crypts sharing the same mutation.[50] Finally, one of the most striking age-related phenotypes in human colon is the increase in colon cancer, which is one of the commonest causes of cancer death in developed countries. As discussed earlier, mtDNA mutations are found in colon cancer cell lines and in colonic tumors. It is not known if the mtDNA mutations in some way promote the development of cancer but this is an area of active research.

An intriguing insight into the possible role of respiratory defects in colon comes from Drosophila aging.[51] Lifespan of the fly appears to be strongly limited by age-related respiratory deficiency in the intestine. Lifespan can be significantly increased by intestine-specific upregulation of mitochondrial biogenesis. Interesting for this discussion is that a prominent aging phenotype of the digestive tract is that it becomes permeable to small molecules such as food dyes. When fed a blue dye, an old fly becomes all blue, while in young flies, the dye remain restricted to the digestive tract. Upregulation of mitochondrial biogenesis alleviates this defect. Therefore, there is a possibility that, similar to the fly, respiratory-deficient crypts in aged human colon may serve as penetration points where unwanted components may get out of the intestine and distort the whole body homeostasis. The model of "penetration points" is interesting conceptually. Similar to muscle, where one respiratory-deficient zone could potentially compromise in the whole muscle fiber, a few respiratory-deficient crypts could compromise the barrier function of the intestine, and thus relatively low fraction of mtDNA mutations can be detrimental.

5.4. Interplay of clonal expansion and *de novo* generation of mtDNA mutations

Importantly, the curve in Fig. 2.6 showing steep increase of *respiratory-deficient cells* does not necessarily imply a steep increase of the *mutational load.* Indeed, lack of respiratory-deficient crypts at young age may be a result of the lack of clonally *expanded* mutations, rather than to lack of mutations *per se*. If clonal expansion is a slow process, as compared to lifespan, then emergence of deficient cells may be limited by expansion of pre-existing mutations, rather than by the *de novo* generation of mutations. These two processes may comprise two synergistic components of the aging process as far as mtDNA mutations are concerned.[18,52] In this way, clonal expansions that cause respiratory deficiency in colonic crypts of a 85-year-old person might have been originally created during early development. This may apply to any other expanded mtDNA mutations,[53] including those observed in muscle fibers and pigmented neurons. In fact, our unpublished data imply that in colon, point mutations comprising expansions in respiratory-deficient crypts may indeed originate at young age.[77] Recent studies have suggested that mtDNA deletions in the brain may also have an early life origin;[54] however, this inference is being debated.[55] A crucial parameter defining interplay between expansion and mutation is the relative rate of expansion.[56] Slow expansion implies early origin of expanded mutations, fast expansion—late, "immediate" origin of expanded mutations. The actual rate of expansion is not known and probably is different in different tissues.

5.5. The "Vicious cycle"

Speaking about the effect of mtDNA mutations in aging, we can't set aside the famous "vicious cycle" concept, which has been influencing the field of mitochondrial mutations for many years. The concept is based on an assumption[57] that mtDNA mutations may cause an increase in ROS production. Unlike the "dominant lethal" mutations model (Section 3.5), "vicious cycle" hypothesis postulates that such increase of ROS does not kill the cell *per se* but instead causes increased generation of mutations, which further increases ROS, resulting in a positive feedback loop. The concept predicts that mtDNA mutations should accumulate exponentially, eventually culminating in an "error catastrophe" identifiable with cellular aging/death. However, the vicious cycle hypothesis is not supported by the data. Most importantly, vicious cycle predicts that a substantial proportion of mutations should be nonclonal, because vicious cycle boosts *de novo* mutation generation rather than expansion of pre-existing mutations. In reality, the fraction of nonexpanded mutations appears to be rather low, at least in case of deletions.[58] Also, a great majority of mutations appear not to increase ROS levels: indeed, there was no ROS increase in heavily mutated tissues of mtDNA mutator mouse.[59] There is no indication that mtDNA mutations increase mutation rates. So if "vicious cycle mutations" (i.e., mtDNA mutations causing increased mtDNA mutation rate) exist, they should be rare, but if so, then a great majority of extra mutations generated because of the presence of the vicious cycle mutations will be of the nonvicious cycle type, and thus, the ability of the cycle to sustain itself is questionable. The vicious cycle concept that dominated in the field for many years is apparently being put to rest.[52,60]

In conclusion, the effects of somatic mtDNA mutations are maximal in limited areas such as pigmented neurons of substantia nigra, respiration-deficient zones of muscle fibers and colonic crypts. It is likely that if mtDNA mutations are somehow involved in the aging process, they act through these critical areas, rather than entire tissues or the whole body.

6. EVOLUTIONARY CONSIDERATIONS AND INTERSPECIES COMPARISONS

Mutational and biochemical analysis of aging tissues described in the preceding text draws a provocative picture of substantially detrimental effects of somatic mitochondrial mutations. Yet, the evidence is not sufficient to conclude whether mtDNA mutations are significant for the aging process. Additional insight into the problem has been gained from evolutionary studies and interspecies comparison.

6.1. Is mitochondrial genome too small?

Detrimental changes caused by mtDNA mutations in aged tissues described in the preceding sections lead to a conceptual problem. mtDNA is a very small genome, about five orders of magnitude (100,000 ×) smaller than the nuclear genome. Because the number of mutations acquired by a DNA segment is expected proportional to its size, the total load of mtDNA somatic mutations is expected to be negligible compared to that of the nuclear genome. If so, why would one even think of any possible impact of mtDNA mutations? Conventional explanation is that mtDNA sustains much higher mutational rates, because it is located close to the RC, one of the major sources of ROS in the cell, lacks protective histones, and has an inferior DNA repair than nuclear DNA.[61] Furthermore, in contrast to nuclear DNA, mtDNA contains almost no nonessential sequences, and therefore, a random mutation in mtDNA is more likely to affect a critical cellular function.[62] The question still remains why do all these factors add up so precisely to allow the tiny mitochondrial genome not to be completely overshadowed by the enormously larger nuclear genome?

The disproportional collective impact of mtDNA mutations (relative to the small mtDNA size) can be explained from the evolution perspective. Note that protective mechanisms that control the rate and/or impact of somatic mutations in mtDNA evolve independently from those of nuclear DNA. For example, enzymes responsible for DNA replication and repair, for "detoxification" of the milieu, and the milieu itself are not shared between the mitochondrial and the nuclear compartments. Other protective mechanisms, such as compensatory upregulation of mtDNA copy number, are also mtDNA specific. Natural selection will only improve mechanisms protecting against mutations if mutations pose a sufficient selective disadvantage, so these mechanisms are expected to be just efficient enough to prevent measurable disadvantage associated with somatic mtDNA mutations. Same is true for nuclear DNA mutations. However, mtDNA maintenance can be much less stringent (in terms of errors per nucleotide) than nuclear and still, the resulting very high mutation rates will pose no disadvantage—because its small genome size *allows* for this luxury. Even with high mutant rates (per nucleotide!), total number of mutations in a small genome can be small and tolerable. In keeping with this, in plants, mitochondrial genomes are much larger than in animals, and, as expected, their mtDNA mutational rates are much lower,[63] as if plants have chosen a different approach than animals: to reduce the mutation rates in their mtDNA rather than reduce mtDNA size.

6.2. Are somatic mtDNA mutations under longevity-related selective pressure?

As we have seen, mutant rates in mtDNA are very high because, from the evolution standpoint, there is no need to make them any lower. If so, why has the burden of mtDNA mutations not evolved to be even higher? Apparently, higher mutation rates would have resulted in sufficient selective disadvantage, which prevents further deterioration of the mtDNA maintenance systems. Intuitively, this alleged disadvantage is expected to be related to aging, because, as we have seen, the impact of somatic mtDNA mutations increases with age. However, this expectation is not necessarily correct. mtDNA mutations can also cause some aging-unrelated problems, e.g., lead to a mutational meltdown due to progressive accumulation of mildly detrimental mutations in successive generations (Mueller's ratchet). While this effect depends on germ line mutations rather than somatic mutations, the two rates may be related because they share mtDNA maintenance machinery. The pressure to avoid Mueller's ratchet might have kept mutations at levels below those relevant to aging. Thus mere fact that mtDNA somatic mutational rates are limited does not mean that mtDNA mutations are of importance for aging.

6.3. MtDNA "mutator" mice: Do they confirm or disprove the mtDNA hypothesis of aging?

A direct way to test whether there is a pressure to keep somatic mtDNA mutations at low levels would be to explore whether an increase of mutational burden results in any obvious disadvantage. This experiment was carried out by creating a genetically modified mouse with a defective error-prone mitochondrial polymerase gamma (polG) gene. When both alleles in the gene are defective, this "mtDNA mutator" mouse accumulates about two dozen mutations per mtDNA molecule and presents with a profound aging-like phenotype.[59,64,65] Unfortunately, while this mouse is a very useful model for studying somatic mtDNA mutations, the overall level of mutations is so much higher than in normal animals, that this result cannot be used to corroborate the involvement of mtDNA mutations in aging. Obviously, high enough mutant loads are detrimental, but this does not mean that moderate loads characteristic of old age are detrimental, too.[66] Moreover, the "heterozygotic mutator" mouse, which carries only *one* defective polG allele and accumulates "only" about five mutations per mtDNA molecule, shows no accelerating aging. Because this mutational load is still much more than a normal mouse does, this seems to imply that mtDNA point mutations have nothing to do with aging.[4]

In addition to point mutations, mtDNA mutator mouse has increased rates of mtDNA deletions. Interestingly, increased deletions appear to coincide with the premature aging phenotype, i.e., they are increased in homozygous mutator only. Does this mean that deletions are causing aging, while point mutations are irrelevant[67]?

In fact, the data from mutator mouse merely demonstrate the complexity of the problem. First, different species may not age similarly. Indeed, as we have seen before, in humans, colon perhaps is one of the tissues most affected by point mutations with ~15% of affected crypts (Section 5.3). In contrast, in the old normal mouse, colon contains very low number of affected crypts (0.75%).[68] Similarly, murine substantia nigra appears to be much less (about an order of magnitude) affected by mtDNA deletions than human[69]! It is therefore possible that despite increased mutant load, heterozygous mutator mouse may still fail to significantly surpass mutational impact of the aged human. Indeed, there is about 20% RC-deficient crypts in the aged heterozygous mutator mouse,[4] which is close to normal aging human. Perhaps, this makes mutator mouse a good model for studies of effects of age-related RC-deficient crypts in human intestine, but this also implies that lack of a dramatic premature aging phenotype in this mouse does not exclude involvement of point mutations in human aging.[70]

With respect to mtDNA deletions, while their load is indeed relatively increased in the mutator mouse, their absolute overall load is quite low. For example, in the mutator mouse brain, there is about 1 mtDNA deletion per 1000 point mutations.[71] It is highly unlikely therefore that the profound phenotype of the mutator mouse is caused by deletions, rather than by greatly overwhelming point mutations. The example of the mtDNA mutator mouse teaches that extreme caution should be exercised in interpreting seemingly clear-cut data.

6.4. Longevity-related sequence traits in mtDNA

Several features of mtDNA sequence, i.e., GC content and the number of repeats, prominently correlate with species' longevity. Naturally, this has been interpreted in favor of specific *longevity-related* selective pressure on mtDNA mutations and of mtDNA mutational hypothesis of aging. These interpretations, however, are subject to debate.

Perhaps the most intriguing observation is a negative correlation between the species' longevity and the frequency of direct repeats in the mitochondrial genome.[72,73] Most recently, an even stronger correlation has been discovered between the frequency of inverted repeats and species longevity.[74] The most parsimonious explanation of this correlation is that because repeats are "mutagenic," i.e., mtDNA deletions arise most readily at these sites (Section 2),[74] and thus reduction of the number of repeats should reduce the rate of deletions. Fewer repeats, the argument goes, are instrumental to keep deletion burden, which rapidly increase with age, at sustainable level in long-lived species. Therefore mutations are important to the aging process. There may be, however, alternative explanations. For example, repeats may be maintained in mtDNA for some purpose (indeed, some of them are relatively conserved, and the number of repeats in mtDNA well exceeds what is expected by chance). If so, a mutation destroying a repeat can be considered mildly detrimental. Long-lived species typically have smaller populations, where because of stronger genetic drift (Section 4.2) fixation of mildly detrimental mutations is easier than in large populations. This could lead to the observed depletion of repeats in long-lived species.

Another property of the mtDNA that correlates very strongly with longevity is the GC content, of which most variable is the G-content.[75,76] It has been suggested[75] that higher GC content affects aging by increasing thermal stability of DNA. More thermostable DNA is less likely to temporarily open "bubbles" of single-stranded DNA and thus presumably is more resistant to chemical damage (because single-stranded DNA is far more chemically reactive) and, therefore, accumulates fewer damage-driven mutations.

Alternative explanation, however, is that in long-lived species, the chemical environment is likely less aggressive (e.g., because this ensures slower damage accumulation presumably required for slower aging). As we have seen earlier (Section 2), the most pressing damage-related mutation in mtDNA is the G to A change likely resulting from deamination of C. Less aggressive environment may mean lower mutational pressure on the G, which could account for the observed higher GC content in long-lived species. Note that in this explanation, mtDNA content is considered merely a marker of chemical damage not a causative factor.

6.5. Extended human lifespan and growing contribution of somatic mtDNA mutations

As we have seen, many lines of evidence suggest that mtDNA mutations are limiting lifespan, but none is truly convincing. Apparently, somatic mutations in mtDNA are easy for natural selection to keep in check. Indeed, animals in the wild typically do not suffer from age-related degenerative disease of a mtDNA mutational cause within their natural lifespan. Note, however, (Section 5) that mtDNA somatic mutations, and even more so—clonal expansions of these mutations, follow rather steep and accelerating accumulation curve (Figs. 2.5 and 2.6). We thus may expect that mtDNA mutations that are kept in check during our natural lifespan will probably swiftly get out of control as human life expectancy in modern world significantly exceeds its "natural" duration. Of note, in contrast to major old-age killers like cancer and cardiovascular disease, where vast scientific effort results in new efficient treatments, there is little if any progress in treating diseases caused by mtDNA mutations. We therefore expect that as human longevity keeps increasing, the importance of mtDNA mutations as a potential component of the aging process may relatively increase.

ACKNOWLEDGMENTS

The authors are indebted to David Samuels, Konstantin Popadin, and Laura Greaves for their valuable discussions and Catherine MacLean and SaiSai Tao for their critical comments and

Evan Feldman for help with data analysis. K. K. was supported by the Ellison Medical Foundation. D. T. was supported by Newcastle University Centre for Ageing and Vitality (supported by the BBSRC and MRC [G0700718]), The Wellcome Trust Centre for Mitochondrial Research [G906919], and UK NIHR Biomedical Research Centre in Age and Age Related Diseases award to the Newcastle upon Tyne Hospitals NHS Foundation.

REFERENCES

1. Harman D. The biologic clock: the mitochondria? *J Am Geriatr Soc.* 1972;20(4): 145–147.
2. Linnane AW, Marzuki S, Ozawa T, Tanaka M. Mitochondrial DNA mutations as an important contributor to ageing and degenerative diseases. *Lancet.* 1989;1(8639):642–645.
3. Terman A, Gustafsson B, Brunk UT. Autophagy, organelles and ageing. *J Pathol.* 2007;211(2):134–143.
4. Vermulst M, Bielas JH, Kujoth GC, et al. Mitochondrial point mutations do not limit the natural lifespan of mice. *Nat Genet.* 2007;39(4):540–543.
5. DiMauro S, Hirano M, Schon EA. Approaches to the treatment of mitochondrial diseases—invited review. *Muscle Nerve.* 2006;34(3):265–283.
6. Archer SL. Mitochondrial dynamics—mitochondrial fission and fusion in human diseases. *N Engl J Med.* 2013;369(23):2236–2251.
7. Longo DL, Archer SL. Mitochondrial dynamics—mitochondrial fission and fusion in human diseases. *N Engl J Med.* 2013;369(23):2236–2251.
8. Boore JL. Animal mitochondrial genomes. *Nucleic Acids Res.* 1999;27(8):1767–1780.
9. Adams KL, Palmer JD. Evolution of mitochondrial gene content: gene loss and transfer to the nucleus. *Mol Phylogenet Evol.* 2003;29(3):380–395.
10. Kukat C, Wurm CA, Spåhr H, Falkenberg M, Larsson N-G, Jakobs S. Super-resolution microscopy reveals that mammalian mitochondrial nucleoids have a uniform size and frequently contain a single copy of mtDNA. *Proc Natl Sci U S A.* 2011;108(33): 13534–13539.
11. Zheng W, Khrapko K, Coller HA, Thilly WG, Copeland WC. Origins of human mitochondrial point mutations as DNA polymerase gamma-mediated errors. *Mutat Res.* 2006;599(1–2):11–20.
12. Marcelino LA, Andre PC, Khrapko K, Coller HA, Griffith J, Thilly WG. Chemically induced mutations in mitochondrial DNA of human cells: mutational spectrum of N-methyl-N'-nitro-N-nitrosoguanidine. *Cancer Res.* 1998;58(13):2857–2862.
13. Krishnan KJ, Reeve AK, Samuels DC, et al. What causes mitochondrial DNA deletions in human cells? *Nat Genet.* 2008;40(3):275–279.
14. Fukui H, Moraes CT. Mechanisms of formation and accumulation of mitochondrial DNA deletions in aging neurons. *Hum Mol Genet.* 2009;18(6):1028–1036.
15. Guo X, Popadin K, Markuzon N, et al. Repeats, longevity and the sources of mtDNA deletions: evidence from "deletional spectra" *Trends Genet.* 2010;26(8):340–343.
16. Kreutzer DA, Essigmann JM. Oxidized, deaminated cytosines are a source of C → T transitions in vivo. *Proc Natl Acad Sci U S A.* 1998;95(7):3578–3582.
17. Kraytsberg Y, Kudryavtseva E, McKee AC, Geula C, Kowall NW, Khrapko K. Mitochondrial DNA deletions are abundant and cause functional impairment in aged human substantia nigra neurons. *Nat Genet.* 2006;38(5):518–520.
18. Kraytsberg Y, Nekhaeva E, Bodyak NB, Khrapko K. Mutation and intracellular clonal expansion of mitochondrial genomes: two synergistic components of the aging process? *Mech Ageing Dev.* 2003;124(1):49–53.
19. Nakada K, Inoue K, Ono T, et al. Inter-mitochondrial complementation: mitochondria-specific system preventing mice from expression of disease phenotypes by mutant mtDNA. *Nat Med.* 2001;7(8):934–940.

20. Moraes CT, Schon EA. +Detection and analysis of mitochondrial DNA and RNA in muscle by in situ hybridization and single-fiber PCR "Met Enzymol". *Methods Enzymol.* 1996;264:522–540.
21. Rossignol R, Faustin B, Rocher C, Malgat M, Mazat JP, Letellier T. Mitochondrial threshold effects. *Biochem J.* 2003;370(Pt 3):751–762.
22. Gilkerson RW, Schon EA, Hernandez E, Davidson MM. Mitochondrial nucleoids maintain genetic autonomy but allow for functional complementation. *J Cell Biol.* 2008;181(7):1117–1128. http://dx.doi.org/10.1083/jcb.200712101.
23. Herbst A, Pak JW, McKenzie D, Bua E, Bassiouni M, Aiken JM. Accumulation of mitochondrial DNA deletion mutations in aged muscle fibers: evidence for a causal role in muscle fiber loss. *J Gerontol A Biol Sci Med Sci.* 2007;62(3):235–245.
24. Sacconi S, Salviati L, Nishigaki Y, et al. A functionally dominant mitochondrial DNA mutation. *Hum Mol Genet.* 2008;17(12):1814–1820. http://dx.doi.org/10.1093/hmg/ddn073.
25. Bai Y, Attardi G. The mtDNA-encoded ND6 subunit of mitochondrial NADH dehydrogenase is essential for the assembly of the membrane arm and the respiratory function of the enzyme. *EMBO J.* 1998;17(16):4848–4858.
26. Fan W, Waymire KG, Narula N, et al. A mouse model of mitochondrial disease reveals germline selection against severe mtDNA mutations. *Science.* 2008;319(5865):958–962.
27. Dubec SJ, Aurora R, Zassenhaus HP. Mitochondrial DNA mutations may contribute to aging via cell death caused by peptides that induce cytochrome c release. *Rejuvenation Res.* 2008;11(3):611–619.
28. Schwarze SR, Lee CM, Chung SS, Roecker EB, Weindruch R, Aiken JM. High levels of mitochondrial DNA deletions in skeletal muscle of old rhesus monkeys. *Mech Ageing Dev.* 1995;83(2):91–101.
29. Johnston W, Karpati G, Carpenter S, Arnold D, Shoubridge EA. Late-onset mitochondrial myopathy. *Ann Neurol.* 1995;37(1):16–23.
30. Khrapko K. Two ways to make an mtDNA bottleneck. *Nat Genet.* 2008;40(2):134–135.
31. Gross NJ, Getz GS, Rabinowitz M. Apparent turnover of mitochondrial DNA and mitochondrial phospholipids in the tissues of the rat. *J Biol Chem.* 1969;244(6):1552–1562.
32. Michikawa Y, Mazzucchelli F, Bresolin N, Scarlato G, Attardi G. Aging-dependent large accumulation of point mutations in the human mtDNA control region for replication. *Science.* 1999;286(5440):774–779.
33. Wang Y, Michikawa Y, Mallidis C, et al. Muscle-specific mutations accumulate with aging in critical human mtDNA control sites for replication. *Proc Natl Acad Sci U S A.* 2001;98(7):4022–4027.
34. Kowald A, Dawson M, Kirkwood TBL. Mitochondrial mutations and ageing: can mitochondrial deletion mutants accumulate via a size based replication advantage? *J Theor Biol.* 2014;340:111–118. http://dx.doi.org/10.1016/j.jtbi.2013.09.009.
35. Bender A, Krishnan KJ, Morris CM, et al. High levels of mitochondrial DNA deletions in substantia nigra neurons in aging and Parkinson disease. *Nat Genet.* 2006;38(5):515–517.
36. Campbell G, Krishnan KJ, Deschauer M, Taylor RW, Turnbull DM. Dissecting the mechanisms underlying the accumulation of mitochondrial DNA deletions in human skeletal muscle. *Hum Mol Genet.* 2014; http://dx.doi.org/10.1093/hmg/ddu176.
37. Kowald A, Kirkwood TBL. Transcription could be the key to the selection advantage of mitochondrial deletion mutants in aging. *Proc Natl Acad Sci U S A.* 2014;111(8):2972–2977.
38. Kowald A, Kirkwood TBL. Evolution of the mitochondrial fusion-fission cycle and its role in aging. *Proc Natl Acad Sci U S A.* 2011;108(25):10237–10242.

39. Avise J, Neigel J, Arnold J. Demographic influences on mitochondrial DNA lineage survivorship in animal populations. *J Mol Evol*. 1984;20(2):99–105.
40. Coller HA, Bodyak ND, Khrapko K. Frequent intracellular clonal expansions of somatic mtDNA mutations: significance and mechanisms. *Ann N Y Acad Sci*. 2002;959:434–447.
41. Jacobs H. Dinosaurs, dementia and cancer. *Trends Genet*. 2000;16:295.
42. Nekhaeva E, Bodyak ND, Kraytsberg Y, et al. Clonally expanded mtDNA point mutations are abundant in individual cells of human tissues. *Proc Natl Acad Sci U S A*. 2002;99(8):5521–5526.
43. Taylor RW, Barron MJ, Borthwick GM, et al. Mitochondrial DNA mutations in human colonic crypt stem cells. *J Clin Invest*. 2003;112(9):1351–1360.
44. Chinnery PF, Samuels DC. Relaxed replication of mtDNA: a model with implications for the expression of disease. *Am J Hum Genet*. 1999;64(4):1158–1165.
45. Inoue K, Nakada K, Ogura A, et al. Generation of mice with mitochondrial dysfunction by introducing mouse mtDNA carrying a deletion into zygotes. *Nat Genet*. 2000;26(2):176–181.
46. Wanagat J, Cao Z, Pathare P, Aiken JM. Mitochondrial DNA deletion mutations colocalize with segmental electron transport system abnormalities, muscle fiber atrophy, fiber splitting, and oxidative damage in sarcopenia. *FASEB J*. 2001;15(2):322–332.
47. Guzman JN, Sanchez-Padilla J, Wokosin D, et al. Oxidant stress evoked by pacemaking in dopaminergic neurons is attenuated by DJ-1. *Nature*. 2010;468(7324):696–700.
48. Luoma P, Melberg A, Rinne JO, et al. Parkinsonism, premature menopause, and mitochondrial DNA polymerase gamma mutations: clinical and molecular genetic study. *Lancet*. 2004;364(9437):875–882.
49. Bennett DA, Beckett LA, Murray AM, et al. Prevalence of parkinsonian signs and associated mortality in a community population of older people. *N Engl J Med*. 1996;334(2):71–76.
50. Greaves LC, Preston SL, Tadrous PJ, et al. Mitochondrial DNA mutations are established in human colonic stem cells, and mutated clones expand by crypt fission. *Proc Natl Acad Sci U S A*. 2006;103(3):714–719.
51. Rera M, Bahadorani S, Cho J, et al. Modulation of longevity and tissue homeostasis by the Drosophila PGC-1 homolog. *Cell Metab*. 2011;14(5):623–634.
52. Elson J, Samuels D, Turnbull D, Chinnery P. Random intracellular drift explains the clonal expansion of mitochondrial DNA mutations with age. *Am J Hum Genet*. 2001;68(3):802–806.
53. Khrapko K, Nekhaeva E, Kraytsberg Y, Kunz W. Clonal expansions of mitochondrial genomes: implications for in vivo mutational spectra. *Mutat Res*. 2003;522(1–2):13–19.
54. Taylor SD, Ericson NG, Burton JN, et al. Targeted enrichment and high-resolution digital profiling of mitochondrial DNA deletions in human brain. *Aging Cell*. 2014;13(1):29–38.
55. Popadin K, Safdar A, Kraytsberg Y, Khrapko K. When man got his mtDNA deletions? *Aging Cell*. 2014 http://dx.doi.org/10.1111/acel.12231.
56. Khrapko K. The timing of mitochondrial DNA mutations in aging. *Nat Genet*. 2011;43(8):726–727.
57. Bandy B, Davison AJ. Mitochondrial mutations may increase oxidative stress: implications for carcinogenesis and aging? *Free Radic Biol Med*. 1990;8(6):523–539.
58. Nicholas A, Kraytsberg Y, Guo X, Khrapko K. On the timing and the extent of clonal expansion of mtDNA deletions: evidence from single-molecule PCR. *Exp Neurol*. 2009;218(2):316–319.
59. Kujoth GC, Hiona A, Pugh TD, et al. Mitochondrial DNA mutations, oxidative stress, and apoptosis in mammalian aging. *Science*. 2005;309(5733):481–484.

60. Hiona A, Leeuwenburgh C. The role of mitochondrial DNA mutations in aging and sarcopenia: implications for the mitochondrial vicious cycle theory of aging. *Exp Gerontol.* 2008;43(1):24–33.
61. Weissman L, de Souza-Pinto NC, Stevnsner T, Bohr VA. DNA repair, mitochondria, and neurodegeneration. *Neuroscience.* 2007;145(4):1318–1329.
62. Lander ES, Lodish H. Mitochondrial diseases: gene mapping and gene therapy. *Cell.* 1990;61(6):925–926.
63. Adams KL, Daley DO, Qiu YL, Whelan J, Palmer JD. Repeated, recent and diverse transfers of a mitochondrial gene to the nucleus in flowering plants. *Nature.* 2000;408(6810):354–357.
64. Zhang D, Mott JL, Chang SW, Denniger G, Feng Z, Zassenhaus HP. Construction of transgenic mice with tissue-specific acceleration of mitochondrial DNA mutagenesis. *Genomics.* 2000;69(2):151–161.
65. Trifunovic A, Wredenberg A, Falkenberg M, et al. Premature ageing in mice expressing defective mitochondrial DNA polymerase. *Nature.* 2004;429(6990):417–423.
66. Khrapko K, Kraytsberg Y, de Grey ADNJ, Vijg J, Schon EA. Does premature aging of the mtDNA mutator mouse prove that mtDNA mutations are involved in natural aging? *Aging Cell.* 2006;5(3):279–282.
67. Vermulst M, Wanagat J, Kujoth GC, et al. DNA deletions and clonal mutations drive premature aging in mitochondrial mutator mice. *Nat Genet.* 2008;40(4):392–394.
68. Greaves LC, Barron MJ, Campbell-Shiel G, Kirkwood TB, Turnbull DM. Differences in the accumulation of mitochondrial defects with age in mice and humans. *Mech Ageing Dev.* 2011;132(11–12):588–591.
69. Guo X, Kudryavtseva E, Bodyak N, et al. Mitochondrial DNA deletions in mice in men: substantia nigra is much less affected in the mouse. *Biochim Biophys Acta.* 2010;1797(6–7): 1159–1162.
70. Khrapko K, Vijg J. Mitochondrial DNA mutations and aging: a case closed? *Nat Genet.* 2007;39(4):445–446.
71. Kraytsberg Y, Simon DK, Turnbull DM, Khrapko K. Do mtDNA deletions drive premature aging in mtDNA mutator mice? *Aging Cell.* 2009;8(4):502–506.
72. Samuels DC. Mitochondrial DNA, repeats constrain the life span of mammals. *Trends Genet.* 2004;20(5):226–229.
73. Khaidakov M, Siegel ER, Shmookler Reis RJ. Direct repeats in mitochondrial DNA and mammalian lifespan. *Mech Ageing Dev.* 2006;127(10):808–812.
74. Yang J-N, Seluanov A, Gorbunova V. Mitochondrial inverted repeats strongly correlate with lifespan: mtDNA inversions and aging. *PLoS One.* 2013;8(9):e73318.
75. Samuels DC. Life span is related to the free energy of mitochondrial DNA. *Mech Ageing Dev.* 2005;126(10):1123–1129.
76. Lehmann G, Segal E, Muradian KK, Fraifeld VE. Do mitochondrial DNA and metabolic rate complement each other in determination of the mammalian maximum longevity? *Rejuvenation Res.* 2008;11(2):409–417.
77. Laura C Greaves, Marco Nooteboom, Joanna L Elson, Helen AL Tuppen, Geoffrey A Taylor, Daniel M Commane, Ramesh P Arasaradnam, Konstantin Khrapko, Robert W Taylor, Thomas BL Kirkwood, John C Mathers, Douglass M Turnbull. Clonal expansion of early to mid-life mitochondrial DNA point mutations drives mitochondrial dysfunction during human ageing. Under review in Plos Genetics.
78. Khrapko K, Vijg J. Mitochondrial DNA mutations in Aging: devils in the details. *Trends in Genet.* 2009;25:91–98.

CHAPTER THREE

Mathematical Models of Mitochondrial Aging and Dynamics

Axel Kowald, Edda Klipp
Theoretical Biophysics, Institute for Biology, Humboldt-Universität zu Berlin, Berlin, Germany

Contents

Abstract

Research on the role of mitochondria in aging and disease is rapidly growing. Furthermore, in recent years, it also became clear that mitochondria are dynamic structures undergoing constant and rapid cycles of fusion and fission. The involvement of mitochondria in multiple complex processes makes them a prime target for mathematical and computational modeling. This review consists of two parts. In the first (Section 2), we provide a detailed introduction to the underlying concepts of mathematical modeling to help the reader who is not so familiar with these techniques to judge the requirements and results that can be obtained through modeling. In the second part (Section 3), we review existing mathematical and computational models that investigate mitochondrial dynamics and the role of mitochondria for the aging process.

Progress in Molecular Biology and Translational Science, Volume 127
ISSN 1877-1173
http://dx.doi.org/10.1016/B978-0-12-394625-6.00003-9

1. INTRODUCTION

Mitochondria are eukaryotic organelles that originated roughly two billion years ago from free living prokaryotic ancestors.[1,2] A remnant from this event is the observation that mitochondria still contain their own genetic material in form of a small circular DNA, called mtDNA. During the course of evolution, most mitochondrial genes have been transferred to the nucleus, but genes for 13 proteins are still located on the human mtDNA.[3] All of these proteins are involved in the process of oxidative phosphorylation (OXPHOS), which takes place in the inner mitochondrial membrane and generates the majority of the cellular energy in form of ATP. All other proteins that are required for a proper functioning of mitochondria are nowadays encoded in the nucleus, synthesized in the cytoplasm, and imported into the mitochondria by a specialized machinery.

But mitochondria are not only the powerhouses of the cell, they are also involved in other important processes like apoptosis, calcium homeostasis, and heme synthesis. Since mitochondria are involved in such a multitude of biochemical pathways, impaired mitochondrial functioning could in principle have complex and severe consequences for the cell and the whole organism. However, eukaryotic cells contain hundreds or thousands of these organelles and it is therefore not obvious to see how damage to a single mitochondrion can affect the functioning of the whole population. And since mitochondria are constantly degraded via mitophagy (even in postmitotic cells), it was generally believed that damaged mitochondria will be removed via this turnover and that mitochondria are therefore damage resistant.[4] This view has by now radically changed since it has been discovered that mtDNA mutations are at the heart of many human diseases like Leber's Hereditary Optic Neuroretinopathy, Myoclonic Epilepsy associated with Ragged-Red Fibers, Mitochondrial Encephalomyopathy Lactic Acidosis and Stroke-like episodes, Kearns–Sayre Syndrome, or Chronic Progressive External Ophthalmoplegia.[5,6] In these diseases, mtDNA molecules with point mutations, partial deletions, or duplications are present in such large numbers per cell that they cause pathological conditions and important areas of research are the mechanisms that allow these mutations to accumulate to such high numbers and to evade the mitochondrial quality assurance system. Obviously, the sheer number of mtDNA molecules that exists per cell does not prevent the spread of disease variants.

Intriguingly, damage to mtDNA is not only involved in the development of mitochondrial diseases, but it has also been implicated with the aging process and forms the core of the mitochondrial theory of aging.[7–10] Many studies have now shown that damage to the mtDNA accumulates with age in various mammalian species such as rats, monkeys, and humans at the single-cell level.[11–16] The most prominent type of damage seems to be deletion mutations, which can have lost up to 50% of the total mtDNA sequence. Importantly, when early studies were performed on tissue homogenates the overall level of damaged mtDNAs was very low and only when studies were performed at the single-cell level it became clear that some cells harbor a large fraction of mtDNA copies with deletions, while others contain seemingly no mutants at all. Furthermore, those cells that do contain deletion mutants are taken over by a single mutant type. However, this clonally expanded mutant differs for different cells. Similarly, also clonally expanded deleterious point mutations have been shown to accumulate with age in some cell types such as colonic crypt stem cells.[17–19]

The large number of interacting reactions and pathways that involve mitochondria together with the limited amount of experimental knowledge makes it very challenging to understand the dynamics of this complex system. This is a situation in which mathematical modeling can be a useful tool to explore the kinetics and feasibility of the various ideas and hypotheses. In Section 2, we will provide an introduction to the basic concepts of mathematical modeling. This is a useful and necessary prerequisite for those who are not so familiar with the techniques of computer simulations and systems biology. In the remainder, we will then discuss mathematical and computational models that have been developed to describe different aspects of mitochondrial biochemistry and debate their relevance for aging and mitochondrial dynamics.

2. FUNDAMENTALS OF MATHEMATICAL MODELING

The development of computational modeling methods in systems biology is driven by a number of challenging questions. Those questions draw a bow from specific ones—such as: What is the reason for the observed oscillations? What is the consequence of this specific mutation?—to relatively general ones—such as: Do our observations make sense within the big picture? In any case, theoretical models are helpful to make abstractions, to formalize observations and hypotheses, and to make testable predictions. They may also help to formally prove or disprove mechanistic assumptions about the potential behavior of the system under study. Modelers apply a selection of techniques such as logical modeling and Boolean networks, Petri nets, ordinary or partial differential equations, or flux-balance modeling to study molecular networks in cells. Simulation strategies may be deterministic or stochastic. The mentioned approaches vary drastically in their level of difficulty and in the amount of data needed to train the model and estimate its parameters. Choosing the most appropriate method to model a given biological system is often difficult because each method has its particular strengths and may be easier or more complicated to apply to available data and more or less useful to answer a limited and specific set of questions. Below we will discuss the basic principles as well as merits and drawbacks of modeling techniques and how they are applied to research about mitochondria in health and disease.

At this stage, it is also important to recall that every useful and predictive model requires the confrontation with biological data. Thus, the intensive discussion between the theoretician and the experimentalist is often of greater importance than the final streamlined model. The process has been frequently called "the iterative cycle of systems biology," starting with a set of data or biological observations which lead to the formulation of a first model, followed by predictions and experiments to prove the predications, update of the model and so on.

A further important aspect is that the quality of models is largely determined by the available data. Currently, we witness a revolution in the modern techniques of molecular genetics revealing all the time new types of data. Especially interesting are here—omics data and all types of whole-genome data. On the other side, models often require precise information for one detailed mechanism or event—so it is not always easy to find the right set of data that are useful for the envisaged model. For the modeler, there are two major sources of data: one source is the direct communication with the experimentalist investigating the biological object. S/he does not only know the currently measured data, but also all the details about the executed experiments, the used model organism or cell line and the little tricks applied to make the experiments work. These details may be decisive for the accuracy and, hence, the success of the model. The second source are databases of all kinds, including the results of measurements as well as derived information such as kinetic parameters (e.g., BRENDA[20]), network reconstructions (such as Reactome[21]) or already existing models (e.g., Biomodels,[22] JWSonline[23,24]).

2.1. Reconstructing networks

Biological networks reflect the regulatory and the functional interactions between molecular components (genes, proteins, and metabolites) but may also be extended to integrate information on cellular behavior and physiological impact. Examples for networks relevant for mitochondria may be (i) the physical interactions of all proteins contained in the mitochondria; (ii) the metabolic network connecting the metabolites transformed in the tricarboxylic acid cycle, an important pathway in cellular energy metabolism; and (iii) regulatory interactions between the genes encoded on the mitochondrial genome. Clearly, depending on the research question, we could identify many more different networks.

Networks usually present very heterogeneous types of information. For example, the transfer of information through regulatory interactions can be distinguished from the transfer of mass during metabolic reactions. The available information on these interactions varies a lot between different systems. For some well-studied model organisms such as *E. coli*, the yeasts *S. cerevisiae* or *S. pombe*, the plant *A. thaliana* or mice, we have rather comprehensive knowledge, while other organisms are largely unexplored. For the human—which is the interesting object to study under the light of mitochondria and disease—information usually stems from a set of cell lines, which have been kept *in vitro* under specific conditions for ages and which are presumably very different from the *in vivo* cells in their tissue environments. However, these are constraints under which predictive models have to be built.

The construction of networks for modeling may proceed in different ways. The classical way is based on molecular biology testing the relation between cause and effect and then concluding that a certain compound has its role in connecting the two. Thus, if we first stimulate a receptor and measure the activity change of a transcription factor and, second, repeat this experiment after knockout of a kinase that we guessed to be involved and find a different response of the transcription factor, we may conclude a role of this kinase in the pathway. Biochemists and molecular biologists have carried out innumerable experiments like this and have elegantly combined the evidences to eventually define cellular pathways. Still, this picture is not yet complete.

Another means of constructing a network is to list all known pairwise interactions of components. An example is the set of all protein–protein interactions measured, for instance, with the Yeast2Hybrid technique.

Sequencing of whole organisms has revealed information on all potential genes, including those coding for metabolic enzymes. By adding the information about their potential substrates and products, one may arrive at a solid metabolic network. The interaction of regulatory proteins with DNA sequences has revealed a number of potential transcription factors. Combining these transcription factors and their target genes leads eventually to a transcriptional regulation network.

An alternative approach for network reconstruction is systematic literature research. Here, the co-occurrence of compounds in scientific articles, weighted according by frequency, is used as an indication that there are also physical interactions between these compounds. For well-studied pathways, this approach can lead to a quite reliable reproduction of the already known structure.

2.2. Logical models with focus on Boolean modeling

Networks can be constructed as systems composed of multiple nodes and edges between them. Each node is associated with a state characterizing either the abundance of the represented molecules or its activation. State changes depend on the own state and on the states of nodes that are connected to the respective node. Logical models restrict the set of possible states of a node to only a few—typically 2—different possibilities. These can present pairs such as present/absent, active/inactive, ON/OFF, expressed/not expressed, or simply 0/1. Such a choice is especially useful if we have only qualitative experimental data available regarding the activity of, e.g., genes and/or proteins. The update of states in time occurs based on the state of the input nodes and on logic rules combining the information of several input nodes such as AND, OR, NOT, or combinations thereof. Figure 3.1 gives an intuitive example of such a network, its rules and dynamics.

This type of network is known as Boolean network, a special type of binary network. Boolean networks were first presented by Kauffman[25,26] striving for a qualitative description of the concerted action of a group of genes during cellular differentiation. Such models were originally developed as a suitable simplification for the analysis of genetic regulation, and were initially studied exclusively from a statistical point of view[27] due to the lack of biological data on experimentally validated biological networks. More recently, Boolean network models have been developed for a series of biological systems showing the suitability of this methodology to capture key aspects of cellular development and differentiation.[28–37]

A

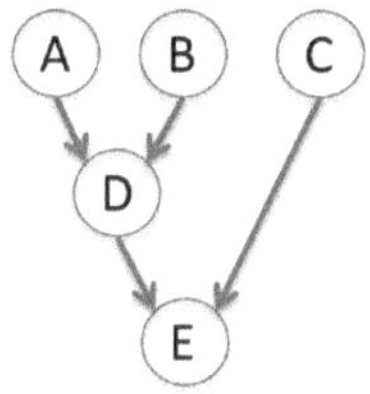

B

A(t+1)=NOT(A(t))
B(t+1)=B(t)
C(t+1)=OFF
D(t+1)=A(t) AND B(t)
E(t+1)=C(t) OR D(t)

C

t	A	B	C	D	E
0	1	1	0	1	1
1	0	1	0	0	1
2	1	1	0	1	0
3	0	1	0	0	1
4	1	1	0	1	0

Figure 3.1 Regulatory network example. (A) The network has six nodes, which could represent activity states of genes. Node A, B, and C are input nodes. A and B together regulate node D. C and D together regulate node E. (B) Boolean rules determining the update of states. Node A is alternatingly ON or OFF (if ON at time *t*, then OFF at time $t+1$). Node B remains at its current state and node C remains OFF. Node D switches ON if both A and B are ON, while node E switches on if C or D (or both) is ON, (C) Temporal evolution of states. The initial state at $t=0$ are given, the following states at $t=1$, $t=2$ and so on are computed from the rules given in (B).

In relation to mitochondria, Sridharan and colleagues developed a Boolean model to conceptualize the accumulated knowledge on the oxidative stress response pathways.[38] By concatenating the oxidative stress response pathway with the PI3-kinase–Akt pathway, they linked oxidative stress to the phenotype of apoptosis. OXPHOS is a metabolic process, which takes place in mitochondria in which ATP is formed as a result of the transfer of electrons from NADH or $FADH_2$ to O_2 by a series of electron carriers and in which 1–2% of electrons are lost during this transfer leading to free radical generation. This ROS production may, in turn, cause genetic instability and cancer.

Boolean models put into focus the influence of the network topology on its dynamical behavior. They provide an easy way to study the impact of positive feedback loops or long paths on the system behavior. These types of network motifs can give rise to interesting phenomena such as multistationarity, damped or sustained oscillations and homeostasis, which are particularly important properties of biological systems.[39–43] The biological interpretation of multistationarity as cellular differentiation, however, goes back already to Delbrück.[44]

Modelers appreciate Boolean networks for their computational tractability and their capability of providing qualitatively correct results. They have, of course, limitations, first of all the discrete description of time and states. In the basic version, states are synchronously updated according to a ticking clock. Since this may not be ensured in the biological counterpart, asynchronous update rules have been introduced. Still the problem of finding the right temporal scheme remains. Moreover, the state space of the whole system becomes much larger compared to synchronous update. In addition to these types of models, Random Boolean models have been introduced, where the update rules for each node at each time are chosen randomly. In Probabilistic Boolean networks, update rules are chosen with a certain probability out of a given set of rules (Ref. 45 with an application presented in Ref. 46).

Boolean models can be considered as a discontinuous step function approximating the smoother dynamics of a continuous system. The other way around, we can also translate Boolean models into continuous models (which are further explained below) by assigning continuous kinetic functions arriving finally at a set of ordinary differential equations (ODEs). These kinetic functions describe the rate of change of activation (or synthesis, or transcription) as a sigmoid function of the state of activation of the controlling input variables. If we normalize the values of the variables such that they are constrained in the range [0,1], then we can compare the continuous dynamics to the dynamics of the binary Boolean simulation.[47] To date, there are a number of tools available allowing for automated analysis of biochemical networks such as Squad (http://www.enfin.org/squad) or CellNOptR.[48]

2.3. Continuous modeling—The basics of network dynamics described with ordinary differential equations

Despite rapid development of new modeling approaches together with newly arising types of data, systems of ODEs are still the most frequently applied techniques for modeling dynamic processes in biological networks.[49] ODEs are employed to describe the temporal changes of molecular concentrations caused by production, degradation, transport, or modification of the modeled substances. Changes of concentrations S_i are expressed as a function of rates v_j of reaction and appropriate stoichiometric coefficients n_{ij}, as described in the following equation:

$$\frac{\mathrm{d}S_i}{\mathrm{d}t} = \sum_{j=1}^{r} n_{ij} v_j \quad \text{for } i = 1, \ldots, m \tag{3.1}$$

These equations are called balance equations, since they present the balance of all processes leading toward and away from a node S_i. Frequently, also a matrix formulation is used, summarizing all concentrations into a vector $\mathbf{S}$, all rates into a vector $\mathbf{v}$, and all stoichiometric coefficients into a matrix $\mathbf{N}$, resulting in

$$\frac{\mathrm{d}\mathbf{S}}{\mathrm{d}t} = \mathbf{N} \cdot \mathbf{v} \tag{3.2}$$

When considering the mitochondrial tricarboxylic acid cycle as presented in Fig. 3.2, an appropriate set of ODEs would read as follows:

$$\begin{aligned}
&\frac{\mathrm{d}}{\mathrm{d}t}\text{citrate} = v_1 - v_2\\
&\frac{\mathrm{d}}{\mathrm{d}t}\textit{cis}\text{-aconitase} = v_2 - v_3\\
&\frac{\mathrm{d}}{\mathrm{d}t}\text{isocitrate} = v_3 - v_4\\
&\frac{\mathrm{d}}{\mathrm{d}t}\text{oxalosuccinate} = v_4 - v_5\\
&\frac{\mathrm{d}}{\mathrm{d}t}\alpha\text{-ketoglutarate} = v_5 - v_6\\
&\frac{\mathrm{d}}{\mathrm{d}t}\text{succinyl-CoA} = v_6 - v_7\\
&\frac{\mathrm{d}}{\mathrm{d}t}\text{succinate} = v_7 - v_8\\
&\frac{\mathrm{d}}{\mathrm{d}t}\text{fumarate} = v_8 - v_9\\
&\frac{\mathrm{d}}{\mathrm{d}t}\text{malate} = v_9 - v_{10}\\
&\frac{\mathrm{d}}{\mathrm{d}t}\text{oxaloacetate} = v_{10} - v_1\\
&\frac{\mathrm{d}}{\mathrm{d}t}\text{acetyl-CoA} = -v_1\\
&\frac{\mathrm{d}}{\mathrm{d}t}\text{CO}_2 = v_6\\
&\frac{\mathrm{d}}{\mathrm{d}t}\text{NAD}^+ = -\frac{\mathrm{d}}{\mathrm{d}t}\text{NADH} = -v_4 - v_6\\
&\frac{\mathrm{d}}{\mathrm{d}t}\text{FAD} = -\frac{\mathrm{d}}{\mathrm{d}t}\text{FADH}_2 = -v_8\\
&\frac{\mathrm{d}}{\mathrm{d}t}\text{GTP} = -\frac{\mathrm{d}}{\mathrm{d}t}\text{GDP} = v_9
\end{aligned} \tag{3.3}$$

where the rates belong to the following reactions: v_1, citrate synthase; v_2 and v_3, aconitase; v_4 and v_5, isocitrate dehydrogenase; v_6, α-ketoglutarate dehydrogenase; v_7, succinyl-CoA-synthetase; v_8, succinate dehydrogenase; v_9, fumarase; v_{10}, malate dehydrogenase. When we sum up all equations except

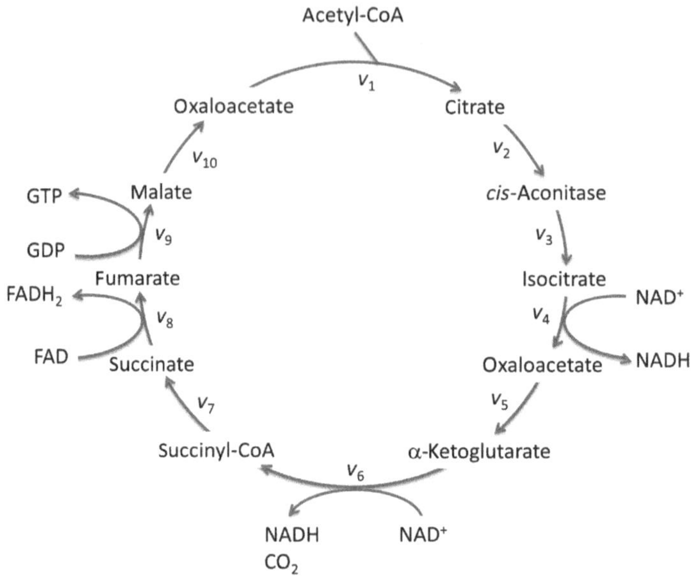

Figure 3.2 The reactions of the tricarboxylic acid cycle as basis for the model represented by the set of equations given in the main text (Eq. 3.3).

of those for NAD^+/NADH, GTP/GDP, FAD/$FADH_2$ as well as acetyl-CoA and CO_2, we find that the sum of the temporal changes is zero, showing that we have a mass conservation in this cycle. While acetyl-CoA must be provided and CO_2 is released to produce the energy rich compounds, the basic skeleton of the cycle is contained (not counting replenishment by other pathways or drain into anabolic pathways).

In order to analyze the dynamics of that system, we need to know the individual rates, v_j (here: $j=1\ldots10$). The reaction rates are not uniquely defined, but have been suggested to follow different types of kinetics such as the mass action law,[50]

$$v_j = k_j \prod_{i \in \{\text{substrates of } v_j\}} S_i \tag{3.4}$$

or the Michaelis–Menten rate law[51,52]

$$v_j = \frac{V_{\max,j} \cdot S_i}{K_{M,ij} + S_i} \tag{3.5}$$

for irreversible reactions where S_i is the substrate, $V_{\max,j}$ denotes the maximal velocity, and $K_{M,ij}$ the Michaelis constant or half-saturation substrate concentration. For reversible reactions with substrate S_{i1} and product S_{i2}, the rate reads:

$$v_j = \frac{V_{\max,j}^{\text{forward}} \cdot \frac{S_{i1}}{K_{M,1j}} - V_{\max,j}^{\text{backward}} \cdot \frac{S_{i2}}{K_{M,2j}}}{1 + \frac{S_{i1}}{K_{M,1j}} + \frac{S_{i2}}{K_{M,2j}}} \tag{3.6}$$

There are even more complicated forms of rate laws to describe some specific kinetics.[53–57] To simplify the choice of kinetics for larger modeling projects, we introduced the convenience kinetics.[56] It abstains from employing individual inhibitions constants for all compounds and it allows for systematic applications of techniques such as parameter balancing.[58]

$$v_j = E_j \frac{k_{\text{cat},j}^{\text{forward}} \cdot \prod_{\text{substrates}} \frac{S_i}{K_{M,ij}} - k_{\text{cat},j}^{\text{backward}} \cdot \prod_{\text{products}} \frac{S_i}{K_{M,ij}}}{\prod_{\text{substrates}} \left(1 + \frac{S_i}{K_{M,ij}}\right) + \prod_{\text{products}} \left(1 + \frac{S_i}{K_{M,ij}}\right) - 1} \tag{3.7}$$

ODE models are frequently used to describe the dynamics of metabolic and signaling networks. When striving to compare their dynamics to real data, we face three levels of uncertainty: (i) is the network structure correct (or should we include or exclude compounds or reactions?), (ii) are the chosen kinetics appropriate, and (iii) what are the values of the kinetic parameters?

The first problem is frequently tackled by experience or by Occam's razor. A more systematic approach is so-called ensemble modeling, where different hypotheses about the model structure are compared with respect to which of them may best fit the experimental data. This approach also helps to tackle the second challenge. Precise *in vitro* kinetic descriptions may be an alternative. Finally, the kinetic parameters of the system are estimated in such a way that the simulated time courses are as close as possible to the measured experimental data.

ODE models have been applied in different flavors to analyze processes relevant for mitochondria in health and disease. Above all, energy conversion in glycolysis, TCA cycle and respiratory chain have been described. Early computational models have focused on the relation between the respiration rate, ATP levels and metabolic demand.[59–61] They analyzed perturbations of the system and interpreted them in the light of mitochondrial dysfunction. Beard[62] has formulated a computational model for the respiratory chain that includes the reactions at complexes I, III, and IV and of the F1Fo ATPase, substrate transporters and cation fluxes over the inner membrane. That model balances the masses and charges. Models describing the dynamics of mitochondrial compounds, protons and cations must address further challenges compared to the plain metabolic model described in Eq. (3.3). First, they must respect volume ratios for transport steps, i.e., a certain amount of cations entering the mitochondrion changes the concentration in the (larger) cytosol less than in the (smaller) mitochondrion. Second, all transport reactions of ions are influenced by the membrane potential and in turn change the membrane potential, $\Delta\Phi$. The change of membrane potential is simply proportional to the sum of ion fluxes (respecting charges), however, the impact of the membrane potential on transport properties is less straight forward and different solutions have been suggested. Figure 3.3 gives an overview over mitochondrial transport steps.

The predictions of the Beard model[62] were compared to data measured on isolated cardiac mitochondria, faithfully reproducing these and other data. A major finding was that the model could only explain the measurements, if one incorporates inorganic phosphate-dependent activation of the dehydrogenase and the electron transport system. A series of models has followed focusing on more and more details and experimental findings,[63,64] such as the role of individual complexes of the respiratory chain for production of reactive oxygen species (ROS).[65] Recent models relate the OXPHOS and its kinetic properties to the work of muscles.[66–68]

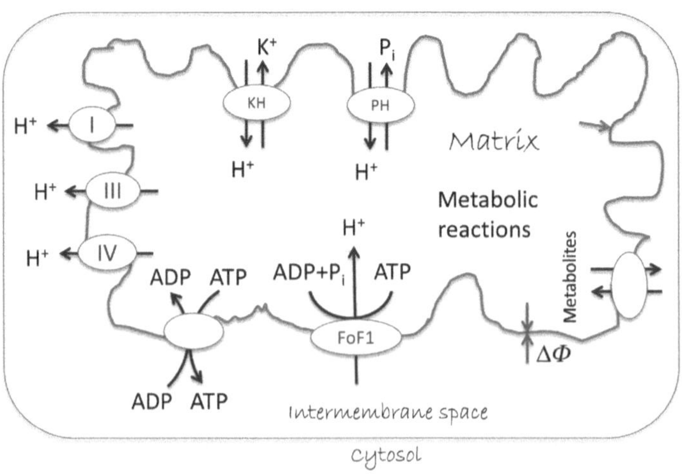

Figure 3.3 Sketch of the mitochondrion with selected transport steps and complexes I, III, and IV of the respiratory chain.

Van Eunen and colleagues have used a metabolic model to study the metabolic syndrome.[69] They have analyzed the dynamics of fatty-acid β-oxidation and subjected it in the model to overload conditions. They found that competition between acyl CoAs of different chain lengths for a set of acyl-CoA dehydrogenases leads to metabolite accumulation in the mitochondria and, hence, unhealthy overload as well as depletion of free CoA.

2.4. Simplifications specific to metabolic networks—Flux-balance analysis

Large metabolic networks are for conceptual simplicity often considered in a so-called steady state. That means to set Eq. (3.2) to zero, i.e.,

$$\frac{\mathrm{d}\mathbf{S}}{\mathrm{d}t} = \mathbf{N} \cdot \mathbf{v} = \mathbf{0} \tag{3.8}$$

The reason is that the detailed kinetic analysis of large network is often complicated or impossible, since there are not enough data to determine all required kinetic parameters. Since **N** and **v** are matrices, Eq. (3.8) defines some dependencies among the rates contained in **v**. However, they are typically not sufficient, to determine all individual v_j. Striving for a modeling approach closer to reality, two further concepts are used: first, all individual reactions are limited by constraints, i.e., upper and lower bounds such as $b_{\mathrm{low}} \leq v_j \leq b_{\mathrm{high}}$. The lower bound b_{low} is often zero flux, the upper bound b_{high} is given by, e.g., the metabolic capacity or physical considerations. This approach is called constraint-based modeling.[70] Second, it is assumed that the metabolic network as a whole fulfills some optimality requirement, for example, maximizing the production of biomass or minimizing the sum of all fluxes needed to produce a compound or alike.[71]

2.5. Agent-based models

An approach that is conceptually different to the hitherto presented network-based methods is that of agent-based models. Those models describe the interaction among multiple autonomous agents. Here, every agent or type of agent has its own set of rules according to which it acts and it perceives information about the action of the other agents. They can move within the simulated space (e.g., membrane, cytoplasm, or mitochondrial matrix). In this way, they are different to cellular automata, where players are restricted to actions on a predefined grid. Agent-based models are used to study complex phenomena and emergent dynamics using populations of agents with simple rules. A popular implementation is the tool NetLogo,[72] which has been used, for example, to study host–pathogen interactions of *Candida* cells growing on an epithelium and provoking defense by macrophages. Another example is the formation of granules in lung epithelia upon infection with *Mycobacterium tuberculosis*.[73] Impairment of mitochondrial quality control can lead to neurodegenerative diseases, diabetes, and cancer. Mitochondrial dynamics comprising growth and mitophagy as well as fusion and fission is important to ensure appropriate mitochondrial membrane potential and prevent the accumulation of damaged material. Figge and colleagues[74] could show that cycles of fusion and fission and mitophagy are essential for ensuring a high average quality of mitochondria, even if there is some damage. To this end, they divided the mitochondrial population into groups and assigned them quality criteria resulting in active and nonactive mitochondria. These groups were then shuffled by fission and fusion or mitophagy leading to dynamics of the overall quality.

3. MODELS OF MITOCHONDRIAL AGING AND DYNAMICS

As mentioned earlier, the accumulation of defective mitochondria, respectively mtDNA mutants in single cells of aged organisms is one of the main suspects in the search for biochemical processes that might be at the heart of the aging process.[7–10] Over the years, several processes were discovered or proposed that have the potential to either promote or prevent such an accumulation. Figure 3.4 provides an overview of these processes and the studies that used mathematical/computational models to investigate them. Please note that these are not necessarily the first publications to mention the respective ideas.

3.1. Early models

The first mathematical models of mitochondrial aging appeared in the early 1990s and investigated the idea that a vicious cycle might drive the accumulation of mtDNA mutations.[86] In this scenario, the ROS that are generated as byproduct during respiration, cause damage to the mtDNA, which results in mitochondrial mutants that produce even more ROS (and then again more damage). The authors constructed a model consisting of seven ODEs that describes the following list of features:

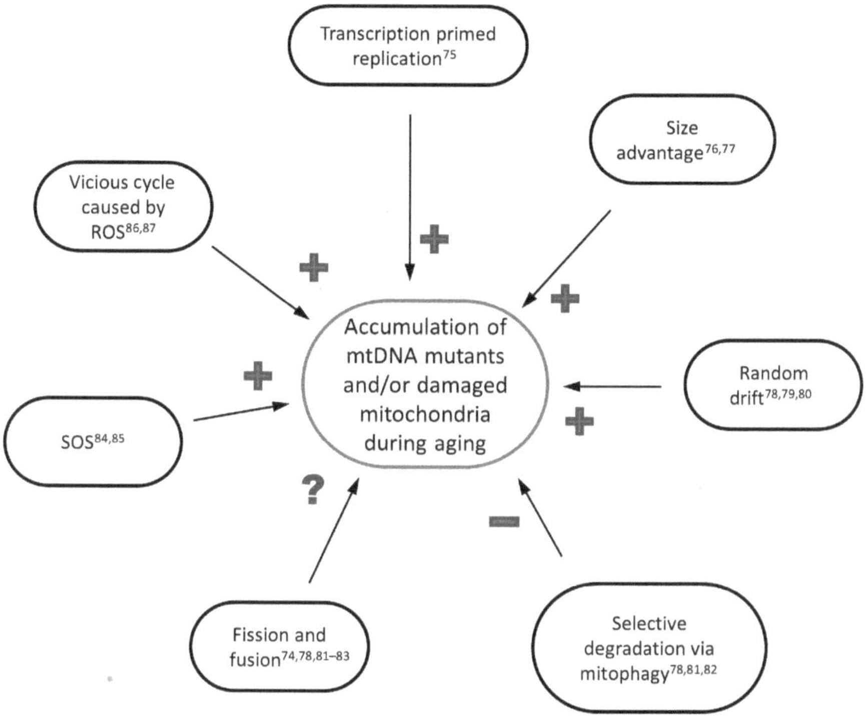

Figure 3.4 Overview of various biochemical processes that have been modeled mathematically to study the accumulation of mtDNA mutants and functionally impaired mitochondria during aging. For each of the processes, literature references are attached, which are discussed in the main text. Plus and minus symbols indicate whether the processes are believed to stimulate or suppress the accumulation of mtDNA mutants. In case of fission/fusion, the consequences are uncertain.

- Mitochondria produce oxygen radicals
- Mitochondria suffer DNA damage caused by the radicals
- Mitochondria produce energy in form of ATP
- Mitochondria produce more radicals and less ATP after DNA damage
- The growth rate of individual mitochondria is controlled by their ATP production
- There is a constant turnover of mitochondria
- Mitochondria are separate entities containing on average five mtDNA molecules, leading to six damage classes (with zero to five mtDNAs being damaged).

The main insight from this model is that the mitochondrial population can either reach a stable steady state or collapse after some time through the unlimited accumulation of mitochondria with damaged mtDNAs. The critical parameters that control the fate of the population are the net growth rate (growth rate—degradation rate) and the mutation rate (which is influenced by the radical production). If the mutation rate exceeds the net growth rate the system collapses, otherwise it is stable.

This model was a first step toward understanding the fate of damaged mitochondria, but it had several shortcomings. For instance, did it not contain any sink for radicals and thus all calculations are based on the radical production rates and not on their actual concentrations. Furthermore, for degradation of mitochondria a simple first-order decay function is used, although it is known that proteins and mitochondria are turned over by special degradation systems. The authors therefore extended their model to include antioxidant enzymes that detoxify ROS, "scavenger" enzymes that degrade all components of the model (proteins as well as mitochondria), and ribosomes that synthesize the different protein types, including themselves.[87] This "Network Theory of Ageing" consists of more than 30 differential equations and represents an early version of a system biological model by integrating several biochemical pathways into a single mathematical description. Like the earlier model also this version shows that the system can be either stable, or collapse after a certain time. This model generated several new results that were only possible after the integration of the different network components. For instance, the model showed that radical damage to the scavenger enzymes resulted in an age-related increase in protein half-life, which agrees with experimental findings.[88–90] It also predicts such an increase in half-life for mitochondria and mtDNA, a point that still awaits experimental investigation. Furthermore, the final breakdown is caused by an interaction of mitochondrial and cytoplasmic reactions, the mitochondria being responsible for a long-term gradual change, which eventually triggers a short-lived cytoplasmic error loop.

However, as interesting as these results are, it is now clear that the accumulation of mtDNA mutants cannot be caused by a vicious cycle, since this mechanism predicts that a single cell should contain a large number of different mtDNA mutants, which has by now been ruled out experimentally.[11–16] Thus, it seems that a selection advantage is required for mutant mtDNAs that allows them to expand clonally once they have been created. An original idea has been put forward by de Grey,[84] who proposes that the selection advantage actually stems from a delayed degradation of damaged organelles, called "survival of the slowest (SOS)." He assumes that damaged mitochondria have a reduced respiratory chain activity, leading to a reduced generation of ROS and ROS-induced membrane damage. If the degradation rate is linked to the level of membrane damage, it then follows that defective mitochondria are degraded at a slower rate than intact ones. To investigate this hypothesis via *in silico* simulations, Kowald and Kirkwood[85] implemented this idea as a mathematical model consisting of nine ODEs. The simulations confirm that such a delayed degradation can indeed lead to an accumulation of defective mitochondria and if it is assumed that the damaged mitochondria grow slower than intact ones (caused by diminished ATP and $\Delta\Psi$) a system with a stable heteroplasmy of mutant and wild type can emerge. Interestingly, this modeling study also investigated the effects of cell division on the stability of the mitochondrial population and found that the growth spurt following division (to regain the normal cell size) shifts the selection advantage toward wild-type mitochondria. This would nicely fit with the observation that defective mitochondria are mainly found in postmitotic cell types. But the SOS idea has its problems. First, it predicts that damaged mitochondria should be degraded at a reduced pace, while experimentally the opposite seems to be true,[91–93] and second, it assumes that mitochondria are separate entities (to establish a link between mutated genotype and its phenotype), while it is now clear that mitochondria are dynamic structures that constantly undergo cycles of fusion and fission.[91,94]

3.2. Mitochondrial dynamics

And this mitochondrial dynamics might have important consequences for the accumulation but also for the removal of mtDNA variants. While mitochondria are fused they can exchange matrix and membrane components like proteins and mtDNA molecules.[95,96] It has been proposed that this mixing acts as a complementation mechanism to maintain the function of mitochondria with defective mtDNAs[97,98] and another idea is that fusion allows the mitochondria to act as cellular power cables, transporting energy via a proton gradient to distant parts of the cell.[99,100]

An important experimental finding is that mitochondrial fragments generated by fission sometimes display a lowered membrane potential[91] and these fragments seem to be preferentially degraded.[91–93] Additionally, it has also been observed that a membrane potential is required for the mitochondrial fusion process.[91] This points to the possibility that mitochondrial dynamics is an essential components of a cellular quality assurance system for removing damaged mitochondria.[91,101]

Mouli *et al.*[81] developed a first computational model to investigate the consequences of repeated cycles of fusion, fission, and autophagy for the stability of a mitochondrial population that is subject to a low, but constant rate of damage. For this purpose, the simulation consisted of 300 mitochondria per cell that underwent pairwise fusion followed by fission. Each mitochondrion contained 10 mtDNAs (called "functional hereditary units") that can be exchanged during the fused state and which are either intact or damaged. Depending on the number of intact mtDNAs mitochondria possess a certain activity level and unfused mitochondria with activity levels at or below three are removed by autophagy. The main results of this study are that the process of selective degradation (e.g., via detecting a lowered membrane potential) is by itself sufficient to stabilize a population of mitochondria despite ongoing damage processes. Furthermore, fusion might reinforce this stabilization since the mixing and redistribution of mtDNAs increases the variance of the activity levels in the mitochondrial population (i.e., two mitochondria with four damaged mtDNAs each, can result in mitochondria with two and six damaged mtDNAs), which in turn improves the efficiency of the selective degradation process. Finally, the study also showed that selective fusion (i.e., only mitochondria above a certain activity level can fuse) boosts the stabilizing effect even further.

Other authors studying mitochondrial dynamics were more interested in the possibility that fission and fusion might be deleterious by their own.[74,83] Also in their model mitochondria belong to different quality states (damage classes) and fusion/fission events can redistribute mitochondria from one quality state into another. The model also assumes a constant source of damage like the model of Mouli *et al.*,[81] but in addition the authors introduce a new type of damage called "infectious damage." The name is somewhat misleading since it is actually damage that is assumed to be directly connected with the fission process itself. The authors motivate this new type of damage with reference to the drop of membrane potential that has sometimes been observed after fission events.[91] This is unusual since the widely accepted interpretation is that this drop merely reflects the successful isolation of a damaged mitochondrial fragment via the fission process. However, from this assumption it consequently follows that too high fusion/fission rates are actually detrimental and the authors propose that it would be beneficial for the cell to decelerate their level of mitochondrial dynamics at advanced ages.

The models of mitochondrial dynamics that were discussed so far all treat the cell, containing the mitochondrial population, as a well-mixed compartment. This simplifies simulations, but is of course a simplification since the spatial location of mitochondria constraints the fusion process insofar that only neighboring mitochondria can fuse. The model of Tam, Gruber, Halliwell, and Gunawan[78] takes this into account by describing the cell as a 2D compartment partitioned into 16 spatial sections. Every cell contained 80 mitochondria, each consisting of four mtDNAs that were initially placed randomly in the different sections. The model was then simulated stochastically involving fusion/fission, degradation, and replication. Only mitochondria in the same or neighboring sections were allowed to fuse and exchange mtDNAs. This not only results in mixing of mitochondrial content, but also in mitochondrial movement, since one of the fusing partners is removed during this step of the simulation. The model does not include *de novo* mutations, but instead all simulations were started with 10% of mutant mtDNAs. The main insight of the model is that a slow fusion and fission rate results in a more uneven distribution of wild-type and mutant mtDNAs across the spatial sections of the cell, accelerating the random segregation of cells into those homoplasmic for wild type of mutant mtDNAs. However, if the degradation probability of mutant mtDNAs was only slightly higher than that of wild type, no clonal expansion occurred and the mitochondrial population eventually became homoplasmic for wild type.

The last result is again a demonstration of the powerful protective capabilities of selective degradation of defective mitochondria. This requires of course that the cell can actually differentiate between intact and defective mitochondria and it is often assumed that this happens via sensing a reduced membrane potential.[81,101] However, this idea can only work if there is a link between genotype and phenotype. If mtDNAs freely diffuse within the mitochondrial network (via mitochondrial fusion) without connection to their gene products, selective degradation cannot work because mitochondria with the selected phenotype (diminished proton gradient) will not necessarily contain the corresponding genotype (mutant mtDNA). Kowald and Kirkwood[82] presented a solution to this problem by proposing that mtDNA molecules are connected to their own gene products (i.e., respiration complexes) either by a direct physical link or because they are confined to the same cristae.[96,102] The proposed link is of course not as tight as the classical connection (DNA inside cell) since some proteins might diffuse away before or after they are inserted into the membrane. To investigate if selection could also work for such a leaky link between genotype and phenotype the authors developed a mathematical model

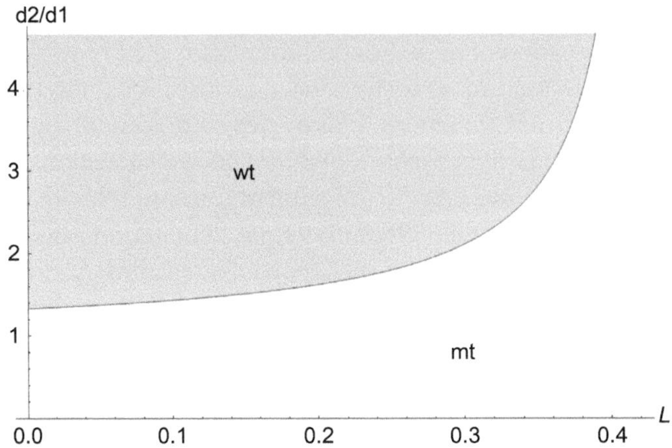

Figure 3.5 Selective degradation can protect a wild-type population of mtDNAs from faster replicating mutant mtDNAs, even if the link between genotype and phenotype is leaky 96. The degree of leakiness, *L*, and the specificity of selective degradation, d2/d1, are important factors controlling the system behavior.

based on differential equations. The analysis shows that selection can indeed maintain a wild-type population of mtDNAs in the presence of a leaky link (Fig. 3.5). Important parameters influencing the outcome of the selection are the amount of leakiness and the degree of selectivity of the preferential degradation process.

3.3. Accumulation of deletion mutants

While mitophagy and preferential degradation seem to be an important cellular components for the removal of defective mitochondria, the question what mechanism drives their accumulation is still wide open. A popular and mechanically very simple explanation is the random drift hypothesis, which states that random fluctuations during replication and degradation of mtDNA molecules are sufficient to explain the sporadic occurrence of high levels of mutated mtDNAs in single cells. It is a very minimalistic idea since it requires neither a special phenotype to be attached to mutated mtDNAs nor a selection advantage.

A first computer simulation of this idea was presented by Chinnery and Samuels.[79] In their study, they simulated the fate of cells containing 1000 mtDNA molecules that could be of two types, wild type or mutant, and these mtDNAs were randomly degraded and replicated. The model contained no *de novo* mutations, but simulations were always started with a certain fraction of mutant mtDNAs. The results basically showed that within a few years random drift can indeed cause gross changes in the fraction of mutant and wild-type mtDNA inside individual cells. The computational study of random drift was continued by a work of Elson, Samuels, Turnbull, and Chinnery,[103] who extended the original computer model by implementing mtDNA mutations (to allow for *de novo* mutations) and by discriminating between different types of mutants. This made it possible to investigate the degree of heteroplasmy that is predicted by the random drift hypothesis. The authors simulated a pool of 600 cells, each initialized with 1000 wild-type mtDNAs that were degraded such that a half-life of 10 days resulted. Degraded mtDNAs were replaced by copying one of the existing ones and this replication process was attached with a certain mutation rate. The simulations were performed for 120 years and the fraction of cells was recorded that contained more than 60% of mutant mtDNAs. It is assumed that such a level of mutant mtDNAs leads to cytochrome *c* oxidase (COX) negative cells, such as those seen in old muscle tissue. The calculations demonstrated that the COX negative cells that were present after 120 years are indeed almost homoplasmic for a single mutant mtDNA, which agrees with experimental observations.

However, the question is if such a process can also work for short-lived species, like rodents, which show a very similar accumulation of defective mtDNAs as in humans.[11,15] To address this question, Kowald and Kirkwood[80] performed a further simulation study, investigating random drift in species with life spans ranging from 3 to 120 years. The results clearly show that random drift is insufficient to explain the accumulation of mitochondrial mutants in short-lived animals, because the level of heteroplasmy dramatically increases for short life spans. For rats with a life span of ca. 3 years, random drifts results in more than 30 different mutant types per COX negative cell, completely incompatible with experimental evidence. We have already discussed mitochondrial dynamics and selective degradation, which might act as cellular pruning mechanism for defective mtDNAs. The existence of such a quality control process poses further serious problems for the random drift hypothesis. The study of Tam, Gruber, Halliwell, and Gunawan,[78] which we already discussed in connection with mitochondrial fission and fusion, showed that selective degradation efficiently prevents the accumulation of mutated mtDNAs in a random drift scenario. This adds further evidence to the argument that a selection advantage is required to explain the accumulation of mtDNA deletion mutants.

Several years ago, it has been realized that the reduced size of mtDNA might provide such a selection advantage, because the deletion mutants would have a shorter replication time.[104,105] Deletions of more than 50% of the genome size have been observed,[11] which could result in an equal reduction of replication time. However, the idea had also been questioned since the replication time of mtDNA is only 1–2 h[106,107] and thus much shorter than the half-life of mtDNA, which is in the order of 1–3 weeks.[108–110] It is therefore difficult to see how under those conditions mtDNA replication can be the rate-limiting step for mitochondrial growth.[103,111] An early modeling study showed that computer simulations based on this idea can indeed very well reproduce experimentally observed deletion size distributions.[76] But a major deficit of the study was that it only contained a very rough representation of the molecular details of the replication process. The increase of the number of mtDNA molecules was simply inversely proportional to the size of the mtDNA, giving an advantage to shorter molecules. Thus, in this model the half-life of mtDNAs has no influence on the selection advantage. Recently, a biochemically more detailed model has been presented by Kowald, Dawson, and Kirkwood[77] that is based on a set of delay differential equations. It assumes that mtDNAs can be in a "free" state in which they can respond to replication signals or in a "busy" state, in which replication takes place. After a certain replication time (which is shorter for deletion mutants) mtDNAs return from the "busy" into the "free" pool together with the newly synthesized daughter mtDNA. The simulations show that the ratio of mtDNA half-life to mtDNA replication time is of crucial importance for the resulting selection advantage. Using accepted literature values (replication time 2 h, half-life 10 days), it takes roughly 95 years until wild-type mtDNAs are replaced by a deletion mutant that has lost 50% of its genome. The critical insight gained from this model is that the smaller size of the deletion mutant represents only an advantage in the busy state via a faster exit time. But a long half-life reduces the fraction of replicating mtDNAs (molecules in the busy state) and consequently the selection advantage dwindles away with increasing ratio of half-life to replication time.

Thus, the various simulation studies have shown that none of the current ideas can convincingly explain the accumulation of mitochondrial deletion mutants across the whole range of life spans found among mammalian species. A new idea to attempt this feat has been proposed by Kowald and Kirkwood[75] and is based on the fact that in metazoan mitochondria the primer required for replication is provided by processed mRNA transcripts of the L-strand promoter.[112] A product inhibition feedback is proposed that downregulates the transcription rate if enough proteins are available. A deletion mutant that has lost the genes for proteins involved in such a feedback mechanism will no longer downregulate its transcription rate, resulting in a higher frequency of priming and replication. Experimentally this idea is backed up with the finding that the deletion spectra in several species identify a genomic region (around ND4/ND5) that is affected by almost all deletions.[75] The authors also developed a deterministic model based on ODEs to describe the competition of mutant and wild-type mtDNAs inside a single cell, as well as a set of stochastic computer simulations to describe the consequences of the idea at the tissue level. The main conclusions are that inside a single cell a mtDNA deletion can replace the wild type in a matter of a few months after the mutation has occurred. Species specific life spans ranging from a few years up to many decades are then a consequence of different mutation rates. Also the resulting degree of mtDNA heteroplasmy in COX negative cells agrees well with experimental findings. Figure 3.6

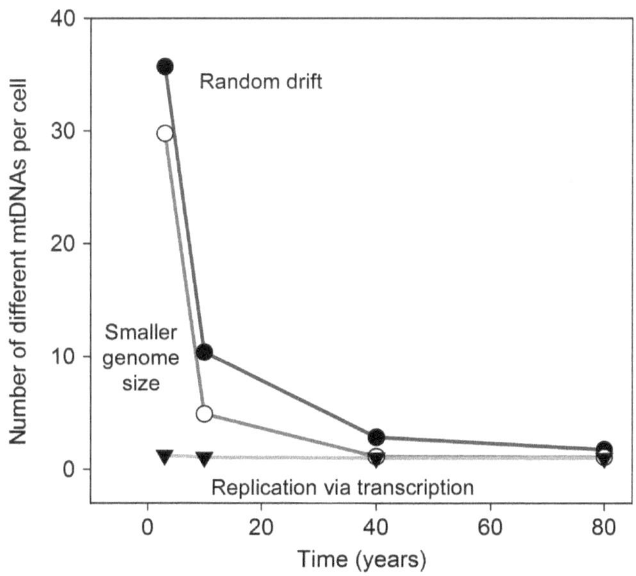

Figure 3.6 Comparison of mtDNA heteroplasmy predicted by different mechanism that have been proposed to explain the accumulation of mtDNA mutants during aging. Simulations were performed for life spans of 3, 10, 40, and 80 years and each data point shows the average number of different mtDNA types per COX negative cell. *Figure taken from Kowald and Kirkwood.*[75] (See the color plate.)

shows a comparison of the predicted degree of heteroplasmy for three of the ideas that have been discussed in this review. Especially for short life spans, the differences are striking and only the "replication via transcription" idea has the potential to explain the accumulation of mtDNA deletion mutants over a wide range of species life spans.

4. CONCLUSIONS AND PERSPECTIVES

Mitochondria are involved in a surprisingly large number of biochemical pathways that are essential for the functioning of eukaryotic cells. Research in recent years has especially highlighted their involvement in aging and disease, but has also revealed new and surprising details of their dynamic behavior, which is characterized by frequent fission and fusion events. Their involvement in multiple complex processes makes mitochondria a prime candidate for mathematical and computational modeling. Such models require that ideas and hypotheses are spelled out in detail and thus help to identify gaps in concepts and current knowledge.

In the first part (Section 2) of this review, we provided a detailed introduction to mathematical modeling techniques to enable the reader to judge which problems can be tackled by mathematical models and what type of results and insights can be gained. In the second part (Section 3), we reviewed publications that use this type of model to investigate the fission and fusion process of mitochondria as well as the accumulation of mitochondrial mtDNA mutants that emerge during the aging process. It can be seen that model development also reflects the progress of biochemical knowledge. While the early models assumed mitochondria to be completely separate entities, the discovery of mitochondrial dynamics made it necessary to develop new models that take the consequences of this process (e.g., content mixing) into account.

But modeling also provides feedback and answers that are only difficult to obtain by experiment. The accumulation of mtDNA deletion mutants in single postmitotic cells is a hallmark of aging that has been observed in several mammalian species. Multiple hypotheses have been proposed to explain this phenomenon, but the lack of experimental techniques to manipulate mtDNA *in vivo*, hamper their validation. Recent modeling studies now provide strong indications that some of them (i.e., random drift and reduced genome size) are very unlikely solutions to this problem.[77,80] This way modeling can help to focus resources and research toward more promising directions.

However, this review also shows that there are important areas of mitochondrial involvement that are not at all or insufficiently investigated by mathematical models. Mitophagy and selective degradation of defective mitochondria emerge as important mechanisms to control the spread of mitochondrial mutants. It would therefore be important to combine these processes with models that study hypotheses regarding the accumulation of mtDNA mutants. The question is no longer how these mutants can accumulate, but how these mutants can accumulate despite an efficient cellular removal process. Another important area of modeling research that is currently completely neglected is the involvement of mtDNA mutations in mitochondrial diseases. This will be an important field of research in the future, since it is currently not clear at all how mtDNA mutations can be responsible for aging as well as diseases, since both maladies have very different phenotypes. An important question in this respect is, why mtDNA mutants seem to replace the wild type completely during aging, while in mitochondrial diseases there appears to be some form of stable heteroplasmy.

REFERENCES

1. Schwartz RM, Dayhoff MO. Origins of prokaryotes, eukaryotes, mitochondria, and chloroplasts. *Science*. 1978;199:395–403.
2. Gray MW, Burger G, Lang BF. Mitochondrial evolution. *Science*. 1999;283(5407): 1476–1481.
3. Anderson S, Bankier AT, Barrell BG, et al. Sequence and organization of the human mitochondrial genome. *Nature*. 1981;290:457–465.
4. Comfort A. The position of aging studies. *Mech Ageing Dev*. 1974;3:1–31.
5. Wallace DC. Mitochondrial DNA, mutations in disease and aging. *Environ Mol Mutagen*. 2010;51(5):440–450.
6. Wallace DC. Mitochondrial diseases in man and mouse. *Science*. 1999;283(5407): 1482–1488.
7. Harman D. Free radical theory of aging: the 'free radical' diseases. *Age*. 1984;7: 111–131.
8. Harman D. The biologic clock: the mitochondria? *J Am Geriatr Soc*. 1972;20(4): 145–147.
9. Miquel J, Economos AC, Fleming J, Johnson Jr JE. Mitochondrial role in cell aging. *Exp Gerontol*. 1980;15(6):575–591.
10. Linnane AW, Marzuki S, Ozawa T, Tanaka M. Mitochondrial DNA mutations as an important contributor to ageing and degenerative diseases. *Lancet*. 1989;1(8639):642–645.
11. Cao Z, Wanagat J, McKiernan SH, Aiken JM. Mitochondrial DNA deletion mutations are concomitant with ragged red regions of individual, aged muscle fibers: analysis by laser-capture microdissection. *Nucleic Acids Res*. 2001;29(21):4502–4508.
12. Brierley EJ, Johnson MA, Lightowlers RN, James OF, Turnbull DM. Role of mitochondrial DNA mutations in human aging: implications for the central nervous system and muscle. *Ann Neurol*. 1998;43(2):217–223.

13. Khrapko K, Bodyak N, Thilly WG, et al. Cell by cell scanning of whole mitochondrial genomes in aged human heart reveals a significant fraction of myocytes with clonally expanded deletions. *Nucleic Acids Res.* 1999;27(11):2434–2441.
14. Gokey NG, Cao Z, Pak JW, et al. Molecular analyses of mtDNA deletion mutations in microdissected skeletal muscle fibers from aged rhesus monkeys. *Aging Cell.* 2004;3(5):319–326.
15. Herbst A, Pak JW, McKenzie D, Bua E, Bassiouni M, Aiken JM. Accumulation of mitochondrial DNA deletion mutations in aged muscle fibers: evidence for a causal role in muscle fiber loss. *J Gerontol A Biol Sci Med Sci.* 2007;62(3):235–245.
16. McKiernan SH, Colman R, Lopez M, Beasley TM, Weindruch R, Aiken JM. Longitudinal analysis of early stage sarcopenia in aging rhesus monkeys. *Exp Gerontol.* 2009;44(3):170–176.
17. McDonald SA, Preston SL, Greaves LC, et al. Clonal expansion in the human gut: mitochondrial DNA mutations show us the way. *Cell Cycle.* 2006;5(8):808–811.
18. Greaves LC, Preston SL, Tadrous PJ, et al. Mitochondrial DNA mutations are established in human colonic stem cells, and mutated clones expand by crypt fission. *Proc Natl Acad Sci USA.* 2006;103(3):714–719.
19. Taylor RW, Barron MJ, Borthwick GM, et al. Mitochondrial DNA mutations in human colonic crypt stem cells. *J Clin Invest.* 2003;112(9):1351–1360.
20. Schomburg I, Chang A, Ebeling C, et al. BRENDA, the enzyme database: updates and major new developments. *Nucleic Acids Res.* 2004;32(Database issue):D431–D433.
21. Matthews L, Gopinath G, Gillespie M, et al. Reactome knowledgebase of human biological pathways and processes. *Nucleic Acids Res.* 2009;37(Database issue):D619–D622.
22. Le Novere N, Bornstein B, Broicher A, et al. BioModels database: a free, centralized database of curated, published, quantitative kinetic models of biochemical and cellular systems. *Nucleic Acids Res.* 2006;34(Database issue):D689–D691.
23. Olivier BG, Snoep JL. Web-based kinetic modelling using JWS Online. *Bioinformatics.* 2004;20(13):2143–2144.
24. Snoep JL, Olivier BG. JWS online cellular systems modelling and microbiology. *Microbiology.* 2003;149(Pt. 11):3045–3047.
25. Glass L, Kauffman SA. The logical analysis of continuous, non-linear biochemical control networks. *J Theor Biol.* 1973;39(1):103–129.
26. Kauffman SA. Metabolic stability and epigenesis in randomly constructed genetic nets. *J Theor Biol.* 1969;22(3):437–467.
27. Kauffman SA. *The Origins of Order: Self-Organization and Selection in Evolution.* New York: Oxford University Press; 1993.
28. Albert R, Othmer HG. The topology of the regulatory interactions predicts the expression pattern of the segment polarity genes in Drosophila melanogaster. *J Theor Biol.* 2003;223(1):1–18.
29. Davidich MI, Bornholdt S. Boolean network model predicts cell cycle sequence of fission yeast. *PLoS One.* 2008;3(2):e1672.
30. Faure A, Naldi A, Chaouiya C, Thieffry D. Dynamical analysis of a generic Boolean model for the control of the mammalian cell cycle. *Bioinformatics.* 2006;22(14): e124–e131.
31. Gupta S, Bisht SS, Kukreti R, Jain S, Brahmachari SK. Boolean network analysis of a neurotransmitter signaling pathway. *J Theor Biol.* 2007;244(3):463–469.
32. Huang S, Ingber DE. Shape-dependent control of cell growth, differentiation, and apoptosis: switching between attractors in cell regulatory networks. *Exp Cell Res.* 2000;261(1):91–103.
33. Kervizic G, Corcos L. Dynamical modeling of the cholesterol regulatory pathway with Boolean networks. *BMC Syst Biol.* 2008;2(1):99.

34. Li F, Long T, Lu Y, Ouyang Q, Tang C. The yeast cell-cycle network is robustly designed. *Proc Natl Acad Sci USA*. 2004;101(14):4781–4786.
35. Mendoza L, Thieffry D, Alvarez-Buylla ER. Genetic control of flower morphogenesis in Arabidopsis thaliana: a logical analysis. *Bioinformatics*. 1999;15(7–8):593–606.
36. Saez-Rodriguez J, Simeoni L, Lindquist JA, et al. A logical model provides insights into T cell receptor signaling. *PLoS Comput Biol*. 2007;3(8):e163.
37. Samal A, Jain S. The regulatory network of E. coli metabolism as a Boolean dynamical system exhibits both homeostasis and flexibility of response. *BMC Syst Biol*. 2008;2:21.
38. Sridharan S, Layek R, Datta A, Venkatraj J. Boolean modeling and fault diagnosis in oxidative stress response. *BMC Genomics*. 2012;13(Suppl 6):S4.
39. Gouzé JL. Positive and negative circuits in dynamical systems. *J Biol Syst*. 1998;6:11–15.
40. Plahte E, Mestl T, Omholt SW. Feedback loops, stability and multistationarity in dynamical systems. *J Biol Syst*. 1995;3:409–413.
41. Snoussi EH. Necessary conditions for multistationarity and stable periodicity. *J Biol Syst*. 1998;6:3–9.
42. Thomas R. Boolean formalization of genetic control circuits. *J Theor Biol*. 1973;42(3):563–585.
43. Thieffry D, Snoussi EH, Richelle J, Thomas R. Positive loops and differentiation. *J Biol Syst*. 1995;3:457–466.
44. Delbrück M. Discussion. *Unités Biologiques Douées de Continuité Génétique*. 8 Lyon: CNRS; 1949:33–35.
45. Shmulevich I, Dougherty ER, Kim S, Zhang W. Probabilistic Boolean networks: a rule-based uncertainty model for gene regulatory networks. *Bioinformatics*. 2002;18(2):261–274.
46. Flottmann M, Scharp T, Klipp E. A stochastic model of epigenetic dynamics in somatic cell reprogramming. *Front Physiol*. 2012;3:216.
47. Mendoza L, Xenarios I. A method for the generation of standardized qualitative dynamical systems of regulatory networks. *Theor Biol Med Model*. 2006;3:13.
48. Terfve C, Cokelaer T, Henriques D, et al. CellNOptR: a flexible toolkit to train protein signaling networks to data using multiple logic formalisms. *BMC Syst Biol*. 2012;6:133.
49. Klipp E. Modelling dynamic processes in yeast. *Yeast*. 2007;24:943–959.
50. Guldberg CM, Waage P. Concerning chemical affinity. *Erdmann's J Prakt Chem*. 1879;127:69–114.
51. Briggs GE, Haldane JB. A note on the kinetics of enzyme action. *Biochem J*. 1925;19(2):338–339.
52. Michaelis L, Menten M. Die Kinetik der Invertinwirkung. *Biochem Z*. 1913;49:333–369.
53. Cornish-Bowden A, Cardenas ML, Letelier JC, Soto-Andrade J, Abarzua FG. Understanding the parts in terms of the whole. *Biol Cell*. 2004;96(9):713–717.
54. Klipp E, Herwig R, Kowald A, Wierling C, Lehrach H. *Systems Biology in Practice*. Weinheim: Wiley; 2005.
55. Koshland Jr DE, Nemethy G, Filmer D. Comparison of experimental binding data and theoretical models in proteins containing subunits. *Biochemistry*. 1966;5(1):365–385.
56. Liebermeister W, Klipp E. Bringing metabolic networks to life: convenience rate law and thermodynamic constraints. *Theor Biol Med Model*. 2006;3:41.
57. Monod J, Wyman J, Changeux JP. On the nature of allosteric transitions: a plausible model. *J Mol Biol*. 1965;12:88–118.
58. Lubitz T, Schulz M, Klipp E, Liebermeister W. Parameter balancing in kinetic models of cell metabolism. *J Phys Chem B*. 2010;114(49):16298–16303.

59. Korzeniewski B, Zoladz JA. A model of oxidative phosphorylation in mammalian skeletal muscle. *Biophys Chem*. 2001;92(1–2):17–34.
60. Korzeniewski B. Regulation of ATP supply in mammalian skeletal muscle during resting state–>intensive work transition. *Biophys Chem*. 2000;83(1):19–34.
61. Korzeniewski B, Froncisz W. An extended dynamic model of oxidative phosphorylation. *Biochim Biophys Acta*. 1991;1060(2):210–223.
62. Beard DA. A biophysical model of the mitochondrial respiratory system and oxidative phosphorylation. *PLoS Comput Biol*. 2005;1(4):e36.
63. Nazaret C, Heiske M, Thurley K, Mazat JP. Mitochondrial energetic metabolism: a simplified model of TCA cycle with ATP production. *J Theor Biol*. 2009;258(3):455–464.
64. Wu F, Yang F, Vinnakota KC, Beard DA. Computer modeling of mitochondrial tricarboxylic acid cycle, oxidative phosphorylation, metabolite transport, and electrophysiology. *J Biol Chem*. 2007;282(34):24525–24537.
65. Bazil JN, Vinnakota KC, Wu F, Beard DA. Analysis of the kinetics and bistability of ubiquinol:cytochrome c oxidoreductase. *Biophys J*. 2013;105(2):343–355.
66. Korzeniewski B. Regulation of oxidative phosphorylation during work transitions results from its kinetic properties. *J Appl Physiol*. 2014;116(1):83–94.
67. Korzeniewski B, Zoladz JA. Slow VO2 off-kinetics in skeletal muscle is associated with fast PCr off-kinetics—and inversely. *J Appl Physiol*. 2013;115(5):605–612.
68. Zoladz JA, Grassi B, Majerczak J, et al. Training-induced acceleration of O(2) uptake on-kinetics precedes muscle mitochondrial biogenesis in humans. *Exp Physiol*. 2013;98(4):883–898.
69. van Eunen K, Simons SM, Gerding A, et al. Biochemical competition makes fatty-acid beta-oxidation vulnerable to substrate overload. *PLoS Comput Biol*. 2013;9(8): e1003186.
70. Price ND, Reed JL, Palsson BO. Genome-scale models of microbial cells: evaluating the consequences of constraints. *Nat Rev Microbiol*. 2004;2(11):886–897.
71. Varma A, Palsson BO. Stoichiometric flux balance models quantitatively predict growth and metabolic by-product secretion in wild-type Escherichia coli W3110. *Appl Environ Microbiol*. 1994;60(10):3724–3731.
72. Sklar E. NetLogo, a multi-agent simulation environment. *Artif Life*. 2007;13(3):303–311.
73. Kirschner DE, Young D, Flynn JL. Tuberculosis: global approaches to a global disease. *Curr Opin Biotechnol*. 2010;21(4):524–531.
74. Figge MT, Reichert AS, Meyer-Hermann M, Osiewacz HD. Deceleration of fusion-fission cycles improves mitochondrial quality control during aging. *PLoS Comput Biol*. 2012;8(6):e1002576.
75. Kowald A, Kirkwood TB. Transcription could be the key to the selection advantage of mitochondrial deletion mutants in aging. *Proc Natl Acad Sci USA*. 2014;111(8):2972–2977.
76. Kowald A, Jendrach M, Pohl S, Bereiter-Hahn J, Hammerstein P. On the relevance of mitochondrial fusions for the accumulation of mitochondrial deletion mutants: a modelling study. *Aging Cell*. 2005;4(5):273–283.
77. Kowald A, Dawson M, Kirkwood TB. Mitochondrial mutations and ageing: can mitochondrial deletion mutants accumulate via a size based replication advantage? *J Theor Biol*. 2014;340(7):111–118.
78. Tam ZY, Gruber J, Halliwell B, Gunawan R. Mathematical modeling of the role of mitochondrial fusion and fission in mitochondrial DNA maintenance. *PLoS One*. 2013;8(10):e76230.
79. Chinnery PF, Samuels DC. Relaxed replication of mtDNA: a model with implications for the expression of disease. *Am J Hum Genet*. 1999;64(4):1158–1165.

80. Kowald A, Kirkwood TB. Mitochondrial mutations and aging: random drift is insufficient to explain the accumulation of mitochondrial deletion mutants in short-lived animals. *Aging Cell.* 2013;12(4):728–731.
81. Mouli PK, Twig G, Shirihai OS. Frequency and selectivity of mitochondrial fusion are key to its quality maintenance function. *Biophys J.* 2009;96(9):3509–3518.
82. Kowald A, Kirkwood TB. Evolution of the mitochondrial fusion-fission cycle and its role in aging. *Proc Natl Acad Sci USA.* 2011;108(25):10237–10242.
83. Figge MT, Osiewacz HD, Reichert AS. Quality control of mitochondria during aging: is there a good and a bad side of mitochondrial dynamics? *Bioessays.* 2013;35(4): 314–322.
84. de Grey ADNJ. A proposed refinement of the mitochondrial free radical theory of aging. *Bioessays.* 1997;19(2):161–166.
85. Kowald A, Kirkwood TB. Accumulation of defective mitochondria through delayed degradation of damaged organelles and its possible role in the ageing of post-mitotic and dividing cells. *J Theor Biol.* 2000;202(2):145–160.
86. Kowald A, Kirkwood TB. Mitochondrial mutations, cellular instability and ageing: modelling the population dynamics of mitochondria. *Mutat Res.* 1993;295(3):93–103.
87. Kowald A, Kirkwood TB. A network theory of ageing: the interactions of defective mitochondria, aberrant proteins, free radicals and scavengers in the ageing process. *Mutat Res.* 1996;316(5–6):209–236.
88. Young V, Munro H. N γ-methylhistidine and muscle protein turnover. *Fed Proc.* 1978;37:2291–3000.
89. Dice JF. Altered intracellular protein degradation in aging: a possible cause of proliferative arrest. *Exp Gerontol.* 1989;24(5–6):451–459.
90. Lavie L, Reznick AZ, Gershon D. Decreased protein and puromycinyl-peptide degradation in livers of senescent mice. *Biochem J.* 1982;202(1):47–51.
91. Twig G, Elorza A, Molina AJ, et al. Fission and selective fusion govern mitochondrial segregation and elimination by autophagy. *EMBO J.* 2008;27(2):433–446.
92. Kim I, Lemasters JJ. Mitophagy selectively degrades individual damaged mitochondria after photoirradiation. *Antioxid Redox Signal.* 2011;14(10):1919–1928.
93. Suen DF, Narendra DP, Tanaka A, Manfredi G, Youle RJ. Parkin overexpression selects against a deleterious mtDNA mutation in heteroplasmic cybrid cells. *Proc Natl Acad Sci USA.* 2010;107(26):11835–11840.
94. Duvezin-Caubet S, Jagasia R, Wagener J, et al. Proteolytic processing of OPA1 links mitochondrial dysfunction to alterations in mitochondrial morphology. *J Biol Chem.* 2006;281(49):37972–37979.
95. Legros F, Malka F, Frachon P, Lombes A, Rojo M. Organization and dynamics of human mitochondrial DNA. *J Cell Sci.* 2004;117(Pt. 13):2653–2662.
96. Wilkens V, Kohl W, Busch K. Restricted diffusion of OXPHOS complexes in dynamic mitochondria delays their exchange between cristae and engenders a transitory mosaic distribution. *J Cell Sci.* 2013;126(Pt. 1):103–116.
97. Youle RJ, van der Bliek AM. Mitochondrial fission, fusion, and stress. *Science.* 2012;337(6098):1062–1065.
98. Westermann B. Mitochondrial fusion and fission in cell life and death. *Nat Rev Mol Cell Biol.* 2010;11(12):872–884.
99. Skulachev VP. Mitochondrial filaments and clusters as intracellular power-transmitting cables. *Trends Biochem Sci.* 2001;26(1):23–29.
100. Skulachev VP. Power transmission along biological membranes. *J Membr Biol.* 1990;114(2):97–112.
101. Twig G, Hyde B, Shirihai OS. Mitochondrial fusion, fission and autophagy as a quality control axis: the bioenergetic view. *Biochim Biophys Acta.* 2008;1777(9):1092–1097.

102. Busch K, Kowald A, Spelbrink JN. Quality matters. How does mitochondrial network dynamics and quality control impact on mtDNA integrity? *Philos Trans R Soc Lond B Biol Sci.* 2014;369(1646):20130442.
103. Elson JL, Samuels DC, Turnbull DM, Chinnery PF. Random intracellular drift explains the clonal expansion of mitochondrial DNA mutations with age. *Am J Hum Genet.* 2001;68(3):802–806.
104. Wallace DC. Mitochondrial genetics: a paradigm for aging and degenerative diseases? *Science.* 1992;256(5057):628–632.
105. Lee CM, Lopez ME, Weindruch R, Aiken JM. Association of age-related mitochondrial abnormalities with skeletal muscle fiber atrophy. *Free Radic Biol Med.* 1998;25(8): 964–972.
106. Clayton DA. Replication of animal mitochondrial DNA. *Cell.* 1982;28(4):693–705.
107. Berk AJ, Clayton DA. Mechanism of mitochondrial DNA replication in mouse L-cells: asynchronous replication of strands, segregation of circular daughter molecules, aspects of topology and turnover of an initiation sequence. *J Mol Biol.* 1974;86(4):801–824.
108. Gross NJ, Getz GS, Rabinowitz M. Apparent turnover of mitochondrial deoxyribonucleic acid and mitochondrial phospholipids in the tissues of the rat. *J Biol Chem.* 1969;244(6):1552–1562.
109. Huemer RP, Lee KD, Reeves AE, Bickert C. Mitochondrial studies in senescent mice—II. Specific activity, bouyant density, and turnover of mitochondrial DNA. *Exp Gerontol.* 1971;6:327–334.
110. Korr H, Kurz C, Seidler TO, Sommer D, Schmitz C. Mitochondrial DNA synthesis studied autoradiographically in various cell types in vivo. *Braz J Med Biol Res.* 1998;31(2):289–298.
111. de Grey ADNJ. *The Mitochondrial Free Radical Theory of Aging.* Austin: R.G. Landes Company; 1999.
112. Shadel GS, Clayton DA. Mitochondrial DNA maintenance in vertebrates. *Annu Rev Biochem.* 1997;66:409–435.

CHAPTER FOUR

Mitochondrial Dynamics in Aging and Disease

Jürgen Bereiter-Hahn
Institute for Cell Biology and Neurosciences, Goethe University Frankfurt am Main, Frankfurt am Main, Germany

Contents

Abstract

Mitochondria are self-replicating organelles but nevertheless strongly depend on supply coded in nuclear genes. They serve many physiological demands in living cells. Supply of the cytoplasm with ATP and engagement in Ca^{2+} regulation belong to the main functions of mitochondria. In large eukaryotic cells, in particular in neurons, with their long dendrites and axons, mitochondria have to move to the sites of their action. This trafficking involves several motor molecules and mechanisms to sense the sites of requirements of mitochondria. With aging and as a consequence of some diseases, mitochondrial components may be rendered dysfunctional, and mtDNA mutations arise during the course of replication and by the action of reactive oxygen species. Mutants in motor molecules engaged in trafficking and in the machinery of fusion and fission are causing severe deficiencies on the cellular level; they support neurodegeneration and, thus, cause many diseases. Frequent fusion and fission events mediate the elimination of impaired parts from mitochondria which finally will be degraded by autophagosomes. Extensive fusion provides a basis for functional complementation. Mobility of proteins and small molecules within the mitochondria is necessary to reach the functional goals of fusion and fission, although cristae and a large

Progress in Molecular Biology and Translational Science, Volume 127
ISSN 1877-1173
http://dx.doi.org/10.1016/B978-0-12-394625-6.00004-0

fraction of proteins of the respiratory complexes proved to be stable for hours after fusion and perform slow exchange of material.

ABBREVIATIONS

AD Alzheimer's disease
ALS amyotrophic lateral sclerosis
CMT Charcot–Marie–Tooth disease type II
Drp1 dynamin-related protein 1
ER endoplasmic reticulum
GED GTPase effector domain
HR hydrophobic repeats
HTT huntingtin
IMM inner mitochondrial membrane
MAP microtubule associated protein
MICOS and MINOS mitochondrial contact site and mitochondrial inner membrane organizing system
Miefl mitochondrial elongation factor
mitoPLD mitochondrial phospholipase D
Mfn mitofusin
MMP mitochondrial membrane potential
NGF nerve growth factor
OPA1 optic atrophy 1
OMM outer mitochondrial membrane
PARL presenilin-associated rhomboid like
PINK1 PTEN-induced putative kinase protein 1
ROS reactive oxygen species

1. INTRODUCTION

"Where life is, there is motion" was the title of an exciting film showing how dynamic structures in the living cell are. All living matter, organelles and cytoplasm, is exposed to continuous movement and exchange of components. Thus, cytoplasm and organelles are dynamic, as long as a cell or an organism is alive. In the case of mitochondria, different types of dynamical behavior are closely related to their broad spectrum of functions ranging from providing metabolic energy, intermediates for syntheses to the regulation of calcium homeostasis, and control of apoptosis and aging.[1–3] Impaired dynamics is inevitably linked to impaired function resulting in aging and disease.[4–6] Mitochondrial dynamics, including trafficking, fusion and fission, and turnover, via biogenesis and mitophagy is the basis for organelle participation during the development and mitochondrial quality control mechanisms.[7–9]

In many cells, the location of mitochondria changes by being dislocated within the cytoplasm. This trafficking process is required, e.g., for the distribution of mitochondria to daughter cells during cell division, for the supply of energy-demanding processes in cells, which are too large to depend on diffusion of metabolites only. Trafficking is mediated by motor molecules and does not rely on diffusion of vesicles and other organelles. This can easily be estimated by assuming diffusion of a particle with an effective diameter of 0.6 μm (such as a mitochondrion) with a hydrodynamic diffusion coefficient of 0.6 $\mu m^2/s$. Pure diffusion would require about 1.5 h to cover 10 μm, while using molecular motors with a speed between 0.5 and 1.5 μm/s requires 20–6 s for the same distance.[10] Because of their size and high energy demand, neurons represent the cell type most sensitive for disturbances in mitochondrial mobility, including distribution within the cytoplasm, fusion and fission, as well as turnover. These activities are also strongly impaired in aged cells, and this explains the age-related appearance of neurodegenerative diseases.[11]

Fusion and fission of mitochondria are of high physiological significance and demonstrate that a single mitochondrion represents a transitory state only within the total chondriome of a cell, which may partially fall apart followed by new assemblies mixing all mitochondrial components,[12] including their DNA,[13] which is organized in nucleoids[14] and continuously looses integrity during aging.[15–20] In addition, mitochondria can elongate, shorten, and branch. Interruption of these processes results in a variety of diseases, which I will discuss in this review. Fission, fusion and other forms of shape changes require rearrangements of cristae and their components. This overall behavior of a mitochondrial particle has to be related to its molecular basis, the dynamic organization of the respiratory complexes, and matrix constituents, what still is a difficult task.

Many recent reviews address these questions but primarily with the general role of mitochondria in aging and disease, not specifically focused to the role of mitochondrial dynamics, as is the goal of the present contribution. Thus, I will concentrate on the role of mitochondrial dynamics in diseases, including disturbed development of cells and organs and its impairment during aging. Neurodegenerative diseases, heart failure, and cancer are the disorders with the highest risk by aging. Therefore, the sensitivity of this type of disorders to impaired mitochondrial dynamics is in the center of this review. The molecular complexes involved in displacements of mitochondria and driving fusion and fission will only be touched and treated with reference to some recent reviews.

2. MITOCHONDRIAL TRAFFICKING AND LOCALIZATION WITHIN CELLS

2.1. Localization of mitochondria within cells

Intracellular localization of mitochondria strongly corresponds to the sites of energy consumption. Obvious examples are the basal labyrinth, e.g., in mammalian proximal kidney tubule epithelial cells, immune synapses,[21] active growth cones, and the juxtaparanodal zones of the nodes of Ranvier with very high active transport rates in neuronal axons.[22,23] The close proximity of mitochondria to myofibrils in skeletal muscle and to cardiac myofibrils and the smooth cardiac endoplasmic reticulum (ER) with its high calcium transport activities controlling cardiomyocyte contraction are further examples. After reaching their destination during ontogeny, these mitochondria seem to remain stationary. In case of growth cones and synaptic bulbs in neurons, microtubule tracks terminate nearby, promoting disembarkation of mitochondria, probably followed by F-actin based trapping.

Locomoting lymphocytes gather their mitochondria in the uropod region,[24] but only if their size is small. Preventing fission resulted in an extensive network of mitochondria impairing cell polarization to follow chemotactic gradients.[24] The localization at the uropod obviously is essential for ATP supply for the movements. This is in favor of the hypothesis that actin and myosin-based contractions in the rear part of the lymphocytes provide the driving force for the movement.[25] Corresponding significance of mitochondrial localization to the direction and persistence of locomotion has been found for epithelial cancer cells where mitochondria are located in front of the nucleus.[26] In aged cells in culture, mitochondria elongate and may form an interconnected reticulum. These changes reduce trafficking and also other forms of mobility and they improve stress resistance.[5,18]

The developmental potential of maturing mouse oocytes is also related to F-actin-mediated redistribution of mitochondria from the cell periphery to perinuclear clustering.[27]

Most of the mitochondria in neurons and cells in culture are in close proximity of the nucleus.[28] This might have several reasons and functional consequences. For animal cells in culture, the simplest explanation is that cytoplasm is the thickest around the nucleus and houses most of the membranous organelles, and most of the microtubules originate from the microtubule-organizing centers close to the nucleus. Because of its geometry, the perikaryon exerts the least resistance against incoming organelles.

In addition, close proximity to microtubule organizing centers (MTOCs) assures equal distribution to daughter cells during cytokinesis.[29] A mutual interaction between mitochondria and mitotic spindle has been described for fission yeast[30] where spindle rotation is reduced by nucleus-associated mitochondria.[31]

Functionally the import of proteins synthesized in the cytoplasm into mitochondria might be most pronounced in this area acting as a "recreation zone" for mitochondria with impaired compounds. Although mitochondria, which have reduced or lost their membrane potential, may be transported into the perinuclear area, most of the mitochondria reveal high membrane potential. Whether it differs between cell periphery and perinuclear area may vary from cell to cell and still is a matter of dispute, depending on the cell type and the fluorochrome used to reveal the membrane potential.[32] However, because in animal cells, mitochondria move along microtubules, their chance to get aligned the proper way to leave the perinuclear area might be low. Furthermore, many seem to be interconnected to a reticulum around the nucleus.[33] Also the number of binding sites to the ER, as characterized by the frequency of mitochondrial inner membrane-organizing system and mitochondrial contact site (MINOS and MICOS), is the highest in proximity to the nucleus as compared with the cell periphery as was shown by superresolution microscopy.[34] Hypoxia signaling by mitochondrial relocalization was identified as an unexpected signaling pathway between mitochondria and gene expression: In pulmonary artery endothelial cells, hypoxia induced retrograde trafficking of mitochondria and increased reactive oxygen species (ROS) production. This resulted in a local increase of ROS in the nucleus-triggering sequence-specific DNA-oxidation that modified the VEGF promoter.[35,36] The mechanism by which the direction of hypoxia-induced retrograde movement is achieved remains unclear. Still the mechanisms controlling the intracellular distribution of mitochondria and their significance are not fully understood.

2.2. Trafficking of mitochondria within cells

Mitochondria in vertebrate cells cultured on a solid substratum (e.g., glass slide) exhibit vivid movements between cell center and periphery, which were recognized first by Lewis and Lewis in[37] primary cells in culture. These cells are well spread with large peripheral cytoplasmic areas around the central nuclear area. It is the size which requires and allows for extensive dislocations of organelles. Neurons are the cell types with the most extreme extensions (axons and dendrites) emanating from the perinuclear cell body (perikaryon). Neuronal tissue consumes about 20% of the total metabolic energy in humans although it comprises only 2% of body mass.[38] Therefore, neurons are exceptionally depending on intact mitochondria and sensitive against disturbances in mitochondrial functions and trafficking. All microtubules within neuronal processes are well oriented with their minus ends in the perikaryon and plus ends toward the periphery. Mitochondria, as do other vesicular organelles, move in both directions along microtubules in axons and dendrites as well as in other animal cells.[39–43]

As was shown by Morris and Hollenbeck[39] for sympathetic chicken neurons in culture, the speed of mitochondrial movements along microtubules can be accelerated, if microfilaments are destroyed by cytochalasin. This corresponds to a higher branching activity in cytochalasin-treated endothelial cells[33] pointing to viscoelastic resistivity of the microfilament network caging mitochondria,[10] which has to be overcome by moving mitochondria. Changing the filamentous network by RhoA activation in cooperation with formin was found to influence anchorage of mitochondria to actin microfilaments and to control mitochondrial motility.[44] In neuronal processes, devoid of microtubules, mitochondria are also moving in both directions, although at reduced speed and with strong preference for retrograde migration, showing that F-actin can support the movement of these organelles.[39] Thus, microfilaments as well as microtubules influence mitochondrial function and location: in chick sensory neurons, cortical F-actin at the nerve growth factor (NGF)-binding sites immobilizes mitochondria[45] and enhances the sensitivity to apoptotic stimuli,[46] while G-actin protects against apoptosis by closing the voltage-dependent anion channel.[47] Because actin is required to recruit the fission protein dynamin-related protein 1 (Drp1) to the outer mitochondrial membrane (OMM), disruption of F-actin attenuates mitochondrial fission[48] while microtubule-disrupting drugs[49] and vimentin disruption[50,51] induce mitochondrial fragmentation.

The driving forces for slow and fast microtubule-based transports are provided by the members of the dynein and kinesin motor molecule families. Dyneins primarily drive retrograde (minus end directed) transport, and kinesins primarily drive anterograde transport. In fact, the motor molecules are parts of the large protein complexes. They are able to interact in a complicated manner[42] and members of both superfamilies may support movement in either directions.

Dyneins are multiprotein complexes containing two heavy chains with large C-terminal motor heads and microtubule- and nucleotide-binding sites. The intermediate and light chains interact with the dynactin complex and are supposed to bind the cargo.[52]

Among the kinesins, members of the kinesin-1 (KIF5), -2 (KIF3), -3 (KIF1), -4 (KIF4), and -13 (KIF1) families participate in axonal transport.[52] These are heterotetramers consisting of two kinesin light chain (KLC) and two kinesin heavy chain (KHC) subunits. Microtubule- und nucleotide-binding sites are located on the KHCs in their N-terminal motor domains. Cargo-binding is located on the KLC domains, in addition they are involved in the regulation of transport. KIF5A and KIF5C are exclusively found in neuronal tissue, while KIF5B and KLC2 are ubiquitously expressed. KIF1ß mutations have been found closely linked to Charcot–Marie–Tooth disease type II (CMT2A),[53] a finding which could not be reproduced in another set of CMT2A-affected families.[54] These mutations act in concert with mutations of the fusion protein mitofusin (Mfn2).[54,55] Dyneins and kinesins moving along microtubules continuously are facing microtubule-associated proteins (MAPs). Single molecule studies on the MAP tau on these movements were performed using tau-decorated microtubules.[55] When meeting patches of bound tau dynein tended to reverse its direction, while kinesin tended to detach. This reaction was 10 times more sensitive to tau than the inhibition of dynein. Thus, different motor molecule activities are differentially modulated which might be a basis for spatial regulation of microtubule-dependent axonal transport.[56]

After release from microtubules, mitochondria may be either trapped by F-actin or move along F-actin propulsed by myosinV family members.[42] More recently myo19 has been identified as a motor protein-connecting mitochondria to actin.[57] In budding yeast and plant cells, mitochondrial movements are driven via actin cables (for review see, e.g., Ref. 29) using the members of the myoV family as the corresponding motor molecules for anterograde movement. Both the anterograde and retrograde movements in yeast require a protein complex within the outer membrane consisting of the proteins Mdm10p, Mdm12p, and Mmm1p.[58] Propulsion might be provided also by forces developed through actin polymerization or retrograde actin cable flow.[59] The description of the molecular basis of the driving machinery is beyond the scope of this paper and has recently been reviewed.[4,52] The machinery may differ even among relatively closely related fungal species as was summarized, e.g., by Bereiter-Hahn and Jendrach.[60]

A generally accepted hypothesis is that organelles are transported to the cell periphery, where they serve their function and after some time, the organelles age, lipids, and proteins become oxidized via free radicals produced within the mitochondrial respiratory chain or in peroxisomes. Then they are brought back to the central cytoplasmic area, which is densely packed with ribosomes, ER, and lysosomes, thus all the organelles are present needed, either to provide new constituents to impaired organelles or to disintegrate them. However, experiments generating cumulative damage of mtDNA, via a mutation of mitochondrial DNA polymerase, which in humans causes progressive diseases like Alper's syndrome, external ophthalmoplegia, and Parkinson´s disease (PD), lead to an increase in peripheral mitochondrial density in *Drosophila* axons.[61] A dominance of synthetic activities in the perikaryon of a neuron does not exclude mtDNA synthesis or proliferation by fission in peripheral areas of axons and dendrites.[62] Obviously the factors determining intracellular distribution of mitochondria in large asymmetric cells are not fully understood.

Because neurons are the cell type with the longest extensions (in large mammals in the range of meters), which have to be supported with molecules and vesicles generated in the perikaryon, they are particularly prone to study mitochondrial behavior and distribution. Within axons and dendrites, a large fraction of mitochondria is stationary (87% in mouse intercostals nerve explants[63]) while others move in anterograde as well as in retrograde directions (2/3, respectively, 1/3 of the mobile population[63]). Their speed is in the range of 0.4–1.5 μm/s.[10,64] Along isolated ovarian, nutritive tubes up to 7 μm/s have been found.[65] This high speed was probably reached, because the hindering F-actin network was destroyed. A single mitochondrion may move, come to a halt and then continue moving in either directions. In nerves, mitochondria may become trapped at growth cones, the juxtaparanodal zone of the nodes of Ranvier, and they are closely associated with synaptic bulbs.[66,67] Synapse formation[68] and function[69,70] depend on the interaction with mitochondria, the speed of mitochondria near the nodes of Ranvier is significantly reduced, when electric activity raises and becomes accelerated at low axonal electric activity.[70,71] These observations underline the hypothesis of codistribution of energy-requiring and energy-delivering structures. Because of its significance, these associations are secured by several mechanisms, including docking, retention, and transport. Docking and retention at axonal synapses is mediated by syntaphilin which can bind microtubules and associates strongly with stationary mitochondria[72,73] as well as with the dynein light chain LC8. These proteins seem to determine the fraction of stationary mitochondria in neurons.[72,73]

The mechanism of mitochondrial translocation provides some cue to the functional implications. Retrograde movements occur independent of the presence of mitochondrial membrane potential (MMP), while only mitochondria with high MMP move anterograde.[74,75] Therefore, at least for the anterograde movements, mitochondria may directly deliver the ATP required to operate the motor molecules. This explains why granular mitochondria lacking membrane potential accumulate in the perinuclear area of neurons, they also lack the ability to fuse with other mitochondria and, therefore, their fate is either to become degraded or probably, to be "refreshed" by import of nucleus coded proteins. This corresponds to increased retrograde transport in neurons treated with the electron transport inhibitor antimycin A,[61] in *Xenopus* heart endothelial cells, however, inhibition of the respiratory chain by cyanide stops mitochondrial trafficking. Also the inner membrane uncoupler CCCP inhibited movements into both directions within neurons.[61] The inverse relationship between MMP and cytoplasmic ATP-content is in favor of this interpretation.[76] Mitochondria lacking MMP cannot provide the ATP required for their transport from electron transport chain activity, but as shown for vesicle transport in *Drosophila* neurons, ATP may be produced via glycolysis by enzymes bound to the cargo (vesicles),[77,78] resembling microcompartmentation of glycolytic enzymes by actin also allowing to shift ATP supply for mobility from respiratory to glycolytic origin.[79]

2.2.1 Neurodegenerative disorders related to impaired mitochondrial trafficking

The significance of axonal transport for functional integrity of neurons becomes evident from the degeneration of neurons defective in axonal transport of mitochondria (reviewed by Saxton and Hollenbeck[4]). This was shown in *Drosophila* by the inhibition of dynactin components.[80] In humans, dynactin mutations are involved in amyotrophic lateral sclerosis.[81] Formation and reorganization of synapses depend on the delivery of presynaptic components via axonal transport[82] its impairment by dysfunctional motor proteins results in a variety of neuropathic diseases, i.e., MCT2 disease[53] and muscular atrophy.[83]

Axonal transport defects are strongly related to the generation of dysfunctional mitochondria.[11,23,53,84–86] Thus impaired neuronal transport of mitochondria in concert with fusion and fission disturbances (see below) is a main reason for many neurodegenerative disorders, such as Alzheimer's disease (AD), PD and Huntington´s disease (HD), and amyotrophic lateral sclerosis (ALS). Neurons of PD patients exhibit a disrupted microtubule network due to mitochondrial deficits, which lead to the accumulation to alpha-synuclein in neurons of PD patients. In a vicious cycle, these aggregates impair mitochondrial function.[87] On the other hand, improving microtubule-dependent traffic restores autophagic flux, reduces alpha-synuclein oligomer content, increases MMP, and decreases mitochondrial ubiquitination levels, thus rescues PD cells.[87] In PD patients the PTEN-induced putative kinase protein 1 (PINK1) is mutated, and it is supposed to interact with mitochondrial transport via complex formation with MIRO.[23] In HD, the related protein huntingtin (HTT) strongly interacts with the motor system for organelle transport. Phosphorylation of wt HTT acts as a molecular switch for anterograde *versus* retrograde movements,[88] a function which is no longer fulfilled by mutant polyglutaminylated HTT does not support the formation of transport complexes, including kinesin and dynein–dynactin binding to the OMM via the HTT-associated protein 1, and thus disrupts the transport of mitochondria along microtubules.[85,89–91] Neurons from ALS patients express mutant superoxide dismutase 1 inhibiting fast axonal transport via KIF5 motors.[92] Whether this effect is a direct cause of ALS remains open.

In AD, the amyloid-ß and the microtubule-associated protein tau represent the key pathogenic factors,[52,93–96] impairing mitochondrial trafficking and enhancing ROS production. In hippocampal neural cultures, amyloid-ß rapidly inhibited axonal motility of mitochondria. This inhibition was not seen in cells with reduced tau[97] while basal trafficking was not influenced by tau reduction. This was a direct proof of tau protein in concert with amyloid-ß aggregates interfering with mitochondrial axonal transport. However, the mechanism underlying these actions still remains obscure, and many other transport phenomena, in addition to the mitochondrial trafficking, are affected.[52]

2.3. How do mitochondria associate with energy requiring structures within cells?

For this purpose, mitochondria require elements sensing concentration gradients of ADP and Ca^{2+}, which then interact with the molecular motors and either inhibit motor molecule activity or support release of the attachment to the cytoskeletal fibers. Concentration gradients of ADP/ATP and of Ca^{2+} can only be established, if their diffusion is restricted. Although direct measurements of intracellular ATP and ADP diffusivities are very rare, experimental evidence for retarded diffusion has been brought forward, i.e., for cardiomyocytes[98] and for well spread, large endothelial cells: HUVEC was injected with ATP or ADP. This evoked strong contrast enhancement and "contraction" (in case of ADP) or swelling loss of contrast (in case of ATP) of the mitochondria seen in phase contrast. Mitochondria being about 50 μm apart from the injection site required about 90 s (for ADP) to react and several minutes, when ATP was injected.[13] Using raster image correlation spectroscopy of Alexa647-ATP in cardiomyocytes,[99] calculated diffusion coefficients were $83 \pm 14\ \mu m^2/s$ for longitudinal and $52 \pm 16\ \mu m^2/s$ for transversal ATP diffusion, which was up to $3.5 \times$ smaller than the diffusion coefficient in the surrounding medium. Binding of ATP to different cellular components and reactions during diffusion may influence the virtual diffusion coefficient in a cell type specific manner, but certainly diffusion of adenine nucleotides is a parameter limiting maximum energy availability for strongly ATP-requiring processes. ATP gradients may be favored as well by intracellular oxygen gradients, which can be induced experimentally[100] and may also develop under natural conditions. Assuming continuous ATP utilization, this gradient would be the inverse of an ADP gradient.

Ca^{2+} sequestering by ER is a highly ATP-consuming process. In addition, mitochondria themselves participate in Ca^{2+} sequestering and by this ensure proper Ca^{2+} buffering in many cell types.[21,101–105] Intimate apposition of mitochondria with ER-membranes[33,103,106,107] is the structural equivalent of this functional entanglement. Microdomains of Ca^{2+} and sequestration have been found in close proximity between the plasma membrane, ER, and mitochondria, which are supposed to enhance local electron transport and, thus, ROS release.[108] Activation of neurons either by local application of NGF[45] or by synaptic stimulation makes mitochondria more likely to stop at the site of NGF release or at postsynaptic sites,[10,40,109] while the block of neuronal activity with tetrodotoxin accelerated mitochondria.[10] However, mitochondria react differently, whether they are located in dendrites or axons, i.e., overactivation of synapses increases the mitochondrial fraction targeted to postsynaptic sites,[110] while in axons this fraction is decreased.[109]

Inhibition of mitochondrial transport along microtubules may be organelle specific, i.e., without effect on that of other organelles as was shown via transmitter-stimulated Ca^{2+}-elevation,[10,111,112] Mfn2 is an OMM protein mediating such specificity.[113] The Ca^{2+}-sensing complex is located within the OMM and consists of the Ca^{2+}-binding Rho-like GTPase Miro,[114,115] which together with Milton binds to the heavy chain of kinesin 1 (Khc1) and by this attaches the motor molecule to mitochondria[116] (Fig. 4.1). Via binding to the EF-hand domains of Miro, Ca^{2+} changes the conformation of

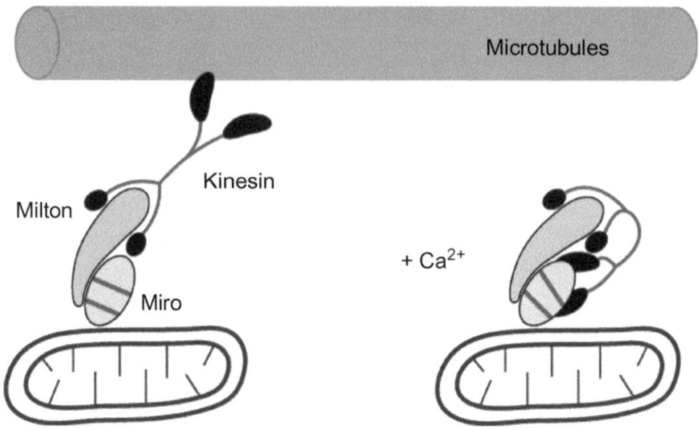

Figure 4.1 Control of mitochondrial trafficking by the Ca^{2+} sensing Miro/Milton complex. This complex binds to the OMM and to the kinesin heavy chain (KHC). At low Ca^{2+} concentrations, kinesin moves along microtubules and by this drives the mitochondrion. Ca^{2+} binds to the two EF-hand domains of Miro. This causes a conformational change in the complex, and KHC loses contact to microtubules. Two alternatives may occur,[22] either KHC remains bound to the Miro/Milton complex in a deactivated form (shown in this scheme) or the kinesin is released from the Miro/Milton complex. In both the cases, driving of mitochondrial movements is interrupted. (See the color plate.)

Miro and thereby blocks Khc1 binding to microtubules. Khc1 localization to the OMM may remain[22,117] (Fig. 4.1).

During antigen-specific T-cell activation, mitochondria localize toward the vicinity of the immune synapse by docking of mitochondria-bound Drp1 to the actin filaments close to the synapse, as was shown by Drp1-silencing abolishing mitochondrial distribution.[118]

3. FUSION AND FISSION REGULATE MITOCHONDRIAL SIZE AND FUNCTIONALITY

In all cell types investigated so far, mitochondria were found to fuse with each other and to undergo fission (Fig. 4.2). These are processes critical for mitochondrial health[121] for embryonic development and to prevent neurodegenerative diseases.[122–124] Fusion of mitochondria with different respiratory deficiencies in hybrids can restore normal function by complementation of mtDNA,[125] thus, fusion provides the possibility to circumvent functional impairments. Mouse knockouts of molecules central to mitochondrial fusion, i.e., Mfn1, Mfn2, and optic atrophy 1 (OPA1), result

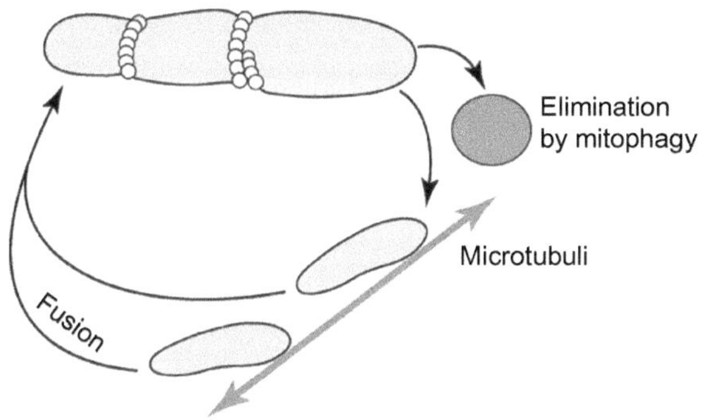

Figure 4.2 Schematic representation of the events shown in Fig. 4.5. Spirals of Drp1 polymers (white circles) establish two fission sites on a tubular mitochondrion with high MMP (bright yellow). The fission products may either retain this high MMP or—in case of dominating dysfunctional components—lose most of the MMP (dark yellow). These mitochondria are not able to fuse while the others can.[119,120] The mitochondria with low MMP become degraded by autophagy (mitophagy). (See the color plate.)

in mitochondrial dysfunction and embryonic lethality.[126–128] Degeneration of retinal glia and ganglia cells leading to the autosomal dominant optic atrophy is caused by the mutations of OPA1.[129,130] The broad range of actions and interactions of molecules related to mitochondrial fission and fusion on the central nervous system still has to be elucidated.[122] The dynamics of chondriome morphology has been modeled by an in-depth mathematical analysis[131] based on the different fusion and fission scenarios known from experimental observations.

Fission and fusion are the main regulators of mitochondrial shape and by this influence dramatically the performance of mitochondria and of the cells. Initiation of apoptosis requires extensive mitochondrial fission. If fission is suppressed by knockout of the fission protein Drp1, no apoptosis takes place, mitochondria become more resistant against stresses. Conservation of the proteins across species, ranging from yeast to mammals, underlines the physiological significance of these processes.

3.1. The fusion and fission machinery and its control

As mitochondria are double-membrane-bound organelles, fusion and fission require the involvement of both membranes (Figs. 4.4 and 4.5). The processes are different at these membranes and may occur asynchronously as was shown for human cells in culture treated with a variety of drugs interacting with mitochondria[132] and on isolated yeast mitochondria fusing *in vitro*.[131] The molecular components interacting in fission and fusion processes were primarily elucidated by studies in yeast,[133–135] but because the processes are evolutionarily conserved, mammalian homologues to the yeast proteins have been identified. We will concentrate on the situation in mammalian cells (Figs. 4.4 and 4.5).

The central group of molecules involved in fission as well as in fusion are evolutionarily well-preserved large GTPases of the dynamin superfamily with a tripartite structure, a GTPase domain, a middle domain, and an assembly or GTPase effector domain (GED) at the C-terminus. In addition to these proteins, Ca^{2+} is involved in the control of mitochondrial fusion and fission. The control mechanisms are not independent from the control of mitochondrial trafficking by Ca^{2+} via the Miro–Milton-complex: Fusion is enhanced at resting state Ca^{2+}-concentrations, while at increased Ca^{2+}-concentrations, fission becomes enhanced[124] due to Drp1 suppression and activation, respectively, by Miro.[114] Also the loss of Miro causes mitochondrial fragmentation and reduced mitochondrial movement (Fig. 4.1), while overexpression of Miro enhances mitochondrial movement and aggregations.[114,136,137]

3.1.1 Fission

In fission, Drp1 is the motor encircling the prospective fission site of a mitochondrion[138] (Fig. 4.2). Drp1 is located in the cytosol and binds locally to a receptor molecule on the OMM. Fis1 for a long time was supposed to mediate this binding as was shown for yeast, however, Fis1 is evenly distributed along a mitochondrion, whereas Drp1 binding is patchy. In mammalian cells, knockdown of Fis1 does not impair fission although overexpression of Fis1 may support fission.[139] In mammalian cells, knockdown studies revealed the mitochondrial fission factor (Mff) as the Drp1 receptor on the OMM[140,141] (Fig. 4.3). Comparable to the endocytotic vesicle membrane constriction by dynamin, Drp1 molecules wrap around the mitochondrion oligomerizing into a higher order spiral structure and constrict the mitochondrial tubule by a GTP hydrolysis-dependent mechanism.[135] This is a highly regulated process involving posttranslational modifications of Drp1. Drp1 translocation to the OMM and GTPase activation are provoked by a variety of stressors from outside and inside the cell inducing mitochondrial fragmentation as an early process in apoptosis or mitophagic clearance of dysfunctional parts. Also during mitosis, enhanced fragmentation has been reported for several cells[143] and activated by Cdk1/cyclinB and aurora A-mediated phosphorylation at S^{616} (in human Drp1),[144] another Drp1-phosphorylating small GTPase is RALA-activating mitotic mitochondrial

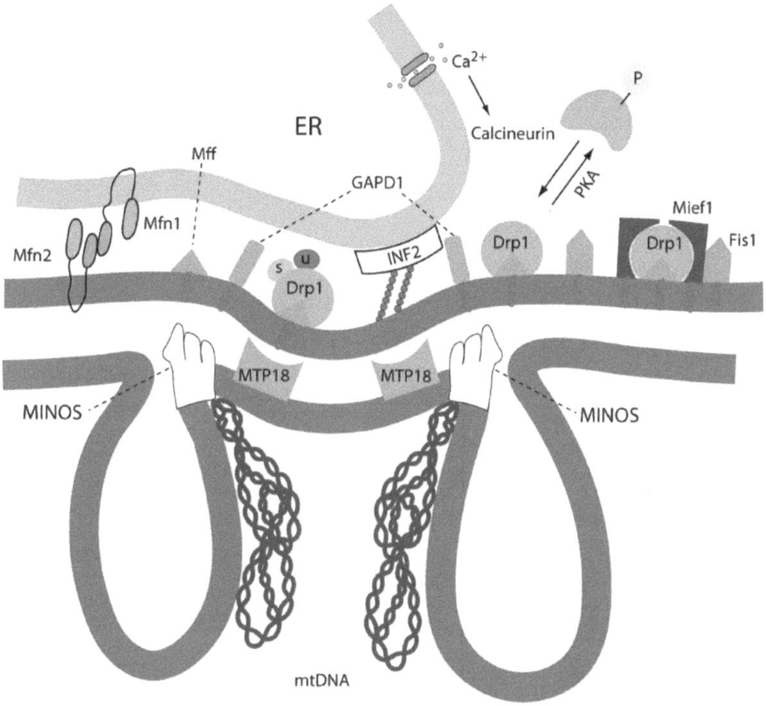

Figure 4.3 Schematic view of the main players in mitochondrial fission. The fission site is close to an ER tubule which is tethered to the OMM by a large protein complex (ERMION), only outlined by the heterotypic Mfn1/Mfn2 interactions. Inverted formin (INF2) bound to the ER membrane organizes actin fibers (red), which start the formation of a fission furrow. The main fission activity is performed by Drp1, which becomes activated either by phosphorylation (e.g., at S^{616}) or by calcineurin-mediated dephosphorylation (at S^{637}) in the presence of high Ca^{2+} released from the ER or influx via the plasma membrane and then binds to the OMM protein Mief1 intensified by Fis1. Ubiquitination (u) or sumoylation (s) of Drp1 contributes to full activation (Drp1 in fission furrow); GAPD1 supports Drp1 polymerization (shown in Fig. 4.2) and wrapping around the mitochondrion. GTPase activity of OMM-bound Drp1 can be inhibited by Mief1, which also binds to Fis1 and by this reduces fission. On the IMM, two protein complexes dominate the behavior. MTP18 interacts with Drp1 from the inner side of the mitochondrion and binds to the MINOS (MITOS/MICOS) complex within the IMM. The MINOS complex (MW in the Megadalton range) contains six IMM proteins, including mitofilin, and is a part of the cristae junctions. It forms a cristae-organizing center linking protein complexes of the OMM, including ER-OMM-binding elements and porin, to the IMM and mtDNA. These multiple partners and their binding are not fully known on a structural basis and, therefore, not included in the scheme. Their entirety was termed ERMIONE.[142] This complex guarantees mtDNA distribution to both the sides of the fission furrow. (See the color plate.)

fission in human HEK cells.[145] Fission is inhibited by protein kinase A-based phosphorylation of a serine within the GED (S^{637} for human, S^{656} for rat Drp1) and by this inhibits GTPase activation and probably also the recruitment of Drp1 to the OMM,[146] however, this action may be cell type specific.[147] Depending on calcium levels, calcineurin dephosphorylation counteracts this type of phosphorylation and supports Drp1 targeting to the OMM.[148,149]

Further, posttranslational modifications are nitrosyslation at Cys^{644}[150] and sumoylation, which both promote mitochondrial fission and pave the way to apoptosis.[151,152] Abnormal mitochondrial elongation can be caused by the downregulation of the Drp1-ubiquitinylating E3 ubiquitin ligase March5 (also known as MITOL).[153] F-actin facilitates the recruitment of Drp1 to mitochondria,[48] which may form a complex with the proapoptotic protein Bax.[154] GDAP1 is an OMM-associated protein, which can cause fragmentation without changing mitochondrial membrane potential.[155] Coexpression with Mfn1 or Mfn2 rescues mitochondrial fragmentation, also if truncated GDAP1 causes mitochondrial elongation.[155]

At the inner mitochondrial membrane (IMM), MTP18 acts in concert with Drp1 at the OMM[156] (Fig. 4.3). MTP18 is a downstream effector of PI3K on the cristae junctions; the MitOS/MINOS/MICOS (mitochondrial organizing structure/mitochondrial inner-membrane organizing system/mitochondrial contact site—these terms relate to the same topic but have been described by different authors[142]) complex links the nucleoid position (Fig. 4.3), the IMM cristae junctions with the OMM.[157] The exact mechanism of IMM fission and synchronization of the events at IMM and OMM are not yet known. At the IMM, a small protein named MTGM is highly expressed in human brain and in tumor cells, it is identical with the reactive oxygen species modulator (ROMO1). Overexpression of MTGM results in mitochondrial extensive fission in concert with Drp1 and inhibits cell proliferation by stalling cells in S phase, but does not cause apoptosis.[158] Thus, MTGM links mitochondrial morphology to cell proliferation. ROMO1 is essential for the oligomerization of OPA1 and, therefore, knockout of ROMO1 has similar effects as the loss of OPA1 (loss of cristae, increased sensitivity to cell death stimuli, fragmented mitochondria).[159]

Mitochondria–ER associations are interesting players within the fission process. The "ERMES" complex (ER–mitochondria encounter structure) has been studied in detail in budding yeast,[106] and it determines fission sites and is close to mtDNA: An ER tubular process bound to the OMM wraps around sites of mtDNA location within the matrix and by this determines fission sites[157,160] (Fig. 4.3). The mitochondria-associated ER membrane forms a macromolecular complex with many constituents which has been reviewed recently by Hayashi *et al.*[161] The key event in fission-related ER-engagement seems to be its binding of inverted formin2 which accumulates actin filaments driving an initial constriction which then allows Drp1 to drive a secondary constriction[162] (Fig. 4.3).

3.1.2 Fusion

Mitofusins (Mfn1 and Mfn2) are the motors driving OMM fusion. Their two transmembrane regions anchor to the OMM and thus mediate the tethering of two adjacent mitochondria (Fig. 4.4) either in tip/side direction or in tip/tip direction. Mutant Mfn2 lacking the two transmembrane domains acts as dominant negative mutants and blocks fusion,[163] and it changes mitochondrial morphology, this becomes ovoid or spherical and shows a

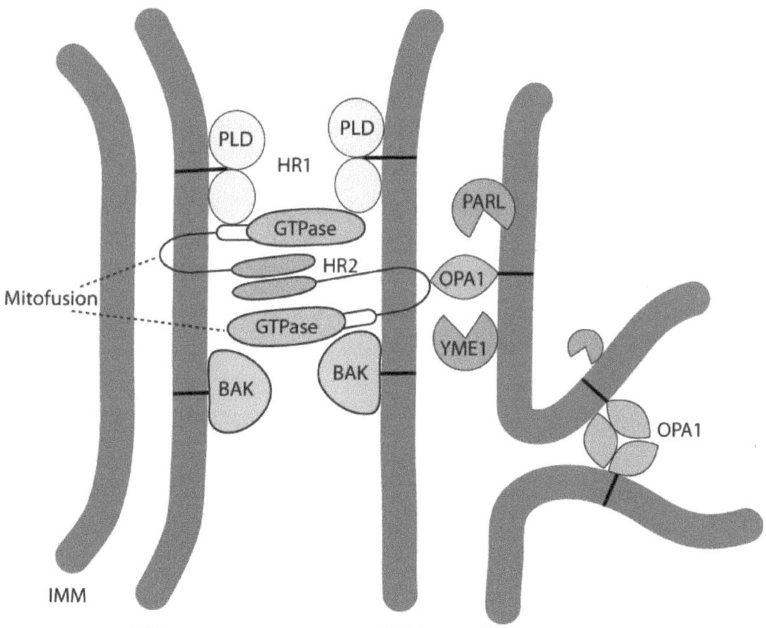

Figure 4.4 Schematic view of the main players in mitochondrial fusion. Mitofusins on the OMM mediate tethering of the two mitochondria via their hydrophobic repeats (HR2). For fusion, GTP splitting is also required. Fusion of membranes is supported by mitochondrial phospholipase (mtPLD). Association of the proapoptotic BAK relates fusion to apoptosis. On the IMM (inner mitochondrial membrane) side, OPA1 interacts with Mfn and forms complexes at the cristae junctions. Posttranslational modification of OPA1 is performed by various proteases (PARL, YME1). (See the color plate.)

"Brownian-type" motility.[126] For fusion, GTPase activity and the interaction between the N- and C-terminal tails are required. Homotypic interaction occurs along hydrophobic reapeats[164] (Fig. 4.4). In addition, Mfn2 mutants may cause the loss of MMP[163] and by this prevent fusion. IMM fusion is mediated through OPA1, which in general controls IMM structure and its remodeling. OPA1 also is a Drp1, which promotes the formation of a branched network of elongated mitochondria but requires the help of mitofusins to evoke this mitochondrial shape.[165] OPA1 is an integral protein of the IMM but faces the intermembrane space[166] and participates in shaping cristae[167] (Fig. 4.4).

Eight alternative splicing forms exist, opening multiple pathways to control mitochondrial fusion altering the spectrum of OPA1 isoforms.[168] Some of them are characteristic for vertebrates and related to apoptotic events.[169] OPA1 is required not only to shape cristae but also to maintain MMP[170] in concert with Mfn1 but not with Mfn2.[171] Posttranslational maturation of OPA1 occurs by proteolytic splitting via members of the rhomboid family—presenilin-associated rhomboid like (PARL)[172]—paraplegin or Yme1 (for review see Refs. 166,173) (Fig. 4.4). Fusion of the two membranes may occur independently as was shown for isolated mitochondria,[174] but IMM fusion requires an intact MMP.[119,120] A further protein in mitochondrial fusion is the mitochondrial phospholipase D (mitoPLD).[175] This dimeric enzyme binds to the OMM and has been shown to generate phosphatidic acid by cardiolipin hydrolysation, what might facilitate membrane fusion.[173] The exact sequence of events and macromolecular interaction during the fusion process of mitochondria in mammalian cells are not known yet.

In mammalian cells, the mitochondrial elongation factor (Mief1; also known as MiD51) promotes elongation and fusion of mitochondria independent of mitofusins. It recruits Drp1 to the OMM but inhibits its activity (Fig. 4.3). Depletion of Mief1 causes mitochondrial fragmentation. Mief1 also interacts with hFis1, which may reverse the Mief1 induced phenotype.[176]

3.2. The physiological significance of fusion and fission

3.2.1 Ubiquitous fusion and fission

The only cells which have long been supposed to be excluded from mitochondrial fusion and fission are adult invertebrate and vertebrate cardiomyocytes. This coincides with the stationary character of these mitochondria. However, recently the existence of a fusion/fission cycle in adult mouse cardiomyocytes, which lasts for about 16 days, has been proven using Mfn1/Mfn2 ablation.[177] Also electron microscopy of rat cardiomyocytes supported the assumption of fusion and fission taking place.[178] Gene trapping of mitofusins or OPA1 arrested mouse heart development and impaired differentiation of cardiomyocytes by interaction with Notch signaling.[179] In *Drosophila*, MARF RNAi-induced cardiomyopathy was rescued by human mitofusin. This observation is an indirect proof for fusion and fission requirements in *Drosophila* adult cardiomyocytes as well.[180]

The dependence of cardiomyocyte integrity on intact fusion/fission cycles points to the general significance in metabolic regulation. This is further stressed by changes of mitochondrial length in mouse acouti-related peptide (Agrp) neurons in the arcuate nucleus of the hypothalamus (which produce the Agrp and the neuropeptide-Y), which stimulate feeding.[181] Their electric activity is the highest when the whole body is deprived from feeding. In this situation mitochondrial number decreases but their size increases. The opposite reaction is found in the close POMC neurons, which functionally counteract the Agrp neurons. Knocking down mitofusins impaired the activity of Agrp neurons during high-fat diet, which can be restored by cell-selective ATP administration,[181] showing that Mfn1 and Mfn2 in Agrp neuron are part of the system-regulating body weight.

Like in cardiomyocytes also in other cell types, the fusion/fission ratio is well balanced,[18] but the frequency depends on the cell type (relatively low in neurons) and decreases drastically at the onset of proliferative senescence.[18] Following the hypothesis of Kowald and Kirkwood that fusion of mitochondria evolved as a consequence of gene transfer to the nucleus, when a single set of genes has to be active only serving all the mitochondria with very different demands[182] instead of multiple copies spread within the chondriome. Then equal distribution of nuclear-coded proteins is achieved by fusion. If continuous fusion is the evolutionary requirement, this can only be reached if balanced by fission.

Mitochondrial networks are almost immobile and, thus, can serve their physiological role only in cells and cellular areas, where trafficking is not required, e.g., in proximity to the nucleus.[33,40,183] The short tubular or spherical mitochondria represent the trafficking fraction, they are the mobile ones being transported to the sites of energy requirements or needs in Ca^{2+} homeostasis (see earlier).[40] In neurons, mitochondria trapped at dendritic spines of axonal buttons neither fuse nor divide as long as they remain stationary.[66] Shifts in the fusion/fission ratio in general are related to cell fate. Relative increase of fission yields fragmented mitochondria. Exceeding fission occurs in the preparation of mitosis, or as a reaction to external stresses (e.g., exposure to elevated ROS levels,[184,185] or uncouplers of oxidative phosphorylation[170]) often as an initiation of apoptotic cell death, including neuronal degenerations or premature senescence.[186] If fission results in mitochondrial fragments with different physiological performance, i.e., one with high and the other with reduced MMP (Fig. 4.5), those with low MMP are no longer able to fuse[8,187] and are prone to autophagic degradation[188] (as indicated in Fig. 4.2). By this, fission is an important step in getting rid of dysfunctional mitochondria or mitochondrial components. However, unbalanced fission of the mitochondrial network in skeletal muscle can activate a FoxO3-dependent atrophy program resulting in muscle atrophy.[189]

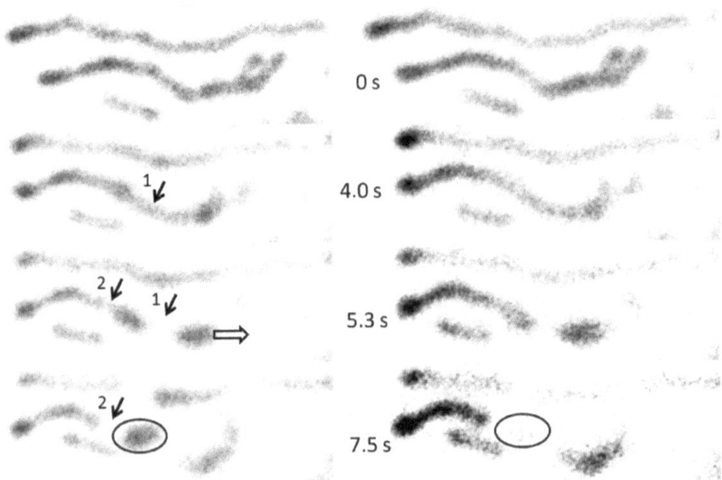

Figure 4.5 Mitochondria in HeLa cells double stained by GFP-CI as a marker for an IMM protein and TMRE to reveal MMP. Fluorescence images are shown as negatives (bright fluorescence appears black). Two subsequent fissions occur, marked with 1 and 2, which result in three mitochondria, two still shows TMRE fluorescence (intensity is lower than at the beginning because of photobleaching), while the middle one (encircled) lost TMRE fluorescence at 7.5 s. After the first fission (at 5.3 s) the smaller part on the right slightly moves away (open arrow). Aggregation of IMM at the site of the future dysfunctional mitochondrion becomes visible in the 4.0 s GFP-frame, no indication in the TMRE set because of its fast diffusion within the mitochondrial matrix, TMRE does not show local effects, the general reduction of TMRE fluorescence is due to photobleaching. Fission takes place despite pretreatment of the cells with nocodazol (25 μM, 30 min) and cytochalasin D (2 μg/mL, 20 min). (This mitochondrial event has been published recently.[60])

Mutants of Drp1 interfere with its fission activity causing extremely elongated mitochondria and reveal the function of some of the molecular domains, i.e., the GTP-binding defective mutant K38A acts as a dominant negative mutant inhibiting the localization on the mitochondrial fission sites.[138,190] The loss of fission ability protects against apoptotic cell death and considerably delays aging in the filamentous fungus *Podospora anserina* and in budding yeast *Saccharomyces cerevisiae*[191] and it delayed mitochondrial cytochrome *c* release and subsequent apoptosis.[192] On the other hand, HeLa cells with mitochondria of reduced size overexpressing Drp1 were more resistant against Ca^{2+}-induced apoptosis than wt cells.[193] These results are not contradictory, because life span will not be limited by Ca^{2+}-related apoptosis. In particular in mammalian cells, a long senescence phase precedes cell death. The role of Drp1 as that of other proteins involved in mitochondrial fission and fusion by far exceeds its function as a motor molecule, they are part of a molecular network pervading the cell in its totality. This is also shown by the dominant-negative mutant A395D in Drp1, which leads to abnormal brain development, optic atrophy, and hypoplasia.[194]

3.2.2 Fusion and fission as related to aging

Reduced expression of hFis1 in mammalian cells induced premature senescence as revealed by morphological changes, lower proliferation, and increased ß-galactosidase activity. The depletion of both hFis1 and OPA1 also resulted in extensive mitochondrial fragmentation but rescued cells from reaching the senescent state.[195]

In senescent endothelial cell cultures (postproliferative stage), mitochondria become stationary,[18] elongate and may form a continuous reticulum also in the periphery due to the downregulation of Fis1 and Drp1 expressions and upregulation of Mfn1 and Mfn2.[5,62] Such a reduction of Drp1 together with an increase of the fusion factor Mfn1 may also be induced by the knockdown of MARCH5 which then caused the senescence of Chan cells.[196] On the other hand, increased length and reticulum formation improve the resistance against stresses, i.e., ROS induced loss of MMP by the upregulation of PINK1 expression, when Drp1 becomes reduced.[5,139,197,198] Stress-induced mitochondrial hyperfusion in mouse embryonic fibroblasts may act on the same line.[199] This process is mediated by SLP2 but independent of Mfn2 and thus differs from the senescence-related elongation, because in senescent endothelial cells, SLP2 is strongly downregulated.[5] The physiological advantage of the reduction of the mobile fraction of the chondriome in aged cells can be assumed to be increased resistance against the higher ROS production and the counterbalance of reduced MMP by increasing the mass of mitochondria per cell.[200] Because these cells no longer divide, distribution of mitochondria to the daughter cells does not require small units. Thus, fusion and fission of mitochondria are closely related to aging. In nerve cells, however, a mitochondrial reticulum is not the way out of loss of function in single smaller units, because neurons require trafficking of mitochondria to guarantee Ca^{2+} homeostasis and ATP supply for the peripheral parts of dendrites and axon. The only way neurons can survive and cope with dysfunctional mitochondria is their elimination via autophagy (mitophagy).[201]

ROS production by mitochondria has long been considered a central cause of aging[202,203] and was extensively reviewed.[6,204] During the aging process, all mitochondrial functions decline, and thus, increase the incidence of age-related disorders.[205,206] Dysfunctional mitochondria accumulate with decreased respiration, change in fusion/fission balance and accumulation of mitochondrial degradation products,[207,208] which then increase ROS sensitivity[209] and accelerate further dysfunctions.

Considering that insulin resistance is one of the severe markers for advanced aging, the action of hyperglycemia on mitochondrial dynamics is of interest. At high glucose levels, cells in culture show mitochondrial fragmentation, enhanced respiration, and ROS production.[210] Exposure to free radicals itself stimulates mitochondrial fragmentation and increases ROS formation,[184] thus, a vicious cycle might be induced by hyperglycemia. This view is supported by the observation that starvation triggers not only the autophagic degradation of organelles but also the enlargement of mitochondria by the inhibition of fission via PKA-mediated Drp1 phosphorylation, impairing its binding to the OMM. Electron transport in elongated mitochondria is optimized for ATP production.[211] Because neurons are supposed to encounter more oxidative stress than other tissues, they are prone to neurodegenerative diseases, in particular if counteracting mechanisms do not function properly.[212]

3.2.3 Fusion and fission as related to apoptosis and neuronal degeneration

The loss of OPA1 suddenly activates apoptosis,[213] because it supports cytochrome *c* release. However, OPA1 does not interfere with the mitochondrial apoptosis regulators Bax and Bak, but oligomeric OPA1 keeps the cristae junctions tight in a function genetically and molecularly distinct from its role in mitochondrial fusion and maintenance of cristae shape.[170]

Mutations of OPA1 are associated with the autosomal-dominant optic atrophy. Homozygous mutations in a mouse model with truncated OPA1 are embryonic lethal, but heterozygous mutant mice showed increased mitochondrial fission and slow degeneration of the optic nerve.[128] Regardless which of the fusion involved proteins is mutated, the lack of mitochondrial fusion causes severe defects, poor cell growth, wide heterogeneity of MMP within one cell, and decreased respiration.[9]

In general, inhibiting mitochondrial fusion promotes apoptosis,[214] independent of the underlying molecular reasons. For example, silencing Mfn1 or Mfn2 increases the sensitivity to apoptotic stimuli,[198] while overexpression reduced Bax-activation, cytochrome release, and apoptotic cell death.[215] In addition, Mfn2 affects the ERK/MAPK signaling pathway,[216] this activity seems to be independent from the role in mitochondrial fusion.

Comparable to impaired mitochondrial trafficking, mutations in the main fusion and fission players have all been causally linked to neurodegenerative conditions as was recently reviewed by Oettinghaus *et al.*,[124] therefore, only some examples are listed here. In HD patients, HTT is expanded by polyglutamine (polyQ), which might be proteolytically cleaved into polyQ-containing N-terminal fragments. With aging these mutant fragments accumulate and have a high tendency to misfold and aggregate. Such aggregations impair microtubule-based transport and mediate mitochondrial fragmentation.[89] Mitochondrial fission activation by mutant HTT relies on its interaction with Drp1 and stimulation of its enzymatic activity. Neuronal cell death can be rescued by reducing Drp1 GTPase activity expressing the dominant-negative DRP1 K38A mutant.[90] In myelinated peripheral nerves, fusion/fission equilibrium disturbed by GDAP1 mutations is closely related to CMT2 disease.[155] Also in AD, the fusion/fission balance is impaired by reduced levels of OPA1, Mfn1, and Mfn2, while levels of Fis1 were significantly increased leading to extensive mitochondrial fragmentation and in concert with impaired trafficking to reduced mitochondrial density in the cell periphery.[217] Thus, both fusion and fission seem to be compromised in AD.[218] Mutations in the Mfn2 cause CMT2, this seems to be related to impaired dendritic outgrowth and spine formation of Purkinje cells, which require Mfn2 but not Mfn1.[219]

3.2.4 Linking trafficking, fusion and fission

With the exception of very densely packed mitochondria in some myocytes, only moving mitochondria are likely to meet and to fuse,[188,220] linking the fusion and fission machineries functionally to the trafficking machineries. This, however, does not imply a relationship between mitochondrial size and speed of movements. Only very long mitochondria and mitochondrial reticula remain stationary, but still can develop branches or bend.

In pancreatic ß-cells, fission has been found to occur more frequently after mitochondria fused. In this case fission is a process paired with fusion, while fusion does not depend on fission events.[8] Fission might result in mitochondrial particles with different MMP (Figs. 4.1 and 4.5). Thus, two populations arise: those which are consistently fusing and dividing and those which are fusing no more, which then are prone to become degraded by autophagy.[8,188] Fusion of mitochondria together with fission comprises an effective means for material distribution throughout the chondriome as was shown by spreading of photo-activated mtGFP from a few irradiated mitochondria to the whole chondriome by the spread of photo-activated IMM proteins (ATP-binding cassette promoter B10) or matrix-targeted PA-GFP.[8] In case of spreading of photo-activated fluorescence from labeled to unlabeled mitochondria, decay was measured by fluorescence decay in INS cells and mouse pancreatic ß cells. Decay was 8.1 times faster for matrix-targeted (10–30 min) than for IMM-targeted fluorochromes (half time up to 60–70 min),[221,222] a similar time scale was found for GFP-labeled C1,[223] however, the distribution of fluorescence within the mitochondria was patchy, not homogenous (comp. Fig. 4.6).

3.2.5 The role of fusion and fission for mtDNA integrity

Mitochondria contain their own DNA (mtDNA), which is organized in hundreds of nucleoids per cell. Nucleoids are of varying size and each contains several mtDNA molecules.[224–226] Within a mitochondrion, these nucleoids are almost static, most probably because they are bound to IMM proteins,[225] they do not appear to exchange DNA among each other, but heterologous nucleoids can complement each other apparently by diffusion of mtDNA transcripts.[227] Nucleoids have been observed to divide[160,224,228] and move apart from each other with a speed comparable to separation of cristae.[229] In many mitochondria, nucleoids show preferred distances to each other. This is in favor of the hypothesis that each group of mtDNA serves for a certain region by transcript diffusion surrounding its confined position. With age, mtDNA is subjected to deletions and to mutations, which in part can be eliminated by repair mechanisms (for review see Ref. 230), but in particular a 4977 bp deletion is a common event in aged cells.[20,231] The continuous fusion and fission processes redistribute mtDNA

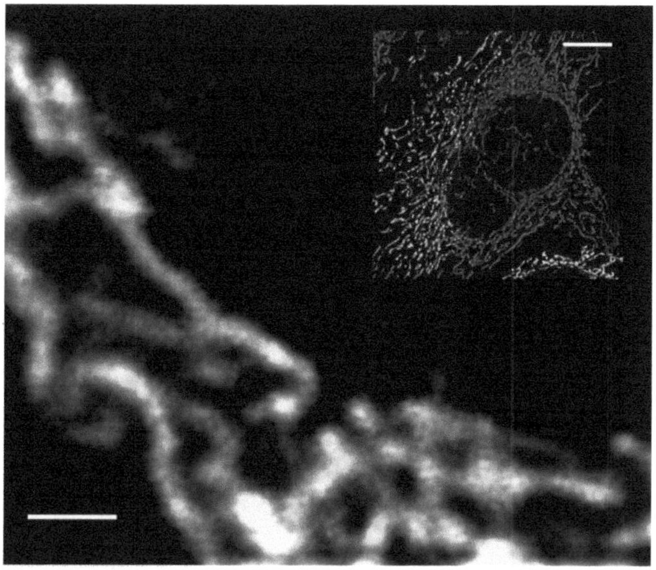

Figure 4.6 Fused mitochondria, 5 h after induction of cell fusion with PEG. CV has been labeled on both the fused cells, in the 3T3 cell with $eGFP_{monomeric}$ and in the HeLa cell with $RFP_{monomeric}$. The resulting hybrid mitochondria show all three possible situations: green and red areas, where the CV complexes remained separated and yellow areas where the CV complexes have mixed. Insert: chimera with differently labeled mitochondria 2 h after fusion. Bar: 2 μm, in insert: 10 μm. *These images have been kindly provided by K. Busch, Osnabrück.* (See the color plate.)

throughout the chondriome. This dynamics has been considered an important factor in delaying senescence due to the delayed accumulation of mutant or truncated mtDNA.[182,232] Tissue-specific dominant mtDNA mutations obviously arise from single mutation expansion and not from ongoing mutational events.[15] The mathematical models describing different scenarios do not explain how a single mutation becomes dominant within a tissue. Fission of mitochondria may produce tubules lacking mtDNA,[224] which in a following fusion process again will be connected to mtRNA supply. In fibroblasts lacking this way, most mitochondria become devoid of nucleoids.[219] Fusion of cells with mitochondria lacking mtDNA with those bearing nucleoids results in the homogenous distribution of nucleoids throughout the chondriome after 12–24 h what has been interpreted as intramitochondrial mobility of nucleoids by the authors,[226] but could just be a result of several fusion and fission cycles. In Drp1-deficient cells or cells undergoing terminal differentiation, nucleoids are enlarged and clustered in hyperfused mitochondria.[224] The same happens if in fission-deficient mitochondria, mtDNA accumulates in bulb-like structures, and apoptotic events are delayed in these cells.[233] ER supports fission close to sites of mtDNA within the matrix resulting in nucleoid separation[157] (Fig. 4.3). Nucleoid separation seems to be a process preceding mitochondrial fission,[160] this is the reason why Drp1 does not colocalize with mtDNA. As discussed earlier, fission can result in very heterogenous mitochondrial tubules (Fig. 4.5). In the case of a confined zone of mtDNA-controlled protein synthesis, deficient DNA could be responsible for the synthesis of dysfunctional proteins which then are removed by fission together with the DNA tethered to IMM proteins. This speculation fits to all the experimental observations but still needs to be proven.

4. DYNAMICS OF PROTEINS WITHIN MITOCHONDRIAL MEMBRANES AND THE MATRIX

Complementation of mitochondrial deficiencies by fusion and separation of dysfunctional components by fission require mobility within the mitochondria. Separation of dysfunctional proteins as shown in Fig. 4.5 would not be possible, if all proteins move faster than the preparation for the fission process can gather these proteins and eliminate them. These movements are all described as diffusive movements although the involvement of active components cannot be excluded.

Small molecules have been shown to diffuse rapidly through the mitochondrial matrix: DsRed diffusion was measured by FRAP and diffused with about 1 μm/s,[28] GFP diffusion in the mitochondrial matrix of fibroblasts, liver, skeletal muscle, and epithelial cell lines as measured by FRAP after spot photobleaching (0.8 μm spot diameter) showed >90% of the fluorescent protein being mobile with diffusion coefficients $2–3 \times 10^{-7}$ cm^2/s.[234] This is 3–4-fold less than the diffusion of GFP in water.[235] Previous studies based on the fluorescence anisotropy of fluorescein in isolated rat liver mitochondria either in orthodox or in condensed conformation and in different osmotic environments revealed much higher values for viscosity and less diffusion,[236] probably because of some immobilization of the dye molecules by the attachment to proteins and by the fact that fluorescence polarization does not measure real diffusivity rather than viscosity in the immediate molecular environment. On the other hand, enzymes of the tricarboxylic acid cycle marked with GFP exhibited low diffusion (ca. 4×10^{-8} cm^2/s), and 45% of the different enzymes were mobile.[237] Considering the very similar values for the four enzymes, this was interpreted as an indication for multienzyme complex formation. The main shortcoming of these measurements, however, is that the bleaching area in these FRAP experiments, exceeded single mitochondria by far according to the images shown. Therefore, the measurements might have revealed mixing of mitochondrial contents by fusion rather than diffusion.

Observing short time contact of fusing mitochondria, which divide several seconds later (kiss-and-run contacts), mixing of matrix material within the whole fusion construct was apparent.[8,220] For IMM proteins, the situation is different as is exemplified by the patchy appearance of the fusion products (Fig. 4.6).[238] The Harlekin view of labeled IMM proteins in long mitochondria may result from a series of consecutive fission and fusion events.[239] No such diffusion restrictions exist for OMM and matrix proteins.[239] The experiments using fusion of cells with differently labeled cells last several hours, and even during this long time still some cristae may preserve their original protein pattern,[239] while others mix (Fig. 4.6). This corresponds to the persistence of cristae. Only a few of them may show lateral displacements, disappear or newly formed cristae may appear.[229,239] Calculations of diffusivity of proteins of the respiratory complexes (mCherry-COX8a and EGFP-$F_1F_oATPase$, γ-subunit) according to the FRAP measurements revealed an almost immobile fraction with 0.007 $\mu m^2/s$ for COX and 0.5×10^{-11} cm^2/s for $F_1F_oATPase$ in HeLa cells. The mobile fraction showed 0.3–0.4 × 10¯8 cm^2/s.[240] The ratio between mobile and immobile fractions might depend on the degree of overexpression of the proteins and the relative amount of labeled proteins integrated in the respiratory complex.[241] Diffusivity values have been calculated on the assumptions made by different models. Sukhorukov and Bereiter-Hahn proved that the two fraction model was well suited[242] while Appelhans *et al.* found anomalous diffusion describing the behavior of single molecules best.[243] Diffusivities of the OMM proteins hFis1 and TOM7 were 0.6 and 0.7×10^{-8} cm^2/s, respectively, and 93% proved to be mobile.[240] Using superresolution microscopy, following the movements of single molecules within mitochondrial cristae revealed three subpopulations of membrane protein mobility, a mobile, a less mobile, and an immobile one. Diffusivity of $F_1F_oATPase$ (CV) in HeLa mitochondria was $D_{app} = 0.082 \times 10^{-8}$ cm^2/s, the second with $D_{app} = 0.019 \times 10^{-8}$ cm^2/s,[239] and the immobile fraction (15%) $D_{app} = 0.004 \times 10^{-8}$ cm^2/s. Complex 2 proved to be more mobile than CV with $D_{app} = 0.039$ for the mobile fraction and $D_{app} = 0.006 \times 10^{-8}$ cm^2/s for the immobile one.[239] Among OMM proteins, Fis1 showed almost unhindered mobility, the D_{app} varying with mitochondrial shape was in the range of $D_{app} = 0.41 \times 10^{-8}$ cm^2/s corresponding to the mobility of TOM20 ($D_{app} = 0.422 \times 10^{-8}$ cm^2/s), which as part of a larger complex also showed less mobile and apparently immobile fractions (0.215 and 0.002×10^{-8} cm^2/s, respectively). This corresponds to a better mixing of CII than CV in fused mitochondria.[223]

ACKNOWLEDGMENTS

I am very grateful to Barbara Tarazi from the group of Prof. Osiewacz for preparing Figs. 4.1–4.3 and to Prof. Osiewacz for critical reading of the paper. I am very indebted to Prof. Dr. Karin Busch for Fig. 4.6.

REFERENCES

1. McBride HM, Neuspiel M, Wasiak S. Mitochondria: more than just a powerhouse. *Curr Biol.* 2006;16(14):R551–R560.
2. Heath-Engel HM, Shore GC. Mitochondrial membrane dynamics, cristae remodelling and apoptosis. *Biochim Biophys Acta.* 2006;1763(5–6):549–560.
3. Spät A, Szanda G, Csordás G, Hajnóczky G. High- and low-calcium-dependent mechanisms of mitochondrial calcium signalling. *Cell Calcium.* 2008;44(1):51–63.
4. Saxton WM, Hollenbeck PJ. The axonal transport of mitochondria. *J Cell Sci.* 2012;125(Pt 9):2095–2104.
5. Mai S, Klinkenberg M, Auburger G, Bereiter-Hahn J, Jendrach M. Decreased expression of Drp1 and Fis1 mediates mitochondrial elongation in senescent cells and enhances resistance to oxidative stress through PINK1. *J Cell Sci.* 2010;123 (Pt 6):917–926.
6. Bereiter-Hahn J. Do we age because we have mitochondria? *Protoplasma.* 2014;251 (1):3–23.
7. Tatsuta T, Langer T. Quality control of mitochondria: protection against neurodegeneration and ageing. *EMBO J.* 2008;27(2):306–314.
8. Twig G, Elorza A, Molina AJA, et al. Fission and selective fusion govern mitochondrial segregation and elimination by autophagy. *EMBO J.* 2008;27(2):433–446.
9. Chen H, Chomyn A, Chan DC. Disruption of fusion results in mitochondrial heterogeneity and dysfunction. *J Biol Chem.* 2005;280(28):26185–26192.
10. Mironov SL. Spontaneous and evoked neuronal activities regulate movements of single neuronal mitochondria. *Synapse.* 2006;59(7):403–411.
11. Chaturvedi RK, Flint BM. Mitochondrial diseases of the brain. *Free Radic Biol Med.* 2013;63:1–29.
12. Jezek P, Plecitá-Hlavatá L. Mitochondrial reticulum network dynamics in relation to oxidative stress, redox regulation, and hypoxia. *Int J Biochem Cell Biol.* 2009;41(10):1790–1804.
13. Bereiter-Hahn J, Vöth M. Dynamics of mitochondria in living cells: shape changes, dislocations, fusion, and fission of mitochondria. *Microsc Res Tech.* 1994;27(3):198–219.
14. Hensen F, Cansiz S, Gerhold JM, Spelbrink JN. To be or not to be a nucleoid protein: a comparison of mass-spectrometry based approaches in the identification of potential mtDNA-nucleoid associated proteins. *Biochimie.* 2014;100:219–226.dx.doi.org/10.1016/j.biochi.2013.09.017.
15. Wiesner RJ, Zsurka G, Kunz WS. Mitochondrial DNA damage and the aging process: facts and imaginations. *Free Radic Res.* 2006;40(12):1284–1294.

16. Lin MT, Simon DK, Ahn CH, Kim LM, Beal MF. High aggregate burden of somatic mtDNA point mutations in aging and Alzheimer's disease brain. *Hum Mol Genet.* 2002;11(2):133–146.
17. Hayashi Y, Yoshida M, Yamato M, et al. Reverse of age-dependent memory impairment and mitochondrial DNA damage in microglia by an overexpression of human mitochondrial transcription factor a in mice. *J Neurosci.* 2008;28(34):8624–8634.
18. Jendrach M, Pohl S, Vöth M, Kowald A, Hammerstein P, Bereiter-Hahn J. Morphodynamic changes of mitochondria during ageing of human endothelial cells. *Mech Ageing Dev.* 2005;126(6–7):813–821.
19. Federico A, Cardaioli E, Da Pozzo P, Formichi P, Gallus GN, Radi E. Mitochondria, oxidative stress and neurodegeneration. *J Neurol Sci.* 2012;322(1–2):254–262.
20. Corral-Debrinski M, Horton T, Lott MT, Shoffner JM, Beal MF, Wallace DC. Mitochondrial DNA deletions in human brain: regional variabililty and increase with advanced age. *Nat Genet.* 1992;2:324–329.
21. Quintana A, Schwindling C, Wenning AS, et al. T cell activation requires mitochondrial translocation to the immunological synapse. *Proc Natl Acad Sci U S A.* 2007;104(36):14418–14423.
22. MacAskill AF, Kittler JT. Control of mitochondrial transport and localization in neurons. *Trends Cell Biol.* 2010;20(2):102–112.
23. Sheng Z-H, Cai Q. Mitochondrial transport in neurons: impact on synaptic homeostasis and neurodegeneration. *Nat Rev Neurosci.* 2012;13(2):77–93.
24. Campello S, Lacalle RA, Bettella M, Mañes S, Scorrano L, Viola A. Orchestration of lymphocyte chemotaxis by mitochondrial dynamics. *J Exp Med.* 2006;203(13):2879–2886.
25. Keller H, Zadeh AD, Eggli P. Localised depletion of polymerised actin at the front of Walker carcinosarcoma cells increases the speed of locomotion. *Cell Motil Cytoskeleton.* 2002;53(3):189–202.
26. Desai SP, Bhatia S, Toner M, Irimia D. Mitochondrial localization and persistent migration of epithelial, cancer cells. *Biophys J.* 2013;104(9):2077–2088.
27. Yu Y, Dumollard R, Rossbach A, Lai FA, Swann K. Redistribution of mitochondria leads to bursts of ATP production during spontaneous mouse oocyte maturation. *J Cell Physiol.* 2010;224(3):672–680.
28. Collins TJ, Bootman MD. Mitochondria are morphologically heterogeneous within cells. *J Exp Biol.* 2003;206(12):1993–2000.
29. Peraza-Reyes L, Crider DG, Pon LA. Mitochondrial manoeuvres: latest insights and hypotheses on mitochondrial partitioning during mitosis in *Saccharomyces cerevisiae*. *Bioessays.* 2010;32(12):1040–1049.
30. Vevea JD, Swayne TC, Boldogh IR, Pon LA. Inheritance of the fittest mitochondria in yeast. *Trends Cell Biol.* 2014;24(1):53–60.
31. Krüger N, Tolić-Nørrelykke IM. Association of mitochondria with spindle poles facilitates spindle alignment. *Curr Biol.* 2008;18(15):R646–R647.
32. Wikstrom JD, Twig G, Shirihai OS. What can mitochondrial heterogeneity tell us about mitochondrial dynamics and autophagy? *Int J Biochem Cell Biol.* 2009;41(10):1914–1927.
33. Bereiter-Hahn J, Vöth M, Mai S, Jendrach M. Structural implicatons of mitochondrial dynamics. *Biotechnol J.* 2008;3:765–780.
34. Jans DC, Wurm CA, Riedel D, Wenzel D, Stagge F, Deckers M. STED super-resolution microscopy reveals an array of MINOS clusters along human mitochondria. *Proc Natl Acad Sci U S A.* 2013;110:8936–8941.
35. Al-Mehdi A-B, Pastukh VM, Swiger BM, et al. Perinuclear mitochondrial clustering creates an oxidant-rich nuclear domain required for hypoxia-induced transcription. *Sci Signal.* 2012;5(231):ra47.
36. Murphy MP. Modulating mitochondrial intracellular location as a redox signal. *Sci Signal.* 2012;5(242):e39.

37. Lewis MR, Lewis WH. Mitochondria (and other cytoplasmic structures) in tissue culture. *Am J Anat.* 1914;17:339–401.
38. Erecińska M, Silver IA. Ions and energy in mammalian brain. *Prog Neurobiol.* 1994;43(1):37–71.
39. Morris RL, Hollenbeck PJ. Axonal transport of mitochondria along microtubules and F-actin in living vertebrate neurons. *J Cell Biol.* 1995;131(5):1315–1326.
40. Li Z, Okamoto K-I, Hayashi Y, Sheng M. The importance of dendritic mitochondria in the morphogenesis and plasticity of spines and synapses. *Cell.* 2004;119(6):873–887.
41. Huang B, Jones SA, Brandenburg B, Zhuang X. Whole-cell 3D STORM reveals interactions between cellular structures with nanometer-scale resolution. *Nat Methods.* 2008;5(12):1047–1052.
42. Ligon LA, Steward O. Role of microtubules and actin filaments in the movement of mitochondria in the axons and dendrites of cultured hippocampal neurons. *J Comp Neurol.* 2000;427(3):351–361.
43. Caviston JP, Holzbaur ELF. Microtubule motors at the intersection of trafficking and transport. *Trends Cell Biol.* 2006;16(10):530–537.
44. Minin AA, Kulik AV, Gyoeva FK, Li Y, Goshima G, Gelfand VI. Regulation of mitochondria distribution by RhoA and formins. *J Cell Sci.* 2006;119(Pt 4):659–670.
45. Chada SR, Hollenbeck PJ, Lafayette W. Nerve growth factor signaling regulates motility and docking of axonal mitochondria. *Curr Biol.* 2004;14:1272–1276.
46. Gourlay CW, Ayscough KR. The actin cytoskeleton: a key regulator of apoptosis and ageing? *Nat Rev Mol Cell Biol.* 2005;6(7):583–589.
47. Kusano H, Shimizu S, Koya RC, et al. Human gelsolin prevents apoptosis by inhibiting apoptotic mitochondrial changes via closing VDAC. *Oncogene.* 2000;19(42):4807–4814.
48. De Vos KJ, Allan VJ, Grierson AJ, Sheetz MP. Mitochondrial function and actin regulate dynamin-related protein 1-dependent mitochondrial fission. *Curr Biol.* 2005;15(7):678–683.
49. Heggeness MH, Simon M, Singer SJ. Association of mitochondria with microtubules in cultured cells. *Proc Natl Acad Sci U S A.* 1978;75(8):3859–3862.
50. Tang HL, Lung HL, Wu KC, Le A-HP, Tang HM, Fung MC. Vimentin supports mitochondrial morphology and organization. *Biochem J.* 2008;410(1):141–146.
51. Kedzior J. Changes in physicochemical properties of microtubules lead to the formation of a single spherical structure of mitochondrial assembly enveloping nuclear chromatins. *J Electron Microsc (Tokyo).* 2004;53(6):659–670.
52. Stokin GB, Goldstein LSB. Axonal transport and Alzheimer's disease. *Annu Rev Biochem.* 2006;75:607–627.
53. Zhao C, Takita J, Tanaka Y, et al. Charcot–Marie–Tooth disease type 2A caused by mutation in a microtubule motor KIF1Bbeta. *Cell.* 2001;105(5):587–597.
54. Züchner S, Mersiyanova IV, Muglia M, et al. Mutations in the mitochondrial GTPase mitofusin 2 cause Charcot–Marie–Tooth neuropathy type 2A. *Nat Genet* 2004;36(5):449–451.
55. Cartoni R, Martinou J-C. Role of mitofusin 2 mutations in the physiopathology of Charcot–Marie–Tooth disease type 2A. *Exp Neurol.* 2009;218(2):268–273.
56. Dixit R, Ross JL, Goldman YE, Holzbaur ELF. Differential regulation of dynein and kinesin motor proteins by tau. *Science.* 2008;319(5866):1086–1089.
57. Quintero OA, DiVito MM, Adikes RC, et al. Human Myo19 is a novel myosin that associates with mitochondria. *Curr Biol.* 2009;19(23):2008–2013.
58. Boldogh IR, Nowakowski DW, Yang H, et al. A protein complex containing Mdm10p, Mdm12p, and Mmm1p links mitochondrial membranes and DNA to the cytoskeleton-based segregation machinery. *Mol Biol Cell.* 2003;14:4618–4627.
59. Fehrenbacher KL, Yang H, Gay AC, Huckaba TM, Pon LA. Live cell imaging of mitochondrial movement along actin cables in budding yeast. *Curr Biol.* 2004;14:1996–2004.

60. Bereiter-Hahn J, Jendrach M. Mitochondrial dynamics. *Int Rev Cell Mol Biol.* 2010;284:1–65.
61. Baqri RM, Turner BA, Rheuben MB, Hammond BD, Kaguni LS, Miller KL. Disruption of mitochondrial DNA replication in *Drosophila* increases mitochondrial fast axonal transport in vivo. *PLoS One.* 2009;4(11):e7874.
62. Amiri M, Hollenbeck PJ. Mitochondrial biogenesis in the axons of vertebrate peripheral neurons. *Dev Neurobiol.* 2008;68(11):1348–1361.
63. Misgeld T, Kerschensteiner M, Bareyre FM, Burgess RW, Lichtman JW. Imaging axonal transport of mitochondria in vivo. *Nat Methods.* 2007;4(7):559–561.
64. Bereiter-Hahn J. Intracellular motility of mitochondria: role of the inner compartment in migration and shape changes of mitochondria in XTH-cells. *J Cell Sci.* 1978;30:99–115.
65. Stebbings H, Hunt C. The translocation of mitochondria along insect ovarian microtubules from isolated nutritive tubes: a simple reactivated model. *J Cell Sci.* 1987;88(Pt 5):641–648.
66. Perkins GA, Tjong J, Brown JM. The micro-architecture of mitochondria at active zones: electron tomography reveals novel anchoring scaffolds and cristae structured for high-rate metabolism. *J Neurosci.* 2010;30(3):1015–1026.
67. Ly CV, Verstreken P. Mitochondria at the synapse. *Neuroscientist.* 2006;12(4):291–299.
68. Ishihara N, Nomura M, Jofuku A, et al. Mitochondrial fission factor Drp1 is essential for embryonic development and synapse formation in mice. *Nat Cell Biol.* 2009;11(8):958–966.
69. Lores-Arnaiz S, Bustamante J. Age-related alterations in mitochondrial physiological parameters and nitric oxide production in synaptic and non-synaptic brain cortex mitochondria. *Neuroscience.* 2011;188:117–124.
70. Zhang CL, Ho PL, Kintner DB, Sun D, Chiu SY. Activity-dependent regulation of mitochondrial motility by calcium and Na/K-ATPase at nodes of Ranvier of myelinated nerves. *J Neurosci.* 2010;30(10):3555–3566.
71. Ohno N, Kidd GJ, Mahad D, et al. Myelination and axonal electrical activity modulate the distribution and motility of mitochondria at CNS nodes of Ranvier. *J Neurosci.* 2011;31(20):7249–7258.
72. Kang J-S, Tian J-H, Pan P-Y, et al. Docking of axonal mitochondria by syntaphilin controls their mobility and affects short-term facilitation. *Cell.* 2008;132(1):137–148.
73. Chen Y-M, Gerwin C, Sheng Z-H. Dynein light chain LC8 regulates syntaphilin-mediated mitochondrial docking in axons. *J Neurosci.* 2009;29(30):9429–9438.
74. Miller KE, Sheetz MP. Axonal mitochondrial transport and potential are correlated. *J Cell Sci.* 2004;117(Pt 13):2791–2804.
75. Verburg J, Hollenbeck PJ. Mitochondrial membrane potential in axons increases with local nerve growth factor or semaphorin signaling. *J Neurosci.* 2008;28(33):8306–8315.
76. Brough D, Schell MJ, Irvine RF. Agonist-induced regulation of mitochondrial and endoplasmic reticulum motility. *Biochem J.* 2005;392(Pt 2):291–297.
77. Zala D, Hinckelmann M-V, Yu H, et al. Vesicular glycolysis provides on-board energy for fast axonal transport. *Cell.* 2013;152(3):479–491.
78. Schiavo G, Fainzilber M. Alternative energy for neuronal motors. *Nature.* 2013;495: 178–180.
79. Tillmann U, Bereiter-Hahn J. Relation of actin fibrils to energy metabolism of endothelial cells. *Cell Tissue Res.* 1986;243(3):579–585.
80. Haghnia M, Cavalli V, Shah SB, et al. Dynactin is required for coordinated bidirectional motility, but not for dynein membrane attachment. *Mol Biol Cell.* 2007;18:2081–2089.
81. Puls I, Jonnakuty C, LaMonte BH, et al. Mutant dynactin in motor neuron disease. *Nat Genet.* 2003;33(4):455–456.

82. Cai Q, Pan P-Y, Sheng Z-H. Syntabulin-kinesin-1 family member 5B-mediated axonal transport contributes to activity-dependent presynaptic assembly. *J Neurosci.* 2007;27(27):7284–7296.
83. Piccioni F, Pinton P, Simeoni S, et al. Androgen receptor with elongated polyglutamine tract forms aggregates that alter axonal trafficking and mitochondrial distribution in motor neuronal. *FASEB J.* 2002;16:1418–1420.
84. Beal MF. Mitochondria take center stage in aging and neurodegeneration. *Ann Neurol.* 2005;58(4):495–505.
85. Li X-J, Orr AL, Li S. Impaired mitochondrial trafficking in Huntington's disease. *Biochim Biophys Acta.* 2010;1802(1):62–65.
86. Salinas S, Bilsland LG, Schiavo G. Molecular landmarks along the axonal route: axonal transport in health and disease. *Curr Opin Cell Biol.* 2008;20(4):445–453.
87. Esteves AR, Gozes I, Cardoso SM. The rescue of microtubule-dependent traffic recovers mitochondrial function in Parkinson's disease. *Biochim Biophys Acta.* 2014;1842(1):7–21.
88. Colin E, Zala D, Liot G, et al. Huntingtin phosphorylation acts as a molecular switch for anterograde/retrograde transport in neurons. *EMBO J.* 2008;27(15):2124–2134.
89. Orr AL, Li S, Wang C-E, et al. N-terminal mutant huntingtin associates with mitochondria and impairs mitochondrial trafficking. *J Neurosci.* 2008;28(11):2783–2792.
90. Song W, Chen J, Petrilli A, et al. Mutant huntingtin binds the mitochondrial fission GTPase dynamin-related protein-1 and increases its enzymatic activity. *Nat Med.* 2011;17(3):377–382.
91. Trushina E, Dyer RB, Ii JDB, et al. Mutant huntingtin impairs axonal trafficking in mammalian neurons in vivo and in vitro. *Mol Cell Biol.* 2004;24(18):8195–8209.
92. Bosco DA, Morfini G, Karabacak NM, et al. Wild-type and mutant SOD1 share an aberrant conformation and a common pathogenic pathway in ALS. *Nat Neurosci.* 2010;13(11):1396–1403.
93. Trinczek B, Ebneth A, Mandelkow EM, Mandelkow E. Tau regulates the attachment/detachment but not the speed of motors in microtubule-dependent transport of single vesicles and organelles. *J Cell Sci.* 1999;112:2355–2367.
94. Stamer K, Vogel R, Thies E, Mandelkow E, Mandelkow E-M. Tau blocks traffic of organelles, neurofilaments, and APP vesicles in neurons and enhances oxidative stress. *J Cell Biol.* 2002;156(6):1051–1063.
95. Rui Y, Tiwari P, Xie Z, Zheng JQ. Acute impairment of mitochondrial trafficking by beta-amyloid peptides in hippocampal neurons. *J Neurosci.* 2006;26(41):10480–10487.
96. Eckert A, Schulz KL, Rhein V, Götz J. Convergence of amyloid-beta and tau pathologies on mitochondria in vivo. *Mol Neurobiol.* 2010;41(2–3):107–114.
97. Vossel KA, Zhang K, Brodbeck J, et al. Tau reduction prevents Abeta-induced defects in axonal transport. *Science.* 2010;330(6001):198.
98. Saks VA, Vasil'eva E, Belikova YO, et al. Retarded diffusion of ADP in cardiomyocytes: possible role of mitochondrial outer membrane and creatine kinase in cellular regulation of oxidative phosphorylation. *Biochim Biophys Acta.* 1993;1144(2):134–148.
99. Vendelin M, Birkedal R. Anisotropic diffusion of fluorescently labeled ATP in rat cardiomyocytes determinede ba raster image correlation spectroscopy. *Am J Physiol Cell Physiol.* 2008;295(5):C1302–C1315.
100. Niethammer P, Kueh HY, Mitchison TJ. Spatial patterning of metabolism by mitochondria, oxygen, and energy sinks in a model cytoplasm. *Curr Biol.* 2008;18(8):586–591.
101. Duchen MR, Verkhratsky A, Muallem S. Mitochondria and calcium in health and disease. *Cell Calcium.* 2008;44(1):1–5.
102. Jacobson J, Duchen MR. Interplay between mitochondria and cellular calcium signalling Introduction. *Mol Cell Biochem.* 2004;256/257:209–218.

103. Hajnóczky G, Csordás G, Das S, et al. Mitochondrial calcium signalling and cell death: approaches for assessing the role of mitochondrial Ca2+ uptake in apoptosis. *Cell Calcium.* 2006;40(5–6):553–560.
104. Lebiedzinska M, Szabadkai G, Jones AWE, Duszynski J, Wieckowski MR. Interactions between the endoplasmic reticulum, mitochondria, plasma membrane and other subcellular organelles. *Int J Biochem Cell Biol.* 2009;41(10):1805–1816.
105. Mann ZF, Duchen MR, Gale JE. Mitochondria modulate the spatio-temporal properties of intra- and intercellular Ca^{2+} signals in cochlear supporting cells. *Cell Calcium.* 2009;46(2):136–146.
106. Friedman JR, Lackner LL, West M, DiBenedetto JR, Nunnari J, Voeltz GK. ER tubules mark sites of mitochondrial division. *Science.* 2011;334(6054):358–362.
107. Rafelski SM. Mitochondrial network morphology: building an integrative, geometrical view. *BMC Biol.* 2013;11(1):71.
108. Davidson SM, Duchen MR. Calcium microdomains and oxidative stress. *Cell Calcium.* 2006;40(5–6):561–574.
109. Chang DTW, Honick AS, Reynolds IJ. Mitochondrial trafficking to synapses in cultured primary cortical neurons. *J Neurosci.* 2006;26(26):7035–7045.
110. Sung JY, Engmann O, Teylan MA, Nairn AC, Greengard P, Kim Y. WAVE1 controls neuronal activity-induced mitochondrial distribution in dendritic spines. *Proc Natl Acad Sci U S A.* 2008;105(8):3112–3116.
111. Rintoul GL, Filiano AJ, Brocard JB, Kress GJ, Reynolds IJ. Glutamate decreases mitochondrial size and movement in primary forebrain neurons. *J Neurosci.* 2003;23(21):7881–7888.
112. Rintoul GL, Reynolds IJ. Mitochondrial trafficking and morphology in neuronal injury. *Biochim Biophys Acta.* 2010;1802(1):143–150.
113. Misko A, Jiang S, Wegorzewska I, Milbrandt J, Baloh RH. Mitofusin 2 is necessary for transport of axonal mitochondria and interacts with the Miro/Milton complex. *J Neurosci.* 2010;30(12):4232–4240.
114. Saotome M, Safiulina D, Szabadkai G, et al. Bidirectional Ca^{2+}-dependent control of mitochondrial dynamics by the Miro GTPase. *Proc Natl Acad Sci U S A.* 2008;105(52):20728–20733.
115. Liu X, Hajnóczky G. Ca^{2+}-dependent regulation of mitochondrial dynamics by the Miro–Milton complex. *Int J Biochem Cell Biol.* 2009;41(10):1972–1976.
116. Stowers RS, Megeath LJ, Górska-Andrzejak J, Meinertzhagen IA, Schwarz TL. Axonal transport of mitochondria to synapses depends on Milton, a novel *Drosophila* protein. *Neuron.* 2002;36(6):1063–1077.
117. Wang X, Schwarz TL. The mechanism of Ca^{2+}-dependent regulation of kinesin-mediated mitochondrial motility. *Cell.* 2009;136(1):163–174.
118. Baixauli F, Martín-Cófreces NB, Morlino G, et al. The mitochondrial fission factor dynamin-related protein 1 modulates T-cell receptor signalling at the immune synapse. *EMBO J.* 2011;30(7):1238–1250.
119. Legros F, Lombe A, Frachon P, Rojo M. Mitochondrial fusion in human cells is efficient, requires the inner membrane potential, and is mediated by mitofusins. *Mol Biol Cell.* 2002;13:4343–4354.
120. Lombe A, Frachon P, Rojo M. Mitochondrial fusion in human cells is efficient, requires the inner membrane potential, and is mediated by mitofusins. *Mol Biol Cell.* 2002;13:4343–4354.
121. Chan DC. Fusion and fission: interlinked processes critical for mitochondrial health. *Annu Rev Genet.* 2012;46:265–287.
122. Chen H, Chan DC. Mitochondrial dynamics—fusion, fission, movement, and mitophagy—in neurodegenerative diseases. *Hum Mol Genet.* 2009;18(R2):R169–R176.

123. Bossy-Wetzel E, Petrilli A, Knott AB. Mutant huntingtin and mitochondrial dysfunction. *Trends Neurosci.* 2008;31(12):609–616.
124. Oettinghaus B, Licci M, Scorrano L, Frank S. Less than perfect divorces: dysregulated mitochondrial fission and neurodegeneration. *Acta Neuropathol.* 2012;123(2):189–203.
125. Ono T, Isobe K, Nakada K, Hayashi JI. Human cells are protected from mitochondrial dysfunction by complementation of DNA products in fused mitochondria. *Nat Genet.* 2001;28(3):272–275.
126. Chen H, Detmer SA, Ewald AJ, Griffin EE, Fraser SE, Chan DC. Mitofusins Mfn1 and Mfn2 coordinately regulate mitochondrial fusion and are essential for embryonic development. *J Cell Biol.* 2003;160(2):189–200.
127. Alavi MV, Bette S, Schimpf S, et al. A splice site mutation in the murine Opa1 gene features pathology of autosomal dominant optic atrophy. *Brain.* 2007;130(Pt 4):1029–1042.
128. Davies VJ, Hollins AJ, Piechota MJ, et al. Opa1 deficiency in a mouse model of autosomal dominant optic atrophy impairs mitochondrial morphology, optic nerve structure and visual function. *Hum Mol Genet.* 2007;16(11):1307–1318.
129. Alexander C, Votruba M, Pesch UE, et al. OPA1, encoding a dynamin-related GTPase, is mutated in autosomal dominant optic atrophy linked to chromosome 3q28. *Nat Genet.* 2000;26(2):211–215.
130. Delettre C, Lenaers G, Griffoin JM, et al. Nuclear gene OPA1, encoding a mitochondrial dynamin-related protein, is mutated in dominant optic atrophy. *Nat Genet.* 2000;26(2):207–210.
131. Sukhorukov VM, Dikov D, Reichert AS, Meyer-Hermann M. Emergence of the mitochondrial reticulum from fission and fusion dynamics. *PLoS Comput Biol.* 2012;8(10):e1002745.
132. Malka F, Guillery O, Cifuentes-Diaz C, et al. Separate fusion of outer and inner mitochondrial membranes. *EMBO Rep.* 2005;6(9):853–859.
133. Hoppins S, Lackner L, Nunnari J. The machines that divide and fuse mitochondria. *Annu Rev Biochem.* 2007;76:751–780.
134. Hoppins S, Nunnari J. The molecular mechanism of mitochondrial fusion. *Biochim Biophys Acta.* 2009;1793(1):20–26.
135. Ingerman E, Perkins EM, Marino M, et al. Dnm1 forms spirals that are structurally tailored to fit mitochondria. *J Cell Biol.* 2005;170(7):1021–1027.
136. Fransson A, Ruusala A, Aspenström P. Atypical Rho GTPases have roles in mitochondrial homeostasis and apoptosis. *J Biol Chem.* 2003;278(8):6495–6502.
137. Fransson S, Ruusala A, Aspenström P. The atypical Rho GTPases Miro-1 and Miro-2 have essential roles in mitochondrial trafficking. *Biochem Biophys Res Commun.* 2006;344(2):500–510.
138. Smirnova E, Griparic L, van der Bliek AM. Dynamin-related protein Drp1 is required for mitochondrial division in mammalian cells. *Mol Biol Cell.* 2001;12(8):2245–2256.
139. Lee Y, Jeong S, Karbowski M, Smith CL, Youle RJ. Roles of the mammalian mitochondrial fission and fusion mediators Fis1, Drp1, and Opa1 in apoptosis. *Mol Biol Cell.* 2004;15:5001–5011.
140. Otera H, Wang C, Cleland MM, et al. Mff is an essential factor for mitochondrial recruitment of Drp1 during mitochondrial fission in mammalian cells. *J Cell Biol.* 2010;191(6):1141–1158.
141. Otera H, Mihara K. Mitochondrial dynamics: functional link with apoptosis. *Int J Cell Biol.* 2012;2012:821676.
142. Van der Laan M, Bohnert M, Wiedemann N, Pfanner N. Role of MINOS in mitochondrial membrane architecture and biogenesis. *Trends Cell Biol.* 2012;22(4):185–192.
143. Yamano K, Youle RJ. Coupling mitochondrial and cell division. *Nat Cell Biol.* 2011;13(9):1026–1027.

144. Taguchi N, Ishihara N, Jofuku A, Oka T, Mihara K. Mitotic phosphorylation of dynamin-related GTPase Drp1 participates in mitochondrial fission. *J Biol Chem.* 2007;282(15):11521–11529.
145. Kashatus DF, Lim K-H, Brady DC, Pershing NLK, Cox AD, Counter CM. RALA and RALBP1 regulate mitochondrial fission at mitosis. *Nat Cell Biol.* 2011;13(9):1108–1115.
146. Chang C-R, Blackstone C. Cyclic AMP-dependent protein kinase phosphorylation of Drp1 regulates its GTPase activity and mitochondrial morphology. *J Biol Chem.* 2007;282(30):21583–21587.
147. Otera H, Mihara K. Molecular mechanisms and physiologic functions of mitochondrial dynamics. *J Biochem.* 2011;149(3):241–251.
148. Cribbs JT, Strack S. Reversible phosphorylation of Drp1 by cyclic AMP-dependent protein kinase and calcineurin regulates mitochondrial fission and cell death. *EMBO Rep.* 2007;8(10):939–944.
149. Cereghetti GM, Stangherlin A, Martins de Brito O, et al. Dephosphorylation by calcineurin regulates translocation of Drp1 to mitochondria. *Proc Natl Acad Sci U S A.* 2008;105(41):15803–15808.
150. Cho D-H, Nakamura T, Fang J, et al. S-nitrosylation of Drp1 mediates beta-amyloid-related mitochondrial fission and neuronal injury. *Science.* 2009;324(5923):102–105.
151. Zunino R, Schauss A, Rippstein P, Andrade-Navarro M, McBride HM. The SUMO protease SENP5 is required to maintain mitochondrial morphology and function. *J Cell Sci.* 2007;120(Pt 7):1178–1188.
152. Wasiak S, Zunino R, McBride HM. Bax/Bak promote sumoylation of DRP1 and its stable association with mitochondria during apoptotic cell death. *J Cell Biol.* 2014;177(3):439–450.
153. Karbowski M, Neutzner A, Youle RJ. The mitochondrial E3 ubiquitin ligase MARCH5 is required for Drp1 dependent mitochondrial division. *J Cell Biol.* 2007;178(1):71–84.
154. Karbowski M, Lee Y-J, Gaume B, et al. Spatial and temporal association of Bax with mitochondrial fission sites, Drp1, and Mfn2 during apoptosis. *J Cell Biol.* 2002;159(6):931–938.
155. Niemann A, Ruegg M, La Padula V, Schenone A, Suter U. Ganglioside-induced differentiation associated protein 1 is a regulator of the mitochondrial network: new implications for Charcot–Marie–Tooth disease. *J Cell Biol.* 2005;170(7):1067–1078.
156. Tondera D, Czauderna F, Paulick K, Schwarzer R, Kaufmann J, Santel A. The mitochondrial protein MTP18 contributes to mitochondrial fission in mammalian cells. *J Cell Sci.* 2005;118(Pt 14):3049–3059.
157. Jayashankar V, Rafelski SM. Integrating mitochondrial organization and dynamics with cellular architecture. *Curr Opin Cell Biol.* 2014;26:34–40.
158. Zhao J, Liu T, Jin S-B, et al. The novel conserved mitochondrial inner-membrane protein MTGM regulates mitochondrial morphology and cell proliferation. *J Cell Sci.* 2009;122(Pt 13):2252–2262.
159. Norton M, Ng AC-H, Baird S, et al. ROMO1 Is an essential redox-dependent regulator of mitochondrial dynamics. *Sci Signal.* 2014;7(310):ra10.
160. Murley A, Lackner LL, Osman C, et al. ER-associated mitochondrial division links the distribution of mitochondria and mitochondrial DNA in yeast. *Elife.* 2013;2:e00422.
161. Hayashi T, Rizzuto R, Hajnoczky G, Su T-P. MAM: more than just a housekeeper. *Trends Cell Biol.* 2009;19(2):81–88.
162. Korobova F, Ramabhadran V, Higgs HN. An actin-dependent step in mitochondrial fission mediated by the ER-associated formin INF2. *Science.* 2013;339(6118):464–467.
163. Honda S, Aihara T, Hontani M, Okubo K, Hirose S. Mutational analysis of action of mitochondrial fusion factor mitofusin-2. *J Cell Sci.* 2005;118(Pt 14):3153–3161.

164. Chan DC. Mitochondrial fusion and fission in mammals. *Annu Rev Cell Dev Biol.* 2006;22:79–99.
165. Cipolat S, Martins de Brito O, Dal Zilio B, Scorrano L. OPA1 requires mitofusin 1 to promote mitochondrial fusion. *Proc Natl Acad Sci U S A.* 2004;101(45):15927–15932.
166. Olichon A, Guillou E, Delettre C, et al. Mitochondrial dynamics and disease, OPA1. *Biochim Biophys Acta.* 2006;1763(5–6):500–509.
167. Griparic L, van der Wel NN, Orozco IJ, Peters PJ, van der Bliek AM. Loss of the intermembrane space protein Mgm1/OPA1 induces swelling and localized constrictions along the lengths of mitochondria. *J Biol Chem.* 2004;279(18):18792–18798.
168. Song Z, Chen H, Fiket M, Alexander C, Chan DC. OPA1 processing controls mitochondrial fusion and is regulated by mRNA splicing, membrane potential, and Yme1L. *J Cell Biol.* 2007;178(5):749–755.
169. Olichon A, Elachouri G, Baricault L, Delettre C, Belenguer P, Lenaers G. OPA1 alternate splicing uncouples an evolutionary conserved function in mitochondrial fusion from a vertebrate restricted function in apoptosis. *Cell Death Differ.* 2007;14(4):682–692.
170. Frezza C, Cipolat S, Martins de Brito O, et al. OPA1 controls apoptotic cristae remodeling independently from mitochondrial fusion. *Cell.* 2006;126(1):177–189.
171. Griffin EE, Detmer SA, Chan DC. Molecular mechanism of mitochondrial membrane fusion. *Biochim Biophys Acta.* 2006;1763(5–6):482–489.
172. Ishihara N, Fujita Y, Oka T, Mihara K. Regulation of mitochondrial morphology through proteolytic cleavage of OPA1. *EMBO J.* 2006;25(13):2966–2977.
173. Suen D-F, Norris KL, Youle RJ. Mitochondrial dynamics and apoptosis. *Genes Dev.* 2008;22(12):1577–1590.
174. Meeusen SL, Nunnari J. How mitochondria fuse. *Curr Opin Cell Biol.* 2005;17(4): 389–394.
175. Choi S-Y, Huang P, Jenkins GM, Chan DC, Schiller J, Frohman MA. A common lipid links Mfn-mediated mitochondrial fusion and SNARE-regulated exocytosis. *Nat Cell Biol.* 2006;8(11):1255–1262.
176. Zhao J, Liu T, Jin S, et al. Human MIEF1 recruits Drp1 to mitochondrial outer membranes and promotes mitochondrial fusion rather than fission. *EMBO J.* 2011;30(14):2762–2778.
177. Chen Y, Liu Y, Dorn GW. Mitochondrial fusion is essential for organelle function and cardiac homeostasis. *Circ Res.* 2011;109(12):1327–1331.
178. Fujioka H, Tandler B, Hoppel C. Mitochondrial division in rat cardiomyocytes: an electron microscope study. *Anat Rec.* 2012;295:1455–1461.
179. Kasahara A, Cipolat S, Chen Y, Korn GW, Scorrano L. Mitochondrial fusion directs cardiomyocyte differentiation via calcineuring and Notch signaling. *Science.* 2013;342:734–737.
180. Dorn GW, Clark CF, Eschenbacher WH, et al. MARF and Opa1 control mitochondrial and cardiac function in *Drosophila*. *Circ Res.* 2011;108(1):12–17.
181. Dietrich MO, Liu Z-W, Horvath TL. Mitochondrial dynamics controlled by mitofusins regulate Agrp neuronal activity and diet-induced obesity. *Cell.* 2013;155(1):188–199.
182. Kowald A, Kirkwood TBL. Evolution of the mitochondrial fusion–fission cycle and its role in aging. *Proc Natl Acad Sci U S A.* 2011;108(39):16481.
183. Popov V, Medvedev NI, Davies HA, Stewart MG. Mitochondria form a filamentous reticular network in hippocampal dendrites but are present as discrete bodies in axons: a three-dimensional ultrastructural study. *J Comp Neurol.* 2005;492(1):50–65.
184. Jendrach M, Mai S, Pohl S, Vöth M, Bereiter-Hahn J. Short- and long-term alterations of mitochondrial morphology, dynamics and mtDNA after transient oxidative stress. *Mitochondrion.* 2008;8(4):293–304.

185. Singh M, Sharma H, Singh N. Hydrogen peroxide induces apoptosis in HeLa cells through mitochondrial pathway. *Mitochondrion*. 2007;7(3):67–73.
186. Stöckl P, Zankl C, Hütter E, et al. Partial uncoupling of oxidative phosphorylation induces premature senescence in human fibroblasts and yeast mother cells. *Free Radic Biol Med*. 2007;43(6):947–958.
187. Rojo M, Legros F, Chateau D, Lombès A. Membrane topology and mitochondrial targeting of mitofusins, ubiquitous mammalian homologs of the transmembrane GTPase Fzo. *J Cell Sci*. 2002;115(Pt 8):1663–1674.
188. Mouli PK, Twig G, Shirihai OS. Frequency and selectivity of mitochondrial fusion are key to its quality maintenance function. *Biophys J*. 2009;96(9):3509–3518.
189. Romanello V, Guadagnin E, Gomes L, et al. Mitochondrial fission and remodelling contributes to muscle atrophy. *EMBO J*. 2010;29(10):1774–1785.
190. Yoon Y, Pitts KR, McNiven MA. Mammalian dynamin-like protein DLP1 tubulates membranes. *Mol Biol Cell*. 2001;12(9):2894–2905.
191. Scheckhuber CQ, Erjavec N, Tinazli A, Hamann A, Nyström T, Osiewacz HD. Reducing mitochondrial fission results in increased life span and fitness of two fungal ageing models. *Nat Cell Biol*. 2007;9(1):99–105.
192. Frank S, Gaume B, Bergmann-Leitner ES, et al. The role of dynamin-related protein 1, a mediator of mitochondrial fission, in apoptosis. *Dev Cell*. 2001;1(4):515–525.
193. Szabadkai G, Simoni AM, Chami M, Wieckowski MR, Youle RJ, Rizzuto R. Drp-1-dependent division of the mitochondrial network blocks intraorganellar Ca^{2+} waves and protects against Ca^{2+}-mediated apoptosis. *Mol Cell*. 2004;16(1):59–68.
194. Waterham HR, Koster J, van Roermund CWT, Mooyer PAW, Wanders RJA, Leonard JV. A lethal defect of mitochondrial and peroxisomal fission. *N Engl J Med*. 2007;356(17):1736–1741.
195. Lee S, Jeong S-Y, Lim W-C, et al. Mitochondrial fission and fusion mediators, hFis1 and OPA1, modulate cellular senescence. *J Biol Chem*. 2007;282(31):22977–22983.
196. Park Y-Y, Lee S, Karbowski M, Neutzner A, Youle RJ, Cho H. Loss of MARCH5 mitochondrial E3 ubiquitin ligase induces cellular senescence through dynamin-related protein 1 and mitofusin 1. *J Cell Sci*. 2010;123(Pt 4):619–626.
197. Jahani-Asl A, Cheung ECC, Neuspiel M, et al. Mitofusin 2 protects cerebellar granule neurons against injury-induced cell death. *J Biol Chem*. 2007;282(33):23788–23798.
198. Sugioka R, Shimizu S, Tsujimoto Y. Fzo1, a protein involved in mitochondrial fusion, inhibits apoptosis. *J Biol Chem*. 2004;279(50):52726–52734.
199. Tondera D, Grandemange S, Jourdain A, et al. SLP-2 is required for stress-induced mitochondrial hyperfusion. *EMBO J*. 2009;28(11):1589–1600.
200. Huang H-M, Fowler C, Xu H, Zhang H, Gibson GE. Mitochondrial function in fibroblasts with aging in culture and/or Alzheimer's disease. *Neurobiol Aging*. 2005;26(6):839–848.
201. Kageyama Y, Zhang Z, Roda R, et al. Mitochondrial division ensures the survival of postmitotic neurons by suppressing oxidative damage. *J Cell Biol*. 2012;197(4):535–551.
202. Harman D. Aging: a theory based on free radical and radiation chemistry. *J Gerontol*. 1956;11(3):298–300.
203. Harmann D. The biologic clock: the mitochondria? *J Am Geriatr Soc*. 1972;20:145–147.
204. Wanagat J, Dai D-F, Rabinovitch P. Mitochondrial oxidative stress and mammalian healthspan. *Mech Ageing Dev*. 2010;131(7–8):527–535.
205. Seo AY, Joseph A-M, Dutta D, Hwang JCY, Aris JP, Leeuwenburgh C. New insights into the role of mitochondria in aging: mitochondrial dynamics and more. *J Cell Sci*. 2010;123(Pt 15):2533–2542.
206. Chan DC. Mitochondria: dynamic organelles in disease, aging, and development. *Cell*. 2006;125(7):1241–1252.

207. Terman A, Kurz T, Navratil M, et al. Mitochondrial turnover and aging the mitochondrial-lysosomal axis theory of aging. *Antioxid Redox Signal.* 2010;12(4):503–535.
208. Strecker V, Mai S, Muster B, et al. Aging of different avian cultured cells: lack of ROS-induced damage and quality control mechanisms. *Mech Ageing Dev.* 2010;131(1):48–59.
209. Lin D-T, Wu J, Holstein D, et al. Ca2+ signaling, mitochondria and sensitivity to oxidative stress in aging astrocytes. *Neurobiol Aging.* 2007;28(1):99–111.
210. Yu T, Robotham JL, Yoon Y. Increased production of reactive oxygen species in hyperglycemic conditions requires dynamic change of mitochondrial morphology. *Proc Natl Acad Sci U S A.* 2006;103(8):2653–2658.
211. Gomes LC, Di Benedetto G, Scorrano L. During autophagy mitochondria elongate, are spared from degradation and sustain cell viability. *Nat Cell Biol.* 2011;13(5):589–598.
212. Kumar H, Lim H-W, More SV, et al. The role of free radicals in the aging brain and Parkinson's disease: convergence and parallelism. *Int J Mol Sci.* 2012;13(8):10478–10504.
213. Olichon A, Baricault L, Gas N, et al. Loss of OPA1 perturbates the mitochondrial inner membrane structure and integrity, leading to cytochrome *c* release and apoptosis. *J Biol Chem.* 2003;278(10):7743–7746.
214. Scorrano L. Proteins that fuse and fragment mitochondria in apoptosis: con-fissing a deadly con-fusion? *J Bioenerg Biomembr.* 2005;37(3):165–170.
215. Shen T, Zheng M, Cao C, et al. Mitofusin-2 is a major determinant of oxidative stress-mediated heart muscle cell apoptosis. *J Biol Chem.* 2007;282(32):23354–23361.
216. Chen K, Guo X, Ma D, et al. Dysregulation of HSG triggers vascular proliferative disorders. *Nat Cell Biol.* 2004;6(9):872–883.
217. Wang X, Su B, Lee H, Li X, Perry G, Smith MA. Impaired balance of mitochondria fission and fusion in Alzheimer disease. *J Neurosci.* 2010;29(28):9090–9103.
218. DuBoff B, Feany M, Götz J. Why size matters—balancing mitochondrial dynamics in Alzheimer's disease. *Trends Neurosci.* 2013;36(6):325–335.
219. Chen H, McCaffery JM, Chan DC. Mitochondrial fusion protects against neurodegeneration in the cerebellum. *Cell.* 2007;130(3):548–562.
220. Liu X, Weaver D, Shirihai O, Hajnóczky G. Mitochondrial "kiss-and-run": interplay between mitochondrial motility and fusion-fission dynamics. *EMBO J.* 2009;28(20):3074–3089.
221. Molina AJA, Shirihai OS. Monitoring mitochondrial dynamics with photoactivatable [corrected] green fluorescent protein. *Methods Enzymol.* 2009;457:289–304.
222. Twig G, Liu X, Liesa M, et al. Biophysical properties of mitochondrial fusion events in pancreatic beta-cells and cardiac cells unravel potential control mechanisms of its selectivity. *Am J Physiol Cell Physiol.* 2010;299(2):C477–C487.
223. Muster B, Kohl W, Wittig I, et al. Respiratory chain complexes in dynamic mitochondria display a patchy distribution in life cells. *PLoS One.* 2010;5(7):e11910.
224. Bereiter-Hahn J, Vöth M. Distribution and dynamics of mitochondrial nucleoids in animal cells in culture. *Exp Biol Online.* 1996;1(4):1–17.
225. Spelbrink JN. Functional organization of mammalian mitochondrial DNA in nucleoids: history, recent developments, and future challenges. *IUBMB Life.* 2010;62(1):19–32.
226. Legros F, Malka F, Frachon P, Lombès A, Rojo M. Organization and dynamics of human mitochondrial DNA. *J Cell Sci.* 2004;117(Pt 13):2653–2662.
227. Gilkerson RW. Mitochondrial DNA, nucleoids determine mitochondrial genetics and dysfunction. *Int J Biochem Cell Biol.* 2009;41(10):1899–1906.
228. Garrido N, Griparic L, Jokitalo E, Wartiovaara J, Van Der Bliek AM, Spelbrink JN. Composition and dynamics of human mitochondrial nucleoids. *Mol Biol Cell.* 2003;14:1583–1596.
229. Dikov D, Bereiter-hahn J. Inner membrane dynamics in mitochondria. *J Struct Biol.* 2013;183:455–466.

230. Hebert SL, Lanza IR, Nair KS. Mitochondrial DNA alterations and reduced mitochondrial function in aging. *Mech Ageing Dev.* 2010;131(7–8):451–462.
231. Meissner C, Bruse P, Mohamed SA, et al. The 4977 bp deletion of mitochondrial DNA in human skeletal muscle, heart and different areas of the brain: a useful biomarker or more? *Exp Gerontol.* 2008;43(7):645–652.
232. Kowald A, Jendrach M, Pohl S, Bereiter-Hahn J, Hammerstein P. On the relevance of mitochondrial fusions for the accumulation of mitochondrial deletion mutants: a modelling study. *Aging Cell.* 2005;4(5):273–283.
233. Ban-Ishihara R, Ishihara T, Sasaki N, Mihara K, Ishihara N. Dynamics of nucleoid structure regulated by mitochondrial fission contributes to cristae reformation and release of cytochrome *c*. *Proc Natl Acad Sci U S A.* 2013;110(29):11863–11868.
234. Partikian A, Olveczky B, Swaminathan R, Li Y, Verkman S. Rapid diffusion of green fluorescent protein in the mitochondrial matrix. *J Cell Biol.* 1998;140(4):821–829.
235. Verkman AS. Solute and macromolecule diffusion in cellular aqueous compartments. *Trends Biochem Sci.* 2002;27(1):27–33.
236. Scalettar BA, Abney JR, Hackenbrock CR. Dynamics, structure, and function are coupled in the mitochondrial matrix. *Proc Natl Acad Sci U S A.* 1991;88(18):8057–8061.
237. Haggie PM, Verkman AS. Diffusion of tricarboxylic acid cycle enzymes in the mitochondrial matrix in vivo Evidence for restricted mobility of a multienzyme complex. *J Biol Chem.* 2002;277(43):40782–40788.
238. Busch KB, Bereiter-Hahn J, Wittig I, Schagger H, Jendrach M. Mitochondrial dynamics generate equal distribution but patchwork localization of respiratory complex I. *Mol Membr Biol.* 2006;23(6):509–520.
239. Wilkens V, Kohl W, Busch K. Restricted diffusion of OXPHOS complexes in dynamic mitochondria delays their exchange between cristae and engenders a transitory mosaic distribution. *J Cell Sci.* 2013;126(Pt 1):103–116.
240. Sukhorukov VM, Dikov D, Busch K, Strecker V, Wittig I, Bereiter-Hahn J. Determination of protein mobility in mitochondrial membranes of living cells. *Biochim Biophys Acta.* 2010;1798(11):2022–2032.
241. Dieteren CEJ, Willems PHGM, Vogel RO, et al. Subunits of mitochondrial complex I exist as part of matrix- and membrane-associated subcomplexes in living cells. *J Biol Chem.* 2008;283(50):34753–34761.
242. Sukhorukov VM, Bereiter-Hahn J. Anomalous diffusion induced by cristae geometry in the inner mitochondrial membrane. *PLoS One.* 2009;4(2):e4604.
243. Appelhans T, Richter CP, Wilkens V, Hess ST, Piehler J, Busch KB. Nanoscale organization of mitochondrial microcompartments revealed by combining tracking and localization microscopy. *Nano Lett.* 2012;12(2):610–616.

CHAPTER FIVE

The Retrograde Response: A Conserved Compensatory Reaction to Damage from Within and from Without

S. Michal Jazwinski

Tulane Center for Aging and Department of Medicine, Tulane University Health Sciences Center, New Orleans, Louisiana, USA

Contents

Abstract

The retrograde response was discovered in *Saccharomyces cerevisiae* as a signaling pathway from the mitochondrion to the nucleus that triggers an array of gene regulatory changes in the latter. The activation of the retrograde response compensates for the deficits associated with aging, and thus it extends yeast replicative life span. The retrograde response is activated by the progressive decline in mitochondrial membrane potential during aging that is the result of increasing mitochondrial dysfunction. The ensuing metabolic adaptations and stress resistance can only delay the inevitable demise of the yeast cell. The retrograde response is embedded in a network of signal transduction pathways that impinge upon virtually every aspect of cell physiology. Thus, its manifestations are complicated. Many of these pathways have been implicated in life span regulation quite independently of the retrograde response. Together, they operate in a delicate balance in promoting longevity. The retrograde response is closely aligned with cell quality control, often performing when quality control is not sufficient to assure longevity. Among the key pathways related to this aspect of retrograde signaling are target of rapamycin and ceramide signaling. The retrograde response can also be found

Progress in Molecular Biology and Translational Science, Volume 127
ISSN 1877-1173
http://dx.doi.org/10.1016/B978-0-12-394625-6.00005-2

in other organisms, including *Caenorhabditis elegans*, *Drosophila melanogaster*, mouse, and human, where it exhibits an ever-increasing complexity that may be corralled by the transcription factor NFκB. The retrograde response may have evolved as a cytoprotective mechanism that senses and defends the organism from pathogens and environmental toxins.

1. INTRODUCTION

Since the establishment of an endosymbiotic relationship between a primitive bacterium and a proto-eukaryotic cell, the resulting mitochondrion has ceded its autonomy to the host by transferring the bulk of its genetic information to the cell nucleus. This process, which may have reached its culmination, may be construed as an obvious subordination of function. However, this would be a superficial view of the state of affairs. It is more appropriate to consider it an integration of processes that requires a close coordination of function.

The superficial interpretation of the cellular state of affairs derives from the simplistic view of mitochondria as an energy factory producing ATP in the process of oxidative phosphorylation. However, mitochondria play vital roles in various anabolic and catabolic processes, as well as in calcium and iron homeostasis. They are also key in reactions involving the generation and management of reactive oxygen species (ROS). These facts suggest the more complex view of mitochondria in cellular function that entails two-way communication between this organelle and the nucleus, as well as interactions with other organelles.

The importance of balanced production of proteins destined to form complexes in the mitochondrion is obvious. This is a matter of mitochondrial biogenesis that is regulated at the supply level, as well as at the level of elimination of supernumerary proteins. The degree of coordination at the supply level between the nucleus and the mitochondrion can vary in different organisms, and it involves both gene regulation and control of protein complex assembly. However, the mechanisms at the disposal level are similar. Beyond this obvious cooperation, there are other lines of communication at play.

Retrograde signaling was the first mechanism of communication from the mitochondrion to the nucleus that was described.[1] As its significance for cell physiology became apparent, it was incorporated within the larger context of the retrograde response.[2] This response was first elaborated in the yeast, *Saccharomyces cerevisiae*. However, now it is known to operate in other organisms, including human. Along the way, variations on this theme have been discovered, and we refer to them here broadly as retrograde response. This broadly understood retrograde response is the topic of this chapter. Recently, the evolutionary origin of such a broadly conceived response as a protective reaction to environmental toxins has been proposed,[3] rendering such a treatment of this topic even more attractive.

The retrograde response is entwined in an array of interactions with other organelles, from which it is difficult to separate. In fact, these interactions have important implications for the role(s) of the retrograde response in cell physiology. Some of these interactions fall into the category of cellular quality control, and their integration with the retrograde response allows it to kick in when quality control is not enough. The intersection of these quality control mechanisms with the retrograde response will also be discussed here.

2. THE RETROGRADE SIGNALING PATHWAY

The first hint of an unusual nuclear response to mitochondrial state was obtained through the discovery of accumulation of a heterogeneous array of RNA species in mitochondrial petite yeast cells (ρ^0) by Parikh *et al.*[4] This led to a series of discoveries over the next 20 years that fleshed out an intracellular signaling pathway from the mitochondrion to the nucleus.[5] A variety of mitochondrial defects trigger this pathway, but the most commonly investigated has been the partial or complete loss of the mitochondrial DNA (mtDNA) genome.

The key event in retrograde signaling is the translocation of the Rtg1–Rtg3 transcription factor from the cytoplasm to the nucleus[6–8] (Fig. 5.1). Rtg1 and Rtg3 are both basic, helix-loop-helix transcription factors (bHLH–LeuZip) that require each other to perform their function in a heterodimeric complex.[9,10] Rtg1–Rtg3 binds to the sequence GTCAC, called the R box, which is located in the promoters of retrograde response target genes. The translocation of Rtg1–Rtg3 to the nucleus requires partial dephosphorylation of Rtg3.[8] However, some phosphorylation of Rtg3 must remain for it to retain transactivation activity in complex with Rtg1. Rtg1 is an unusual bHLH–LeuZip transcription factor because of its truncated basic domain and no apparent transcription activation domain. It lies in the top layer of the hierarchical transcription factor regulatory network in yeast, and it is not regulated by any other transcription factor.[11] These top-layer transcription factors are variable in expression from cell to cell,

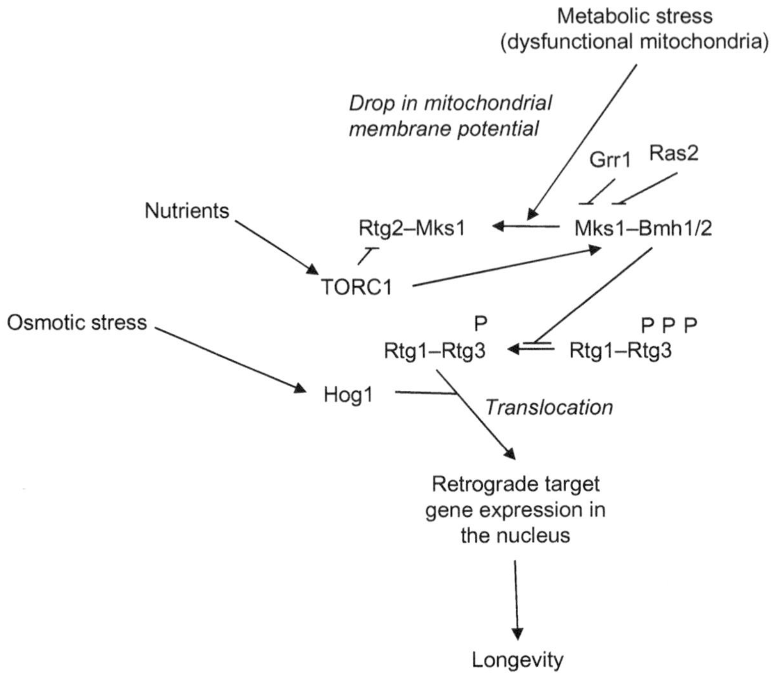

Figure 5.1 The retrograde response signaling pathway. This schematic view summarizes the core elements of the retrograde response. Stress, whether metabolic or osmotic, triggers the pathway, and nutrients modulate the response through TORC1. Rtg2 sequesters Mks1 when a drop in mitochondrial membrane potential signals mitochondrial dysfunction. This prevents the Mks1–Bmh1/2 complex from inhibiting the partial dephosphorylation of Rtg3 in the Rtg1–Rtg3 heterodimer. Partial dephosphorylation results in the translocation of this transcription factor from the cytoplasm to the nucleus, where it activates retrograde response target genes, resulting in an increase in longevity. The Hog1 protein kinase, in response to hyperosmotic shock, facilitates this translocation, as well. Hog1 is also required for chromatin binding of Rtg1–Rtg3 and the transactivation of retrograde response target genes. TORC1 responds to nutrient availability by inhibiting retrograde signaling both upstream and downstream of Rtg2. The core retrograde response depicted here cross-talks with several other signaling pathways (see text). Notably, sphingolipid signaling interacts with the retrograde response modulating the effects of TORC1 on autophagy. A key target of ceramide signaling is the PP2A, Sit4, one of the mediators of TORC1 signaling. These interactions have important implications for cellular quality control. The retrograde response also impacts the mutual interaction of mitochondria and the vacuole, with consequences for yeast longevity.

and they allow at least some cells in a clone to respond to environmental challenges (see Ref. 11 for further discussion). Recent computer modeling has demonstrated the homology of the mammalian Myc–Max transcription factor to Rtg1–Rtg3.[12]

Translocation of Rtg1–Rtg3 requires Rtg2.[8] Rtg2 possesses no homologs in higher organisms, at least as of this writing. It has features common to certain bacterial phosphatases.[13] Thus, it has been presumed to actually dephosphorylate Rtg3.[8] However, more complicated scenarios have been put forth. These include Mks1 and the 14-3-3 proteins Bmh1 and Bmh2. Mks1 negatively regulates retrograde signaling by forming a complex with either Bmh1 or Bmh2, which maintains Rtg3 in a hyperphosphorylated state.[8,14–17] Rtg2 can sequester Mks1 preventing the formation of this complex. In this way, Rtg2 is a positive regulator of retrograde signaling. Another positive regulator of retrograde signaling is Grr1,[18] a ubiquitin ligase component, because it helps remove Mks1.

The target of rapamycin complex 1 (TORC1) negatively regulates retrograde signaling.[19–21] It appears to act both upstream and downstream of Rtg2, although the details are not entirely clear. One of the components of TORC1 is Lst8, which was shown in genetic studies to act both upstream and downstream of Rtg2 as a negative regulator, consistent with the above conclusions. The retrograde response plays an anabolic role in the cell. It is repressed by glutamate,[5] and *RTG* gene mutants are all glutamate auxotrophs.[9,10] Availability of glutamate inside the cell activates TORC1, repressing the retrograde response. Furthermore, the presence of glutamate outside the cell is sensed by the SPS amino acid sensor in the plasma membrane.[22] *LST8* mutants that act upstream of Rtg2 resemble mutations in the SPS sensor.[23]

Mks1 is a negative regulator of the Ras2–cAMP pathway.[24] The deletion of *RAS2* abrogates the retrograde response.[25] However, it is not known whether Ras2 operates in a cAMP-dependent or -independent manner[26] in the retrograde response. Ras2 has a positive effect on replicative life span in yeast,[26] which is measured by the number of times an individual cell divides, but it does so through a cAMP-independent pathway. The cAMP pathway curtails life span. Activation of the retrograde response also extends yeast replicative life span,[25] suggesting that perhaps it is the cAMP-independent pathway that is crucial, but this has not been proven.

There are several other examples of crosstalk between the retrograde response and other signaling pathways in the cell. TORC1-mediated phosphorylation of the AGC protein kinase, Sch9, is downregulated in ρ^0 cells.[27] Phosphorylation of Sch9 directly counteracts the induction of stress responses under the control of the transcription factor Msn2–Msn4, and it also promotes ribosome biogenesis.[28] Deletion of *SCH9* extends yeast replicative life span,[29] as would be expected. Interestingly, phosphorylated

Sch9 also blocks protein kinase A activity.[30] This has the opposite effect on Msn2–Msn4 and the stress responses that depend on it. It also would paradoxically block metabolism and growth in the presence of nutrients. However, protein kinase A feeds back to counteract the effect of Sch9 on its activity. This balances cell growth, metabolism, and stress responses, while preventing an exaggerated response to feedback inhibition of TORC1 by dysfunctional mitochondria.

Osmotic stress results in a transient reduction in Sch9 phosphorylation by TORC1.[28] It also recruits Rtg1–Rtg3.[31] Thus, the retrograde signaling responds to both metabolic stress and to osmotic stress. The latter can also be considered a form of metabolic stress as it stimulates the glycolytic breakdown of glucose to glycerol. The regulation of Rtg1–Rtg3 by the hyperosmolarity-activated, Hog1 protein kinase pathway is complex.[32] Hog1 is necessary for translocation of Rtg1–Rtg3 into the nucleus, its binding to chromatin, and its transcriptional activity.

The signal elicited by dysfunctional mitochondria to activate the retrograde response is a drop in mitochondrial membrane potential.[33] This activation requires Rtg2 and Rtg1–Rtg3. Curiously, the requirement for Rtg2 appears to be by-passed in certain ρ^0 yeast strains.[34] This may be the result of the effect of background mutations. There is a gradual decline in mitochondrial membrane potential as yeasts progress through their replicative life spans.[35] Coupled to this is a commensurate increase in the activation of the retrograde response.[36] This suggests that the retrograde response is activated to compensate for accumulating mitochondrial dysfunction, allowing yeast cells to live as long as they do. Indeed, increasing activation of the retrograde response is matched by an increased life span extension.[37]

Coupling the lack of mtDNA with deletion of *CRD1*, which encodes cardiolipin synthase, leads to a more pronounced phenotype of division delay.[38] This is caused by excessive expression of Swe1, a morphogenesis checkpoint protein. Induction of *SWE1* requires Rtg2 and Rtg3, in a typical retrograde response. Cardiolipin is required for the activity of inositol sphingolipid phospholipase C activity,[39] which requires the translocation of the enzyme, Isc1, to the mitochondrial membrane during diauxic shift.[40] Isc1 is the yeast equivalent of mammalian sphingomyelinase, which generates the intracellular signaling molecule ceramide. Deletion of *ISC1* prevents the induction of a large array of genes that adapt the cell to late-logarithmic phase.[41] This defect is similar to that found in ρ^0 cells. However, the response to *ISC1* deletion does not require the retrograde signaling pathway. Instead, the ceramide signaling triggered by Isc1 may activate a pathway that is parallel to the retrograde signaling pathway.

CRD1 deletion reduced replicative life span at 37 °C.[42] It did not prevent life span extension by calorie restriction, but it prevented extension by mild heat stress or osmotic stress. Cells lacking Crd1 display severe vacuolar defects, including loss of vacuolar acidification due to deficient vacuolar proton pumping.[43] These phenotypes are rescued by deletion of *RTG2*. These results suggest that among the consequences of perturbation of the inner mitochondrial membrane by elimination of cardiolipin is the activation of the retrograde response. They also point to signaling from the mitochondrion to the vacuole. Curiously, signaling in the opposite direction, from the vacuole to the mitochondrion, has been suggested in other studies.[44] The connection between mitochondria and the vacuole becomes more complicated in ρ^- cells, in which an increase in vacuolar pH (loss of the normal acidification) suppresses some of the associated phenotypes.[45] Interestingly, overexpression of the oxysterol-binding protein homolog Osh6 in yeast rescues vacuole-fusion defects, and it also extends replicative life span in wild-type cells.[46]

Isc1 regulates Sit4, which is the catalytic subunit of protein phosphatase 2A (PP2A) known to be activated by ceramide.[47] Thus, the parallel pathway to the retrograde response mentioned above could involve PP2A. PP2A is also regulated by TORC1, a common feature with retrograde signaling. Ceramide was shown many years ago to arrest cells in the G_1 phase of the cell cycle by activating PP2A.[48] The *ISC1* gene interacts with *SWE1*, as it does with other cell-cycle checkpoint genes.[49] Thus, the retrograde response may regulate expression of the *SWE1* gene, as described above, while ceramide signaling may regulate Swe1 protein activity. The importance of the intersection of ceramide signaling and mitochondrial function is further highlighted by the suppression of the reduced proliferation, mitochondrial protein import defect, lower mitochondrial membrane potential, and retrograde signaling in ρ^0 cells by the deletion of PP2A.[50] The mitochondrial dysfunction, oxidative stress sensitivity, and premature chronological aging observed in cells devoid of Isc1 are suppressed by deletion of either *TOR1* or *SCH9*.[51] This is associated with a reduction in the deficit in autophagy exhibited by cells lacking Isc1. The protection afforded by *SCH9* deletion, but not by *TOR1* deletion, is associated with Hog1 hyperphosphorylation.

3. CONSEQUENCES OF RETROGRADE SIGNALING

Gene expression changes in ρ^0 as compared to ρ^+ cells are extensive.[52,53] Of key importance is the activation of the glyoxylate cycle in peroxisomes, which provides the citrate to fuel the synthesis of α-ketoglutarate in the Krebs cycle, by-passing the defect in electron transport in ρ^0 cells. Furthermore, the glyoxylate cycle utilizes acetate for net synthesis, unlike the Krebs cycle which loses a carbon atom as carbon dioxide. The retrograde response also results in activation of anaplerotic reactions. Some of these changes have been documented at the biochemical level and not only at the level of transcription.[54] The generation of α-ketoglutarate is of singular importance, as this is converted to glutamate and glutamine which provide a source of nitrogen in all biosynthesis.

There are 410 genes whose expression is significantly altered in the retrograde response.[52] Many of these genes adapt the cell to withstand stress, in addition to the metabolic stress that directly accompanies the loss of mitochondrial electron transport. One of the activated genes is *PDR5*.[52,55] Pdr5 is a member of the ATP-binding cassette superfamily, and it functions as a multidrug resistance pump in the plasma membrane.[56] The *LAC1* ceramide synthase gene[57–59] is under the same transcriptional control as *PDR5*, and it is coordinately regulated with it.[60] The magnitude and complexity of the gene expression changes elicited in the retrograde response has hampered the identification of those that are necessary and sufficient for replicative life span extension. This may require a systems biology approach to reach resolution.

LAC1 is the ortholog of the yeast ceramide synthase gene *LAG1*, the first longevity-determining gene cloned as such.[61,62] Ceramide signaling in yeast plays an important role in processes such as cell-cycle control and stress resistance.[63] Indeed, it impinges upon virtually every known signaling cascade important for long life in yeast, including the Snf1, protein kinase A, TORC1, and Sch9 pathways.[64] Furthermore, inositol-phosphorylceramide turnover, which is important for resistance to heat stress, involves the enigmatic *RAS1* gene.[65] *RAS1* is also essential for sublethal heat stress-induced replicative life span extension in yeast,[66] and it is a longevity-determining gene in yeast.[26]

The activation of the retrograde response has a curious effect in promoting the accumulation of episomal circles derived from rDNA.[67] These circles, called extrachromosomal ribosomal DNA circles (ERC), increase exponentially in amount during the entire replicative life span of both ρ^+ and ρ^0 cells,[36] and they are deleterious to the cell.[68] This ERC accumulation does not interfere with the increased life span of ρ^0 cells, indicating that the retrograde response alleviates this deleterious effect, at least to an extent. The mechanism underlying the phenomenon of increased longevity coupled to elevated ERC production resides in a dual role of Rtg2. Rtg2 is needed for propagation of the retrograde signal generated by dysfunctional mitochondria, but it also functions in the suppression of ERC production.[36] Its quantity in the cell is not sufficient to fully support both functions, however. As the retrograde response is progressively activated during the yeast life span, Rtg2 is drawn away from its role in suppressing genome instability, in a commensurate manner.

The effect of Rtg2 on ERC production indicates a role for this protein in the nucleus. This protein has yet another nuclear-based role. It is a component of the SAGA-like complex SLIK, in which Rtg2 replaces Spt8.[69] SAGA and SLIK are transcriptional coactivator complexes that contain the histone acetylase Gcn5, and they are conserved among eukaryotes. Rtg2 was detected at the *CIT2* gene promoter, along with SLIK which was required for induction of this gene. Gcn5 is required for activation of the retrograde target gene *CIT2* and for the extension of life span afforded by the retrograde response.[70] The notion that emerges from these studies is that SLIK is essential for the retrograde response. However, the supposed retrograde response in which Rtg2-containing SLIK was discovered was induced by growth on the nonfermentable carbon source acetate. This, by definition, is not a true retrograde response which occurs when mitochondria are dysfunctional and thus the cell cannot grow on acetate. Growth on acetate is more akin to the diauxic shift induced gene expression that requires *ISC1*.[41]

SAGA functions in the regulation of expression of approximately 10% of the yeast genome primarily involved in stress resistance.[71] The activity of SAGA at these loci is counteracted by the histone deacetylases Rpd3 and Hda1. Deletion of *RPD3* or *HDA1* can extend replicative life span in contrast to deletion of *SIR2*, which possesses a different gene silencing spectrum.[72] The effects on life span of Rpd3, Hda1, and Gcn5 are consistent. However, they do not distinguish between SAGA, SLIK, or the more diminutive Gcn5–Ada2 histone acetylase complexes.

4. OTHER RETROGRADE RESPONSES IN YEAST

Another retrograde-like response, termed mitochondrial back-signaling has been described in yeast.[73] This signaling is induced upon deletion of the *AFO1/MRPL25* nuclear gene that encodes a protein of the large subunit of the mitochondrial ribosome. The extension of replicative life span that ensues requires active TORC1 and the transcription factor Sfp1 that activates cytoplasmic ribosomal protein expression. This response is activated only in ρ^0 cells, during growth on glucose, in a strain in which the Rtg-dependent retrograde response is suppressed by this carbon source.

A different perturbation of mitochondrial protein synthesis also extends replicative life span. The deletion of nuclear genes that encode the proteins of the mitochondrial translation control complex results in life span extension in a Sir2-dependent manner.[74] The life span extension and enhanced genomic silencing that result require signaling by protein kinase A and the Msn2–Msn4 transcription factor which elevates Pnc1 expression. This promotes scavenging of the Sir2 inhibitor nicotinamde. This pathway for life extension operates in the absence of mitochondrial respiration, and it converges with reduced TOR signaling to extend longevity. This has most of the features of the retrograde response, and the relationship of this mechanism of life extension to the retrograde response is of great interest. It was shown some time ago that inhibition of mitochondrial protein synthesis by erythromycin extends yeast replicative life span.[75]

5. THE RETROGRADE RESPONSE AND CELL QUALITY CONTROL

Cell quality control consists of a large suite of processes that serve to maintain the fidelity and efficiency of cell function in time and space. This chapter is not directly concerned with cellular quality control. However, I would be remiss in not pointing to some of the intersections of the retrograde response with cell quality control processes.

Mitophagy is a form of selective autophagy. In yeast, Atg32 tags mitochondria for selective degradation in the vacuole.[76,77] Uth1 is an outer mitochondrial membrane protein that modulates mitophagy by altering its efficiency and selectivity.[78] The *UTH1* gene was originally discovered as a gene whose deletion extends replicative life span.[79] The protein phosphatase Aup1, located in the mitochondrial intermembrane space, is required for mitophagy in stationary phase.[80] In this process, the retrograde response is activated in an Rtg3-dependent manner. This activation is required for autophagy, and Rtg3 is dephosphorylated by Aup1. It is worth noting that survival in stationary phase termed chronological life span requires retrograde signaling, dependent on Rtg2, as well.[81]

TORC1 inhibits general (nonselective) autophagy by phosphorylating Atg13.[82] Complex sphingolipids, in yeast, negatively modulate autophagy induced by nitrogen starvation.[83] Ceramide, a substrate for the synthesis of complex sphingolipids, promotes autophagy in mammalian cells.[84] Thus, the primary regulator of autophagy appears to be TORC1, while ceramide signaling may modulate its effects.

TORC2 stimulates ceramide synthase activity via Ypk2, an SGC kinase, which it phosphorylates.[85] TORC2 activity is stimulated by the presence of nutrients, a growth signal. In the presence of active Ipt1 and Skn1, this would accentuate complex sphingolipid synthesis.[83] On the other hand, the stress responsive protein phosphatase calcineurin inhibits ceramide synthase, in a calcium/calmodulin-dependent manner,[85] which would attenuate sphingolipid production. Thus, TORC2 may modulate autophagy, upregulating it during periods of stress and downregulating it during active growth.

The Lag1/Lac1 ceramide synthases are located in the endoplasmic reticulum (ER). There is a physical connection between the ER and the mitochondrion (ERMES), which has been proposed to regulate mitochondrial function.[86] This raises the possibility that ceramide may have an important role in such regulation. Gem1 helps tether the ER to mitochondria in the ERMES.[87] ERMES are important for efficient mitochondrial calcium uptake in mammalian cells,[88] which may affect NFκB activation in cells that are in respiratory stress.[89] Mammalian ρ^0 cells show an increase in cytoplasmic calcium that results in the activation of calcineurin,[90] which would downregulate ceramide synthase and perhaps tip the balance toward autophagy. Furthermore, ERMES may play a more direct role in yeast mitophagy. Degradation of Gem1 would release mitochondria from their tether to the ER, facilitating mitophagy. This scenario gains some plausibility because Miro, the mammalian homolog of Gem1, is a PINK1 substrate that is degraded through Parkin-mediated polyubiquitination.[91] This results in detachment of mitochondria from the ER, and it occurs when mitochondria become dysfunctional through depolarization.[92]

Another mechanism of mitochondrial quality control depends on the prohibitins, Phb1/Phb2. These residents of the inner mitochondrial membrane play the role of chaperones in assembly of protein complexes and in the degradation of proteins by the inner membrane m-AAA protease.[93–95] The prohibitins are induced in the retrograde response.[53] Lack of Phb1 and/or Phb2 dramatically curtails the replicative life span of ρ^0 cells, which can be rescued by deletion of *RAS2*, a gene that plays a role in stimulating mitochondrial biogenesis.[95] This effect is likely the result of reduction in the mitochondrial protein degradation burden to an extent that can be handled by cells lacking prohibitins. This interpretation has gained support in studies involving mitochondrial dysfunction induced by the elimination of the ATP/ADP translocator Aac2,[96] in which the defect resulting from the combination with prohibitin deficit was suppressed by the reduction of protein synthesis in the cytoplasm. Further support for this interpretation has been obtained in calorie restriction studies in yeast.[97]

6. RETROGRADE RESPONSE IN OTHER ORGANISMS

When the life span-extending effect of the retrograde response was discovered, it was proposed that a similar response occurs in *Caenorhabditis elegans*, based on the metabolic changes that occur in that organism during aging.[25] This expectation was fulfilled when it was shown that knockdown of respiratory chain components in the roundworm can extend life span.[98–100] The picture that has emerged since has become more complicated.

The knockdown of *cco-1*, which encodes a subunit of mitochondrial cytochrome oxidase, activates the hypoxia-inducible transcription factor HIF-1 to extend life span in the worm.[101] A retrograde response similar to this has recently been found in the mouse.[102] Interestingly, ceramide synthesized by the worm *hyl-2* encoded ceramide synthase, a homolog of Lag1,[57] is involved in the activation of HIF-1,[103] and elevation of HIF-1 levels extends worm longevity.[104] HYL-2 synthesizes a ceramide homolog that has a prosurvival effect, while the ceramide generated by Lag1 homolog HYL-1 has an apoptotic effect.[103] This is consistent with the life span extension seen in the roundworm on knockdown of *hyl-1*.[105]

The signal that triggers the retrograde response and activates HYL-1 appears to be production of ROS by mitochondria.[101] However, another study demonstrated the involvement of the mitochondrial unfolded protein response, which engaged the transcription factors UBL-5 and DVE-1.[106] This occurred in a cell nonautonomous fashion that suggests the operation of a diffusible mitokine secreted by certain cells, to which other cells respond. In a third study, the mediator of the response to reduced mitochondrial electron transport chain function was shown to be CEH-23, a predicted homeobox transcription factor.[107] In all three studies, the implicated pathways were shown to be both necessary and sufficient for the increased longevity. Thus, their mutual relationship is not clear at present. This picture is further complicated by the fact that disruption of the worm electron transport chain at different places can extend life span but have different associated phenotypes.[108] The same study showed that ROS signaling is involved, but in a distinct mechanism that incorporates bile acid-like mediators which exert hormonal effects.[109]

A mitochondrial unfolded protein response to aggregated ornithine decarboxylase is operative in mammalian cells. However, it involves the CHOP transcription factor, rather than UBL-5 and DVE-1 homologues.[110] Interestingly, CHOP is activated by the ceramide species generated by one of the mammalian homologs of Lag1, Lass6, which protects cells from ER stress-induced apoptosis.[111] In yeast, the highly conserved mitochondrial protein Vms1 recruits Cdc48 upon oxidative stress.[112] Cdc48 is a component of the ubiquitin/proteasome system that is involved in the degradation of ER proteins. The compromise of this mechanism curtails chronological life span. Interestingly, TORC1 and ceramide signaling converge on the Sit4 protein phosphatase to link to the ER unfolded protein response.[113]

The knockdown of any one of several respiratory chain components in the fruit fly, *Drosophila melanogaster*, also elicited a retrograde response, often with associated increased longevity.[114] Furthermore, a mutant in the *sbo* gene that is defective in coenzyme Q synthesis also extended life span.[115] Unfortunately, the signaling pathways that were activated were not identified in these studies.

Reduced activity of the *MCLK1* gene, which is involved in coenzyme Q synthesis in the mouse, and thus curtailment of respiratory chain activity, substantially extended life span without negative impact on growth or fertility.[116] These effects were similar to those found earlier in *C. elegans clk-1* gene mutants. Mitochondrial cytochrome oxidase assembly is disrupted in *SURF1* knockout mice, also resulting in a marked increase in life span.[117] The signaling pathways involved in these "retrograde responses" have not been elucidated thus far. In a different approach, the uncoupler dinitrophenol was provided in the mouse diet.[118] This enhanced tissue respiration, improved the serum glucose, triglyceride, and insulin profiles, decreased reactive oxygen species and tissue DNA and protein oxidation, decreased body weight, as well as increased life span.

Defects in mitochondrial respiration have long been known to alter nuclear gene expression.[119] Calcium and NFκB signaling have been implicated in these "retrograde responses".[1] The conserved master regulator NFκB responds to stress of various sorts, including ROS.[12] It is also involved in mitochondrial biogenesis. The gene expression profiles of ρ^0/ρ^+ pairs of human cells derived from different cell layers were compared.[120] They were quite heterogeneous among the three pairs, likely due to differences in the pathologies induced by mitochondrial dysfunction in the different cell types.

However, there was a common core of a handful of changes in gene expression that reflect adaptation to survival without mitochondrial respiratory chain activity, and they are broadly reminiscent of the physiologic changes induced in the yeast retrograde response. One of these common gene expression changes was the activation of c-Myc, a bHLH–LeuZip transcription factor, which is responsible for the shift to glycolytic metabolism.[121] As mentioned earlier, the mammalian Myc–Max heterodimeric transcription factor is homologous to yeast Rtg1–Rtg3.[12]

Normal human diploid fibroblasts have a limited number of population doublings available to them, undergoing the process of cell senescence in culture.[122] Addition of dinitrophenol to the culture medium, which resulted in mild mitochondrial uncoupling, delayed this replicative senescence.[123] At the same time, it lowered ROS production, reduced telomere shortening, prevented DNA repair foci in the nucleus, and induced gene expression changes. These findings support the operation of a retrograde response in human cells.

We have proposed[124] that NFκB may have evolved as a master regulator along with the appearance of more complex immune systems.[12] Its appearance in phylogeny would have allowed the coordination of the ability to respond to a variety of inputs with an array of outputs, among which is the retrograde response. *C. elegans* does not possess an NFκB homolog, perhaps explaining the conundrum of three different, necessary and sufficient, retrograde responses. NFκB is likely to be upstream of Myc in the mammalian retrograde response, because it has two binding sites in the Myc promoter.[125] During cell senescence, NFκB and Myc activity are suppressed by Cdk2.[126,127] This finally shuts down the compensatory function of NFκB, which is actually activated during aging.

7. EVOLUTION OF THE RETROGRADE RESPONSE AS A CYTOPROTECTIVE MECHANISM

Recently, it has been proposed, based largely on the consideration of genetic analyses of *C. elegans*, that longevity regulation is the result of the evolution of cytoprotective pathways underpinning diverse triggers across species.[3] These pathways encompass disparate functions, and inhibition of each of them induces cytoprotective responses that buffer stress and damage. The cornerstone of these considerations is the retrograde response activated by mitochondrial dysfunction.

Insulin/IGF-1 signaling is one of these pathways. Its inhibition results in the utilization of the well-known survival/dispersal form-generating dauer response that is harnessed in the adult worm to increase resistance to heat, oxidative stress, anoxia, heavy metals, osmotic stress, pathogens, and UV stress. The relationship of longevity extending mechanisms to such use of dispersal form pathways was proposed earlier.[128] This class of mechanisms encompasses the unfolded protein response in both the ER and in mitochondria. It also engages resistance to pathogens. It includes detoxification of xenobiotics and the induction of autophagy. Disruptions of mitochondrial function and of cytoplasmic translation also extend life span in the worm, and they are associated with resistance to diverse toxins, as well as resistance to a variety of other stressors. The cross-protection afforded by exposure to one stressor on the resistance to an array of diverse stressors is typical of a hormetic response. The dose of the stressor that induces a hormetic response is key. This was clearly demonstrated in yeast exposed to heat shock.[66,129] The overall notion here is that exposure to pathogens or environmental toxins, whatever their nature, triggers a common suite of responses. These responses encompass detoxification, antimicrobial defenses, and chaperones and antioxidants to repair damaged cellular structures. These are the same structures that are prone to damage during aging. The sensor of impending damage is the damage itself rather than a detector of the myriad agents that can elicit such damage, and it is this damage that triggers the protective response.

This novel perspective on the evolution of longevity promoting pathways presents new opportunities to discover additional mechanisms associated with life span extension. More importantly, it creates a unifying vision of longevity-determining mechanisms that may prove instrumental in devising useful approaches to modulating longevity and healthy aging, both in model organisms and in humans.

ACKNOWLEDGMENT

The research in the author's laboratory is supported by grants from the National Institute on Aging of the National Institutes of Health (U.S.P.H.S.).

REFERENCES

1. Butow RA, Avadhani NG. Mitochondrial signaling: the retrograde response. *Mol Cell.* 2004;14:1–15.
2. Jazwinski SM. Molecular mechanisms of yeast longevity. *Trends Microbiol.* 1999;7: 247–252.

3. Shore DE, Ruvkun G. A cytoprotective perspective on longevity regulation. *Trends Cell Biol.* 2013;23:409–420.
4. Parikh VS, Morgan MM, Scott R, Clements LS, Butow RA. The mitochondrial genotype can influence nuclear gene expression in yeast. *Science.* 1987;235:576–580.
5. Liu Z, Butow RA. Mitochondrial retrograde signaling. *Annu Rev Genet.* 2006;40: 159–185.
6. Rothermel BA, Shyjan AW, Etheredge JL, Butow RA. Transactivation by Rtg1p, a basic helix–loop–helix protein that functions in communication between mitochondria and the nucleus in yeast. *J Biol Chem.* 1995;270:29476–29482.
7. Rothermel BA, Thornton JL, Butow RA. Rtg3p, a basic helix–loop–helix/leucine zipper protein that functions in mitochondrial-induced changes in gene expression, contains independent activation domains. *J Biol Chem.* 1997;272:19801–19807.
8. Sekito T, Thornton J, Butow RA. Mitochondria-to-nuclear signaling is regulated by the subcellular localization of the transcription factors Rtg1p and Rtg3p. *Mol Biol Cell.* 2000;11:2103–2115.
9. Liao X, Butow RA. RTG1 and RTG2: two yeast genes required for a novel path of communication from mitochondria to the nucleus. *Cell.* 1993;72:61–71.
10. Jia Y, Rothermel B, Thornton J, Butow RA. A basic helix–loop–helix–leucine zipper transcription complex in yeast functions in a signaling pathway from mitochondria to the nucleus. *Mol Cell Biol.* 1997;17:1110–1117.
11. Jothi R, Balaji S, Wuster A, et al. Genomic analysis reveals a tight link between transcription factor dynamics and regulatory network architecture. *Mol Syst Biol.* 2009;5:294.
12. Srinivasan V, Kriete A, Sacan A, Jazwinski SM. Comparing the yeast retrograde response and NF-kappa B stress responses: implications for aging. *Aging Cell.* 2010;9: 933–941.
13. Koonin EV. Yeast protein controlling inter-organelle communication is related to bacterial phosphatases containing the Hsp 70-type ATP-binding domain. *Trends Biochem Sci.* 1994;19:156–157.
14. Sekito T, Liu Z, Thornton J, Butow RA. RTG-dependent mitochondria-to-nucleus signaling is regulated by MKS1 and is linked to formation of yeast prion [URE3]. *Mol Biol Cell.* 2002;13:795–804.
15. Dilova I, Chen CY, Powers T. Mks1 in concert with TOR signaling negatively regulates RTG target gene expression in S. cerevisiae. *Curr Biol.* 2002;12:389–395.
16. Dilova I, Aronova S, Chen JC, Powers T. Tor signaling and nutrient-based signals converge on Mks1p phosphorylation to regulate expression of Rtg1.Rtg3p-dependent target genes. *J Biol Chem.* 2004;279:46527–46535.
17. Liu Z, Sekito T, Spirek M, Thornton J, Butow RA. Retrograde signaling is regulated by the dynamic interaction between Rtg2p and Mks1p. *Mol Cell.* 2003;12:401–411.
18. Liu Z, Spirek M, Thornton J, Butow RA. A novel degron-mediated degradation of the RTG pathway regulator, Mks1p, by SCFGrr1. *Mol Biol Cell.* 2005;16:4893–4904.
19. Komeili A, Wedaman KP, O'Shea EK, Powers T. Mechanism of metabolic control. Target of rapamycin signaling links nitrogen quality to the activity of the Rtg1 and Rtg3 transcription factors. *J Cell Biol.* 2000;151:863–878.
20. Giannattasio S, Liu Z, Thornton J, Butow RA. Retrograde response to mitochondrial dysfunction is separable from TOR1/2 regulation of retrograde gene expression. *J Biol Chem.* 2005;280:42528–42535.
21. Breitkreutz A, Choi H, Sharom JR, et al. A global protein kinase and phosphatase interaction network in yeast. *Science.* 2010;328:1043–1046.
22. Forsberg H, Ljungdahl PO. Sensors of extracellular nutrients in Saccharomyces cerevisiae. *Curr Genet.* 2001;40:91–109.

23. Roberg KJ, Bickel S, Rowley N, Kaiser CA. Control of amino acid permease sorting in the late secretory pathway of Saccharomyces cerevisiae by SEC13, LST4, LST7 and LST8. *Genetics*. 1997;147:1569–1584.
24. Matsuura A, Anraku Y. Characterization of the MKS1 gene, a new negative regulator of the Ras-cyclic AMP pathway in Saccharomyces cerevisiae. *Mol Gen Genet.* 1993;238:6–16.
25. Kirchman PA, Kim S, Lai CY, Jazwinski SM. Interorganelle signaling is a determinant of longevity in Saccharomyces cerevisiae. *Genetics*. 1999;152:179–190.
26. Sun J, Kale SP, Childress AM, Pinswasdi C, Jazwinski SM. Divergent roles of RAS1 and RAS2 in yeast longevity. *J Biol Chem*. 1994;269:18638–18645.
27. Kawai S, Urban J, Piccolis M, Panchaud N, De Virgilio C, Loewith R. Mitochondrial genomic dysfunction causes dephosphorylation of Sch9 in the yeast Saccharomyces cerevisiae. *Eukaryot Cell.* 2011;10:1367–1369.
28. Urban J, Soulard A, Huber A, et al. Sch9 is a major target of TORC1 in Saccharomyces cerevisiae. *Mol Cell.* 2007;26:663–674.
29. Fabrizio P, Pozza F, Pletcher SD, Gendron CM, Longo VD. Regulation of longevity and stress resistance by Sch9 in yeast. *Science*. 2001;292:288–290.
30. Zhang A, Shen Y, Gao W, Dong J. Role of Sch9 in regulating Ras-cAMP signal pathway in Saccharomyces cerevisiae. *FEBS Lett.* 2011;585:3026–3032.
31. Pastor MM, Proft M, Pascual-Ahuir A. Mitochondrial function is an inducible determinant of osmotic stress adaptation in yeast. *J Biol Chem*. 2009;284:30307–30317.
32. Ruiz-Roig C, Noriega N, Duch A, Posas F, de Nadal E. The Hog1 SAPK controls the Rtg1/Rtg3 transcriptional complex activity by multiple regulatory mechanisms. *Mol Biol Cell.* 2012;23:4286–4296.
33. Miceli MV, Jiang JC, Tiwari A, Rodriguez-Quinones JF, Jazwinski SM. Loss of mitochondrial membrane potential triggers the retrograde response extending yeast replicative lifespan. *Front Genet.* 2011;2:102.
34. Woo DK, Phang TL, Trawick JD, Poyton RO. Multiple pathways of mitochondrial–nuclear communication in yeast: intergenomic signaling involves ABF1 and affects a different set of genes than retrograde regulation. *Biochim Biophys Acta.* 2009;1789: 135–145.
35. Lai CY, Jaruga E, Borghouts C, Jazwinski SM. A mutation in the ATP2 gene abrogates the age asymmetry between mother and daughter cells of the yeast Saccharomyces cerevisiae. *Genetics*. 2002;162:73–87.
36. Borghouts C, Benguria A, Wawryn J, Jazwinski SM. Rtg2 protein links metabolism and genome stability in yeast longevity. *Genetics*. 2004;166:765–777.
37. Jazwinski SM. Metabolic control and gene dysregulation in yeast aging. *Ann N Y Acad Sci*. 2000;908:21–30.
38. Chen S, Liu D, Finley Jr RL, Greenberg ML. Loss of mitochondrial DNA in the yeast cardiolipin synthase crd1 mutant leads to up-regulation of the protein kinase Swe1p that regulates the G2/M transition. *J Biol Chem*. 2010;285:10397–10407.
39. Vaena de Avalos S, Su X, Zhang M, Okamoto Y, Dowhan W, Hannun YA. The phosphatidylglycerol/cardiolipin biosynthetic pathway is required for the activation of inositol phosphosphingolipid phospholipase C, Isc1p, during growth of Saccharomyces cerevisiae. *J Biol Chem*. 2005;280:7170–7177.
40. Vaena de Avalos S, Okamoto Y, Hannun YA. Activation and localization of inositol phosphosphingolipid phospholipase C, Isc1p, to the mitochondria during growth of Saccharomyces cerevisiae. *J Biol Chem*. 2004;279:11537–11545.
41. Kitagaki H, Cowart LA, Matmati N, et al. ISC1-dependent metabolic adaptation reveals an indispensable role for mitochondria in induction of nuclear genes during the diauxic shift in Saccharomyces cerevisiae. *J Biol Chem*. 2009;284:10818–10830.

42. Zhou J, Zhong Q, Li G, Greenberg ML. Loss of cardiolipin leads to longevity defects that are alleviated by alterations in stress response signaling. *J Biol Chem.* 2009;284:18106–18114.
43. Chen S, Tarsio M, Kane PM, Greenberg ML. Cardiolipin mediates cross-talk between mitochondria and the vacuole. *Mol Biol Cell.* 2008;19:5047–5058.
44. Hughes AL, Gottschling DE. An early age increase in vacuolar pH limits mitochondrial function and lifespan in yeast. *Nature.* 2012;492:261–265.
45. Garipler G, Dunn CD. Defects associated with mitochondrial DNA damage can be mitigated by increased vacuolar pH in Saccharomyces cerevisiae. *Genetics.* 2013;194:285–290.
46. Gebre S, Connor R, Xia Y, et al. Osh6 overexpression extends the lifespan of yeast by increasing vacuole fusion. *Cell Cycle.* 2012;11:2176–2188.
47. Barbosa AD, Osorio H, Sims KJ, et al. Role for Sit4p-dependent mitochondrial dysfunction in mediating the shortened chronological lifespan and oxidative stress sensitivity of Isc1p-deficient cells. *Mol Microbiol.* 2011;81:515–527.
48. Nickels JT, Broach JR. A ceramide-activated protein phosphatase mediates ceramide-induced G1 arrest of Saccharomyces cerevisiae. *Genes Dev.* 1996;10:382–394.
49. Tripathi K, Matmati N, Zheng WJ, Hannun YA, Mohanty BK. Cellular morphogenesis under stress is influenced by the sphingolipid pathway gene ISC1 and DNA integrity checkpoint genes in Saccharomyces cerevisiae. *Genetics.* 2011;189:533–547.
50. Garipler G, Mutlu N, Lack NA, Dunn CD. Deletion of conserved protein phosphatases reverses defects associated with mitochondrial DNA damage in Saccharomyces cerevisiae. *Proc Natl Acad Sci USA.* 2014;111:1473–1478.
51. Teixeira V, Medeiros TC, Vilaca R, Moradas-Ferreira P, Costa V. Reduced TORC1 signaling abolishes mitochondrial dysfunctions and shortened chronological lifespan of Isc1p-deficient cells. *Microbial Cell.* 2014;1:21–36.
52. Epstein CB, Waddle JA, Hale Wt, et al. Genome-wide responses to mitochondrial dysfunction. *Mol Biol Cell.* 2001;12:297–308.
53. Traven A, Wong JM, Xu D, Sopta M, Ingles CJ. Interorganellar communication. Altered nuclear gene expression profiles in a yeast mitochondrial DNA mutant. *J Biol Chem.* 2001;276:4020–4027.
54. Small WC, Brodeur RD, Sandor A, et al. Enzymatic and metabolic studies on retrograde regulation mutants of yeast. *Biochemistry.* 1995;34:5569–5576.
55. Hallstrom TC, Moye-Rowley WS. Multiple signals from dysfunctional mitochondria activate the pleiotropic drug resistance pathway in Saccharomyces cerevisiae. *J Biol Chem.* 2000;275:37347–37356.
56. Balzi E, Wang M, Leterme S, Van Dyck L, Goffeau A. PDR5, a novel yeast multidrug resistance conferring transporter controlled by the transcription regulator PDR1. *J Biol Chem.* 1994;269:2206–2214.
57. Jiang JC, Kirchman PA, Zagulski M, Hunt J, Jazwinski SM. Homologs of the yeast longevity gene LAG1 in Caenorhabditis elegans and human. *Genome Res.* 1998;8: 1259–1272.
58. Guillas I, Kirchman PA, Chuard R, et al. C26-CoA-dependent ceramide synthesis of Saccharomyces cerevisiae is operated by Lag1p and Lac1p. *EMBO J.* 2001;20: 2655–2665.
59. Schorling S, Vallee B, Barz WP, Riezman H, Oesterhelt D. Lag1p and Lac1p are essential for the Acyl-CoA-dependent ceramide synthase reaction in Saccharomyces cerevisiae. *Mol Biol Cell.* 2001;12:3417–3427.
60. Kolaczkowski M, Kolaczkowska A, Gaigg B, Schneiter R, Moye-Rowley WS. Differential regulation of ceramide synthase components LAC1 and LAG1 in Saccharomyces cerevisiae. *Eukaryot Cell.* 2004;3:880–892.

61. Egilmez NK, Chen JB, Jazwinski SM. Specific alterations in transcript prevalence during the yeast life span. *J Biol Chem*. 1989;264:14312–14317.
62. D'Mello NP, Childress AM, Franklin DS, Kale SP, Pinswasdi C, Jazwinski SM. Cloning and characterization of LAG1, a longevity-assurance gene in yeast. *J Biol Chem*. 1994;269:15451–15459.
63. Dickson RC. Roles for sphingolipids in Saccharomyces cerevisiae. *Adv Exp Med Biol*. 2010;688:217–231.
64. Liu J, Huang X, Withers BR, Blalock E, Liu K, Dickson RC. Reducing sphingolipid synthesis orchestrates global changes to extend yeast lifespan. *Aging Cell*. 2013;12:833–841.
65. Kaibuchi K, Miyajima A, Arai K, Matsumoto K. Possible involvement of RAS-encoded proteins in glucose-induced inositolphospholipid turnover in Saccharomyces cerevisiae. *Proc Natl Acad Sci USA*. 1986;83:8172–8176.
66. Shama S, Lai CY, Antoniazzi JM, Jiang JC, Jazwinski SM. Heat stress-induced life span extension in yeast. *Exp Cell Res*. 1998;245:379–388.
67. Conrad-Webb H, Butow RA. A polymerase switch in the synthesis of rRNA in Saccharomyces cerevisiae. *Mol Cell Biol*. 1995;15:2420–2428.
68. Sinclair DA, Guarente L. Extrachromosomal rDNA circles—a cause of aging in yeast. *Cell*. 1997;91:1033–1042.
69. Pray-Grant MG, Schieltz D, McMahon SJ, et al. The novel SLIK histone acetyltransferase complex functions in the yeast retrograde response pathway. *Mol Cell Biol*. 2002;22:8774–8786.
70. Kim S, Ohkuni K, Couplan E, Jazwinski SM. The histone acetyltransferase GCN5 modulates the retrograde response and genome stability determining yeast longevity. *Biogerontology*. 2004;5:305–316.
71. Huisinga KL, Pugh BF. A genome-wide housekeeping role for TFIID and a highly regulated stress-related role for SAGA in Saccharomyces cerevisiae. *Mol Cell*. 2004;13:573–585.
72. Kim S, Benguria A, Lai CY, Jazwinski SM. Modulation of life-span by histone deacetylase genes in Saccharomyces cerevisiae. *Mol Biol Cell*. 1999;10:3125–3136.
73. Heeren G, Rinnerthaler M, Laun P, et al. The mitochondrial ribosomal protein of the large subunit, Afo1p, determines cellular longevity through mitochondrial back-signaling via TOR1. *Aging*. 2009;1:622–636.
74. Caballero A, Ugidos A, Liu B, et al. Absence of mitochondrial translation control proteins extends life span by activating sirtuin-dependent silencing. *Mol Cell*. 2011;42:390–400.
75. Holbrook MA, Menninger JR. Erythromycin slows aging of Saccharomyces cerevisiae. *J Gerontol A Biol Sci Med Sci*. 2002;57:B29–B36.
76. Kim I, Rodriguez-Enriquez S, Lemasters JJ. Selective degradation of mitochondria by mitophagy. *Arch Biochem Biophys*. 2007;462:245–253.
77. Kissova I, Salin B, Schaeffer J, Bhatia S, Manon S, Camougrand N. Selective and non-selective autophagic degradation of mitochondria in yeast. *Autophagy*. 2007;3:329–336.
78. Kissova I, Deffieu M, Manon S, Camougrand N. Uth1p is involved in the autophagic degradation of mitochondria. *J Biol Chem*. 2004;279:39068–39074.
79. Kennedy BK, Austriaco Jr NR, Zhang J, Guarente L. Mutation in the silencing gene SIR4 can delay aging in S. cerevisiae. *Cell*. 1995;80:485–496.
80. Journo D, Mor A, Abeliovich H. Aup1-mediated regulation of Rtg3 during mitophagy. *J Biol Chem*. 2009;284:35885–35895.
81. Barros MH, Bandy B, Tahara EB, Kowaltowski AJ. Higher respiratory activity decreases mitochondrial reactive oxygen release and increases life span in Saccharomyces cerevisiae. *J Biol Chem*. 2004;279:49883–49888.

82. Kamada Y, Yoshino K, Kondo C, et al. Tor directly controls the Atg1 kinase complex to regulate autophagy. *Mol Cell Biol.* 2010;30:1049–1058.
83. Thevissen K, Yen WL, Carmona-Gutierrez D, et al. Skn1 and Ipt1 negatively regulate autophagy in Saccharomyces cerevisiae. *FEMS Microbiol Lett.* 2010;303:163–168.
84. Lavieu G, Scarlatti F, Sala G, et al. Sphingolipids in macroautophagy. *Methods Mol Biol.* 2008;445:159–173.
85. Aronova S, Wedaman K, Aronov PA, et al. Regulation of ceramide biosynthesis by TOR complex 2. *Cell Metab.* 2008;7:148–158.
86. Kornmann B, Walter P. ERMES-mediated ER-mitochondria contacts: molecular hubs for the regulation of mitochondrial biology. *J Cell Sci.* 2010;123:1389–1393.
87. Kornmann B, Osman C, Walter P. The conserved GTPase Gem1 regulates endoplasmic reticulum-mitochondria connections. *Proc Natl Acad Sci USA.* 2011;108: 14151–14156.
88. de Brito OM, Scorrano L. Mitofusin 2 tethers endoplasmic reticulum to mitochondria. *Nature.* 2008;456:605–610.
89. Biswas G, Tang W, Sondheimer N, Guha M, Bansal S, Avadhani NG. A distinctive physiological role for IkappaBbeta in the propagation of mitochondrial respiratory stress signaling. *J Biol Chem.* 2008;283:12586–12594.
90. Biswas G, Adebanjo OA, Freedman BD, et al. Retrograde Ca^{2+} signaling in C_2C_{12} skeletal myocytes in response to mitochondrial genetic and metabolic stress: a novel mode of inter-organelle crosstalk. *EMBO J.* 1999;18:522–533.
91. Wang X, Winter D, Ashrafi G, et al. PINK1 and Parkin target Miro for phosphorylation and degradation to arrest mitochondrial motility. *Cell.* 2011;147:893–906.
92. Narendra D, Tanaka A, Suen DF, Youle RJ. Parkin-induced mitophagy in the pathogenesis of Parkinson disease. *Autophagy.* 2009;5:706–708.
93. Berger KH, Yaffe MP. Prohibitin family members interact genetically with mitochondrial inheritance components in Saccharomyces cerevisiae. *Mol Cell Biol.* 1998;18: 4043–4052.
94. Nijtmans LG, de Jong L, Artal Sanz M, et al. Prohibitins act as a membrane-bound chaperone for the stabilization of mitochondrial proteins. *EMBO J.* 2000;19: 2444–2451.
95. Kirchman PA, Miceli MV, West RL, Jiang JC, Kim S, Jazwinski SM. Prohibitins and Ras2 protein cooperate in the maintenance of mitochondrial function during yeast aging. *Acta Biochim Pol.* 2003;50:1039–1056.
96. Wang X, Zuo X, Kucejova B, Chen XJ. Reduced cytosolic protein synthesis suppresses mitochondrial degeneration. *Nat Cell Biol.* 2008;10:1090–1097.
97. Schleit J, Johnson SC, Bennett CF, et al. Molecular mechanisms underlying genotype-dependent responses to dietary restriction. *Aging Cell.* 2013;12:1050–1061.
98. Dillin A, Hsu AL, Arantes-Oliveira N, et al. Rates of behavior and aging specified by mitochondrial function during development. *Science.* 2002;298:2398–2401.
99. Lee SS, Lee RY, Fraser AG, Kamath RS, Ahringer J, Ruvkun G. A systematic RNAi screen identifies a critical role for mitochondria in C. elegans longevity. *Nat Genet.* 2003;33:40–48.
100. Cristina D, Cary M, Lunceford A, Clarke C, Kenyon C. A regulated response to impaired respiration slows behavioral rates and increases lifespan in Caenorhabditis elegans. *PLoS Genet.* 2009;5:e1000450.
101. Lee SJ, Hwang AB, Kenyon C. Inhibition of respiration extends C. elegans life span via reactive oxygen species that increase HIF-1 activity. *Curr Biol.* 2010;20:2131–2136.
102. Gomes AP, Price NL, Ling AJ, et al. Declining NAD(+) induces a pseudohypoxic state disrupting nuclear–mitochondrial communication during aging. *Cell.* 2013;155: 1624–1638.

103. Menuz V, Howell KS, Gentina S, et al. Protection of C. elegans from anoxia by HYL-2 ceramide synthase. *Science*. 2009;324:381–384.
104. Mehta R, Steinkraus KA, Sutphin GL, et al. Proteasomal regulation of the hypoxic response modulates aging in C. elegans. *Science*. 2009;324:1196–1198.
105. Tedesco P, Jiang J, Wang JQ, Jazwinski SM, Johnson TE. Genetic analysis of hyl-1, the C. elegans homolog of LAG1/LASS1. *Age*. 2008;30:43–52.
106. Durieux J, Wolff S, Dillin A. The cell-non-autonomous nature of electron transport chain-mediated longevity. *Cell*. 2011;144:79–91.
107. Walter L, Baruah A, Chang HW, Pace HM, Lee SS. The homeobox protein CEH-23 mediates prolonged longevity in response to impaired mitochondrial electron transport chain in C. elegans. *PLoS Biol*. 2011;9:e1001084.
108. Yang W, Hekimi S. Two modes of mitochondrial dysfunction lead independently to lifespan extension in Caenorhabditis elegans. *Aging Cell*. 2010;9:433–447.
109. Liu JL, Desjardins D, Branicky R, Agellon LB, Hekimi S. Mitochondrial oxidative stress alters a pathway in Caenorhabditis elegans strongly resembling that of bile acid biosynthesis and secretion in vertebrates. *PLoS Genet*. 2012;8:e1002553.
110. Zhao Q, Wang J, Levichkin IV, Stasinopoulos S, Ryan MT, Hoogenraad NJ. A mitochondrial specific stress response in mammalian cells. *EMBO J*. 2002;21: 4411–4419.
111. Senkal CE, Ponnusamy S, Bielawski J, Hannun YA, Ogretmen B. Antiapoptotic roles of ceramide-synthase-6-generated C16-ceramide via selective regulation of the ATF6/CHOP arm of ER-stress-response pathways. *FASEB J*. 2010;24:296–308.
112. Heo JM, Livnat-Levanon N, Taylor EB, et al. A stress-responsive system for mitochondrial protein degradation. *Mol Cell*. 2010;40:465–480.
113. Mousley CJ, Tyeryar K, Ile KE, et al. Trans-Golgi network and endosome dynamics connect ceramide homeostasis with regulation of the unfolded protein response and TOR signaling in yeast. *Mol Biol Cell*. 2008;19:4785–4803.
114. Copeland JM, Cho J, Lo Jr T, et al. Extension of Drosophila life span by RNAi of the mitochondrial respiratory chain. *Curr Biol*. 2009;19:1591–1598.
115. Liu J, Wu Q, He D, et al. Drosophila sbo regulates lifespan through its function in the synthesis of coenzyme Q in vivo. *J Genet Genomics*. 2011;38:225–234.
116. Lapointe J, Hekimi S. Early mitochondrial dysfunction in long-lived Mclk1 +/− mice. *J Biol Chem*. 2008;283:26217–26227.
117. Dell'agnello C, Leo S, Agostino A, et al. Increased longevity and refractoriness to Ca(2+)-dependent neurodegeneration in Surf1 knockout mice. *Hum Mol Genet*. 2007;16:431–444.
118. Caldeira da Silva CC, Cerqueira FM, Barbosa LF, Medeiros MH, Kowaltowski AJ. Mild mitochondrial uncoupling in mice affects energy metabolism, redox balance and longevity. *Aging Cell*. 2008;7:552–560.
119. Heddi A, Lestienne P, Wallace DC, Stepien G. Mitochondrial DNA expression in mitochondrial myopathies and coordinated expression of nuclear genes involved in ATP production. *J Biol Chem*. 1993;268:12156–12163.
120. Miceli MV, Jazwinski SM. Common and cell type-specific responses of human cells to mitochondrial dysfunction. *Exp Cell Res*. 2005;302:270–280.
121. Collier JJ, Doan TT, Daniels MC, Schurr JR, Kolls JK, Scott DK. c-Myc is required for the glucose-mediated induction of metabolic enzyme genes. *J Biol Chem*. 2003;278: 6588–6595.
122. Hayflick L, Moorhead PS. The serial cultivation of human diploid cell strains. *Exp Cell Res*. 1961;25:585–621.
123. Passos JF, Saretzki G, Ahmed S, et al. Mitochondrial dysfunction accounts for the stochastic heterogeneity in telomere-dependent senescence. *PLoS Biol*. 2007;5:e110.

124. Jazwinski SM, Kriete A. The yeast retrograde response as a model of intracellular signaling of mitochondrial dysfunction. *Front Physiol.* 2012;3:139.
125. Duyao MP, Kessler DJ, Spicer DB, Sonenshein GE. Binding of NF-KB-like factors to regulatory sequences of the c-myc gene. *Curr Top Microbiol Immunol.* 1990;166:211–220.
126. Semsei I, Ma SY, Cutler RG. Tissue and age specific expression of the myc proto-oncogene family throughout the life span of the C57BL/6J mouse strain. *Oncogene.* 1989;4:465–471.
127. Campaner S, Doni M, Hydbring P, et al. Cdk2 suppresses cellular senescence induced by the c-myc oncogene. *Nat Cell Biol.* 2010;12:54–59, pp. 1–14.
128. Jazwinski SM. Longevity, genes, and aging. *Science.* 1996;273:54–59.
129. Shama S, Kirchman PA, Jiang JC, Jazwinski SM. Role of RAS2 in recovery from chronic stress: effect on yeast life span. *Exp Cell Res.* 1998;245:368–378.

CHAPTER SIX

Mitochondrial Acetylation and Genetic Models of Parkinson's Disease

Georg Auburger, Suzana Gispert, Marina Jendrach
Experimental Neurology, Goethe University Medical School, Frankfurt am Main, Germany

Contents

Abstract

Parkinson's disease (PD) is frequent at old age, leading to atrophy of specific neurons and to early death. Lifespan and healthy aging of organisms depend on growth factor/nutrient signaling and on bioenergetics via mitochondria, all of which regulate downstream nuclear functions through FOXO and SIR proteins. Mammalian SIRtuins include the mitochondrial deacetylase SIRT3, and recently mitochondrial lysine acetylation (AcLys) was found to initiate mitochondrial degradation by autophagy. This mitophagy process is closely regulated by PINK1 and Parkin, two interacting proteins which relocalize to mitochondria with deficient proton gradients, and whose mutations cause autosomal recessive variants of PD. Strong generalized deacetylation of mitochondrial proteins and altered SIRT3 levels occur in rodent models of PD before the onset of toxic aggregate formation. We propose that the development of site-specific AcLys-antibodies and their characterization in patients will have medical value.

Progress in Molecular Biology and Translational Science, Volume 127
ISSN 1877-1173
http://dx.doi.org/10.1016/B978-0-12-394625-6.00006-4

1. INTRODUCTION: LONGEVITY MODULATION BY NUTRIENT AND BIOENERGETIC PATHWAYS

Parkinson's disease (PD) is the second most frequent neurodegenerative syndrome at old age, leading to shortened lifespan, with bioenergetic failure due to mitochondrial dysfunction playing a crucial role in its pathogenesis. Any organism must strive to optimize its life-expectancy and its disease-risks, and the crucial determinant in most cases is the balance between feeding and fasting excesses for each cell. Research into the genetic determinants of longevity and healthy aging in organisms revealed several pathways[1]: Firstly, glucose, insulin, and growth factors controlling the PI3K/AKT/AMPK phosphorylation pathways. Secondly, amino acids regulating the TOR phosphorylation pathway and mRNA translation. Thirdly, mitochondrial respiration and oxidative stress modulating genotoxicity, cell cycle, senescence, and apoptosis. Fourth, the coordination of downstream nuclear functions for metabolic stress adaptation through FOXO transcription factors and diverse SIR proteins (Sirtuins). Sirtuins play a central role in NAD^+-dependent deacetylation and deacylation, thus reflecting the bioenergetic state of the cell by specific biochemical responses, prominently in the mitochondria.

2. THE CENTRAL ROLE OF SIRTUINS

The role of nuclear sirtuins for longevity became evident, when increased gene dosage of Sir2p (silent information regulator 2) in *Saccharomyces cerevisiae* was observed to slow aging and extend lifespan in mother cells,[2] and when increased gene dosage of *sir-2* in *Caenorhabditis elegans* was found to extend lifespan by up to 50%.[3] Yeast Sir2 and its mammalian ortholog SIRT1 were shown to couple NAD^+ hydrolysis to lysine deacetylation of nuclear histone proteins.[4] This hydrolysis yields nicotinamide (NAM), *O*-acetyl-ADP-ribose, and the deacetylated substrate.[5] Small molecule activators of nuclear sirtuins, such as the red wine component resveratrol, lower the Michaelis constant for both the acetylated substrate and NAD^+, and increase cell survival by stimulating SIRT1-dependent deacetylation of the nuclear cell cycle regulator and tumor suppressor p53.[6] The posttranslational modification (PTM) of FOXO family members by nuclear SIRT1/SIR-2.1 as well as by mitochondrial SIRT3 underscores the role of deacetylation for longevity regulation.[7–11] Also the generation of NAM seems to be important, given that NAM, its methylation derivative 1-methylnicotinamide, and overexpression of nicotinamide-N-methyltransferase extend *C. elegans* lifespan independent of sir-2.[12] Thus, NAD^+ plays a central role for lifespan.

Mammals have seven members of the sirtuin gene family, with SIRT1–3 showing strong NAD^+-dependent deacetylation activity, while SIRT4–7 show strong deacetylation and are more accurately described as NAD^+-dependent deacylases (e.g., demalonylases and desuccinylases).[13] SIRT1, SIRT6, and SIRT7 are primarily nuclear proteins; SIRT2 is cytoplasmic; and SIRT3, SIRT4, and SIRT5 are imported into mitochondria.[14] SIRT1 deacetylates many transcription factors, which on one hand control the choice by cells of an oxidative versus glycolytic metabolic strategy, such as the mitochondrial biogenesis modulator PGC-1α, the hypoxia response factor HIF-1α, and the lipid catabolism for ketogenesis inductor PPARα, or on the other hand control the cellular timing that balances repair versus apoptosis during damage responses, such as FOXO, p53, and PARP.[14] SIRT1 also deacetylates several components of the autophagy complex, namely Atg5, Atg7, and Atg8, thus activating autophagy.[15] SIRT6 weakly deacetylates histones at promoters of NF-κB-regulated genes to suppress the proinflammatory pathways,[16] but was recently shown to have strong effects on lysine fatty acylation (such as myristoylation) and thus regulate the secretion of proteins like TNP-α.[17] SIRT7 mediates histone H3K18Ac deacetylation to inhibit tumor suppressor transcription, thus maintaining anchorage-independent growth and preventing contact inhibition.[18] SIRT2 deacetylates the cytoplasmic growth signaling hub Akt to activate it[19] and the microtubule component α-tubulin in correlation with age.[20,21] Among these nonmitochondrial sirtuins, SIRT1 plays an upstream role for mitochondrial SIRT3 expression[22] and SIRT1 is best characterized.

As an intriguing disease aspect, inhibitors of the cytoplasmic SIRT2 rescue the aggregation toxicity in genetic models of human cells and the dopaminergic neuron loss of fly organism, which are mediated by α-synuclein as a trigger for the age-related PD.[23] In primary cortical neurons, PGC-1α or its nuclear activation by SIRT1 compensated the mitochondrial loss induced by mutant α-synuclein.[24] Mutations mediating an α-synuclein gain-of-function result in its co-aggregation with the microtubule component tau in the process of causing early onset autosomal dominant PD,[25] while gene variants at the chromosomal loci of α-synuclein and tau predispose to old-age sporadic PD.[26] Interestingly, α-synuclein displaying N-terminal acetylation has decreased aggregation rates,[27,28] while tau deacetylation was shown to protect from aggregate formation and neurodegeneration.[29] Thus, cytoplasmic acetylation of α-synuclein was already implicated repeatedly in the neuroprotection against PD.

Among the three sirtuins in the mitochondrial matrix, SIRT4 uses NAD^+ for ADP-ribosylation to repress the activity of glutamate dehydrogenase and thus downregulate insulin secretion from pancreatic beta-cells.[30] SIRT5 acts as NAD^+-dependent lysine desuccinylase, demalonylase, and deacetylase to activate carbamoyl phosphate synthase 1, the rate-limiting step of the urea cycle.[31] SIRT3 is the strongest and the best understood among mitochondrial deacetylases. It is translated in the cytoplasm as an enzymatically inactive precursor, then its first 100 amino acids are proteolytically cleaved upon import into mitochondria, generating the mature active SIRT3 of 28 kDa.[32] It acts on many targets during fasting conditions or calorie restriction—for extensive reviews, see[5,32]—to suppress carbohydrate utilization and oxidative stress, while stimulating fatty acid catabolism, amino acid catabolism, ketone body synthesis, and acetate usage in the TCA cycle. SIRT3 is also the only sirtuin for which a human polymorphism (an enhancer in intron 5) has been reported to associate with extreme longevity.[14] Thus, SIRT3 seems to be the main deacetylase counteracting the widespread acetylation of mitochondrial matrix proteins.

An elevated NAD^+/NADH ratio stimulates sirtuin enzymatic activity, but regulates also the expression at least of SIRT3, SIRT1, SIRT4, and SIRT6. The expression of SIRT3 is increased during caloric deprivation (e.g., fasting, calorie restriction, exercise), exposure to cold, or acute high-fat diet[33–38] and decreased during prolonged caloric excess (e.g., high-fat diet over 13 weeks, diabetes mellitus) as well as in aged human tissue.[34,35,37,39,40] SIRT3 expression is controlled at the transcription level by PGC-1α.[41] Also SIRT1 expression is increased during calorie restriction[42,43] and influences downstream SIRT3 induction.[22] The fasting-triggered increase in SIRT1 levels is mediated by PPARα expression induction and occurs in parallel to increasing NAD^+ levels due to enhanced nicotinamide phosphoribosyltransferase (NAMPT) activity.[44] In dependence on NAD^+/NADH levels also additional biomarkers of feeding or fasting excesses, of circadian clock, or of aging change their expression together with SIRT3 and SIRT1.[45,46] A well-established example is NAMPT expression, which also increases during fasting, calorie restriction, and exercise.[43,47,48] In a study of NAD^+ effects during aging, NAMPT and UCP3 levels were found downregulated, and levels of SIRT1, SIRT6, HIF-1α, the SIRT3-interactor OGG1, and VEGF were upregulated.[47]

Conversely, the expression of SIRT4 is suppressed by calorie restriction and induced in diabetes mellitus mouse models.[49,50] Fasting also leads to an increase of the mitochondrial acetyl-CoA synthase (ACeCS2) expression in parallel to decreased cytoplasmic ACeCS1 expression.[51,52] Thus, a battery of expression changes reveals the bioenergetics status of tissues (Table 6.1).

Mitochondrial matrix proteins are subject to extensive acetylation, in particular mitochondrial dehydrogenases, with high conservation from *Drosophila melanogaster* to humans.[53] This was previously thought to occur in a spontaneous manner due to the millimolar concentrations of acetyl-CoA, which can initiate nonenzymatic acetylation reactions.[32] However, general control of amino acid synthesis 5-like 1 (GCN5L1) was recently shown to

Table 6.1 NAD^+-dependent expression changes

	CR	FS	EX	AG	CE	AF	DM	CA
SIRT1	↑	↑		↑				↓
SIRT2								↓
SIRT3	↑	↑	↑		↑	↑	↓	↓
SIRT4	↓						↑	
SIRT5								↓
SIRT6				↑				↓
SIRT7								↓
PPARα		↑						
PGC-1α		↑						
HIF-1α				↑				
NAMPT	↑	↑	↑	↓				
UCP3				↓				
OGG1				↑				
VEGF				↑				
ACeCS1		↓						
ACeCS2		↑						
GCN5L1	?	?	?	?	?	?	?	?
TFEB	?	?	?	?	?	?	?	?

CR, calorie restriction; FS, fasting; EX, exercise; AG, aging; CE, cold exposure; AF, acute high-fat diet; DM, diabetes mellitus or chronic high-fat diet; CA, carcinoma.

counter SIRT3 effects on acetylation and respiration in mitochondria, identifying it as a critical component of the mitochondrial acetyltransferase program.[54] GCN5L1 possesses significant homology to prokaryotic acetyltransferases, is mitochondrially enriched, more highly expressed in oxidative tissues, and promotes acetylation of electron transport chain proteins.[54] Yeast GCN5 is part of a histone acetyltransferase multiprotein complex that is important for transcriptional co-activation and for genome stability.[55] Currently it is unclear whether enzymatic acetylation of mitochondrial proteins occurs in bulk, or shows differential regulation of subsets of lysine sites or proteins in response to metabolic demands. This seems conceivable in view of the selective deacetylation of the SIRT3 target LCAD during fasting of wild-type (WT) mice, while global mitochondrial protein acetylation is increased and SIRT3 expression is also high, and its selective hyperacetylation during high-fat diet, while global mitochondrial protein acetylation is high and SIRT3 expression is low.[39] Thus, mitochondrial acetylation is an active and possibly selective process, which is poorly understood yet.

3. MITOCHONDRIAL DEACETYLATION EFFECTS ON OXIDATIVE STRESS AND CANCER

The acetylation of mitochondrial matrix proteins is high, while cellular energy conversion is dominated by carbohydrate catabolism. SIRT3-induced deacetylation promotes the usage of fuels that might otherwise be stored (fat and amino acids) or ignored (acetate).[14] Fatty acid oxidation produces $FADH_2$ instead of NADH and bypasses respiratory complex I as a source of reactive oxygen species (ROS).[14] SIRT3 also minimizes ROS production at respiratory complex III[56] and enhances the anti-oxidant properties of mitochondrial superoxide dismutase (SOD2).[57,58] The loss-of-function of SIRT3 results in mouse mammary tumors and in tumor-permissive phenotype together with stress-induced genome instability of mouse embryonal fibroblasts, an effect that could be rescued by SOD2.[59] SIRT3 was shown to control the Warburg effect of tumors through the destabilization of HIF-1α, a transcription factor that controls glycolytic gene expression.[60] SIRT3 is also a known interactor of FOXO3a, a stimulator of pro-apoptotic signaling.[57] Decreased SIRT3 expression was observed in breast cancer, in gastric cancer with poor prognosis, in lung cancer cells, and in liver cancer of humans.[60–66] Interestingly, mRNA quantification demonstrated downregulation of *SIRT1*, *SIRT2*, *SIRT3*, *SIRT5*, *SIRT6*, and *SIRT7* in head and neck squamous cell carcinoma.[67] Paradoxically, the expression of SIRT3 and NAMPT was increased in well-differentiated thyroid carcinomas,[68] and the expression of *SIRT3* and *SIRT7* mRNA was increased in node-positive breast cancer,[69] in keeping with a dual role of SIRT3 for tumors.[70] Overall, most observations correlate low SIRT3 levels, which would be expected to result in high-mitochondrial protein acetylation, to increased oxidative stress, genome instability, and cancer.

4. MITOPHAGY REGULATION

Conversely, the restriction of mitochondrial protein acetylation by *GCN5L1* knockdown was recently reported to initiate the degradation of mitochondria by autophagy (a process now termed mitophagy). For a starvation-induced bulk mitophagy process the autophagy mediators Atg5 and p62 are required, while the mitochondrial depolarization triggered role of Parkin for selective mitophagy is not needed.[71] In confirmation, the genetic deletion of the mitochondrial acetyltransferase component GCN5L1 stimulated the master inductor of autophagy transcriptional factor EB (TFEB).[72] The subsequent increase in mitochondrial degradation by autophagy (mitophagy) in these cells was compensated by induction of PGC-1α expression and mitochondrial biogenesis, resulting in enhanced turnover of mitochondria.[72] Since, young mitochondria and glycolysis become more prominent in such cells, it is understandable why their resistance to mitochondrial stressors was increased.[71] In comparison, nuclear protein deacetylation seems to act preferentially on mitochondrial biogenesis, given that SIRT1 overexpression or GCN5 acetyltransferase silencing in primary cortical neurons increased mitochondrial density in cell bodies and axons via PGC-1α.[24] Mitochondrial and nuclear lysine deacetylases influence not only mitophagy and mitochondrial biogenesis, but affect at several points also the mitochondrial fusion–fission balance and the mitochondrial transport, together with their alterations in neurodegenerative diseases.[73]

Macroautophagy can occur in a nonselective manner upon nutrient deprivation to supply cells with essential metabolic building blocks and energy, or as cargo-specific quality-control process under nutrient-rich conditions (pruning) to mediate the removal of superfluous or damaged organelles and protein aggregates that otherwise could be toxic.[74] This process depends on several Atg proteins,[74] as well as on the LC3-/GABARAP-receptors NIX and FUNDC1[75] (see Fig. 6.1). In contrast, the function of the

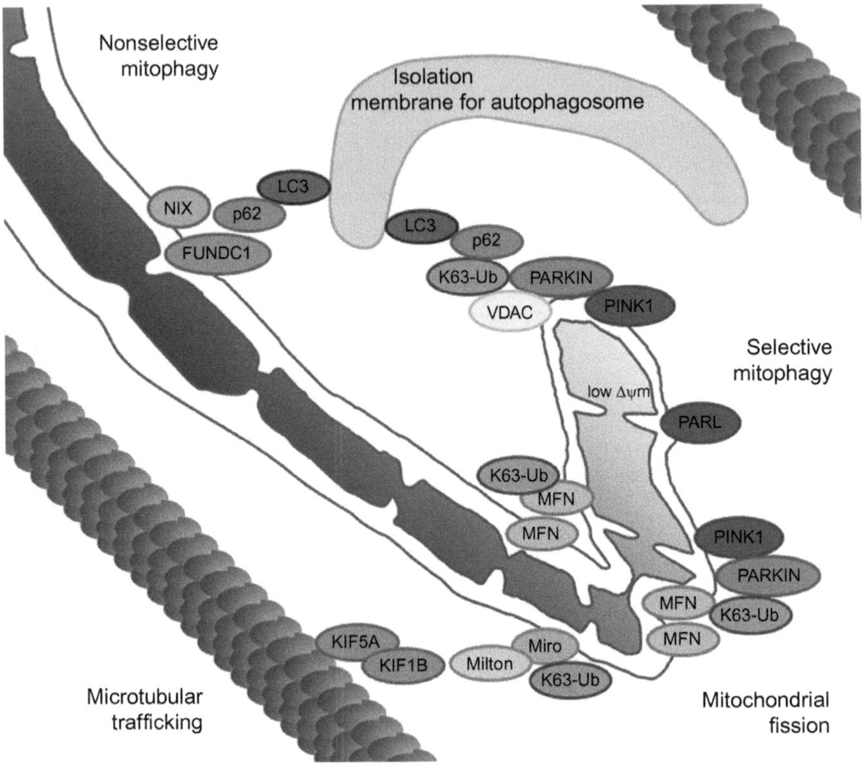

Figure 6.1 Regulation of selective versus nonselective mitophagy. (See the color plate.)

cytoplasmic ubiquitin E3 ligase Parkin in mitophagy is the relocalization to mitochondrial compartments with decreased membrane potential (Δψm) and their selective labeling by K63- and K27-polyubiquitination.[76–79] This is followed by the seeding of a multiprotein complex that includes the Mitofusins (MFN1/2), the LC3-adaptor p62, and the histone deacetylase 6 at the outer mitochondrial membrane (OMM), for removal towards the minus-end of microtubules via Miro, Milton, KIF1B, and KIF5A,[80–82] and the final autophago-lysosomal elimination of such dysfunctional mitochondria.[83,84] The extrusion of ubiquitinated proteins from membranes occurs with the help of the AAA-ATPase p97 well known for its role in ER-associated degradation, the final proteolysis is performed by the cytoplasmic proteasome, and the mitochondrial inner membrane and matrix are degraded by the autophago-lysosomal pathway.[85] Parkin is attracted to dysfunctional mitochondria by the upstream action of PINK1, these factors being co-induced transcriptionally during starvation-triggered autophagy[86] or by decreased levels of mitochondrial fission factors.[87] The N-terminus of PINK1 bears a mitochondrial targeting signal and a putative transmembrane motif, and a C-terminal serine/threonine kinase domain.[88,89] After cytoplasmic translation and import into mitochondria, PINK1 protein is immediately degraded in a voltage-sensitive manner, so that only mitochondrial compartments with decreased membrane potential ($\Delta\psi m$) accumulate it, with its kinase activity signaling from the outer membrane to the cytosol.[90]

5. MITOCHONDRIAL CLEARANCE IN PD PATIENT CELLS

Loss-of-function mutations in Parkin and PINK1 cause autosomal-recessive variants of PD named PARK2 and PARK6, respectively,[88,91] which are characterized by particularly early onset, slow progression, and sustained therapy benefit.[92] The characterization of patient cells soon discovered a susceptibility to mitochondrial apoptosis,[93] deficits in mitochondrial respiration together with oxidative stress,[94–96] alterations in the mitochondrial fusion–fission balance,[97] and a susceptibility to degradation challenges.[98] The analysis of mitochondrial clearance pathology was performed in patient primary skin fibroblasts and in fibroblast-derived induced pluripotent stem-cells reprogrammed as dopaminergic neurons, and confirmed the influence of PINK1 on Parkin, the selective stabilization of PINK1 and enrichment of Parkin on dysfunctional mitochondria, the crucial role of $\Delta\psi m$, the Parkin-mediated degradation of mitofusins and other OMM proteins, and the pathological accumulation of mitochondria in parallel to an upregulation of PGC-1α as an effort to obtain fresh mitochondria.[99–103]

6. AVAILABLE GENETIC ANIMAL MODELS FOR PD-ASSOCIATED MITOCHONDRIAL PATHOLOGY

The mitophagy defect caused by PINK1- and Parkin-deficiency results in dramatic *D. melanogaster* phenotypes selectively in tissues with respiratory stress. In addition to reduced lifespan and locomotor deficits, a selective degeneration of dopaminergic neurons and flight muscles was prominent, with many enlarged and swollen mitochondria detected.[104–106] The corresponding knockout mice for PINK1 and Parkin apparently have much better compensation of the mitochondrial stress, since their lifespan is unaltered and dopaminergic neuron death or mitochondrial morphology deficits have not been observed.[107–115] Mitochondrial morphology problems became prominent when Parkin-deficiency was combined with overexpression of α-synuclein as a Parkinson-specific mitochondrial stressor.[116,117] The gain-of-function of α-synuclein through missense mutations such as A53T or through gene duplication/triplication causes autosomal-dominant early onset PD through toxicity for dopaminergic neurons[118,119] and was shown to inhibit mitochondrial fusion, an effect that could be rescued by PINK1, Parkin, and DJ-1.[120,121] PINK1 deficiency also appears to interact with changes in α-synuclein abundance, because accumulation/aggregation of α-synuclein was observed in brain, neuroblastoma cells, and skin fibroblasts with PINK1 deficiency,[122–124] and because altered α-synuclein levels were apparent in the brainstem of PINK1-deficient mice.[113] As a chaperone component of the SNARE complex,[125] α-synuclein inhibits membrane fusion also in the presynapse in an age-progressive manner in our mouse mutants, with deleterious effects on release of the neurotransmitter dopamine long before the formation of demonstrable protein aggregates or death of dopaminergic neurons.[126–130]

7. PRELIMINARY FINDINGS ON MITOCHONDRIAL ACETYLATION IN OUR PD MOUSE MODEL

To maximize pathology and phenotype, we crossed our PrPmtA mouse line with 1.5-fold overexpression of human A53T-α-synuclein in brain into our PINK1-KO mouse line to obtain doubly homozygous mutants (DM) and aged them in parallel to WT mice derived from common ancestors, as previously described.[113,126] For a hypothesis-free analysis whether lysine acetylation (AcLys) changes are detectable in the nucleus, in the cytoplasm, at the OMM, or in the mitochondrial matrix, whole brains from these mice were subjected to an unbiased screen available commercially (PTMscan® from Cell Signaling). In short, protein extracts underwent trypsin digests, quantifiable immunoprecipitation with a proprietary anti-AcLys-antibody, and mass spectrometry sequencing of the peptide with the relevant PTM. The data obtained—after filtering by technical quality-control criteria and consistency criteria between the three DM versus three WT samples—are shown in Table 6.2. Deacetylation was prominent, with levels mostly beyond −twofold for many mitochondrial proteins, and levels beyond −fivefold for several components of respiratory complex V. The deacetylation for the OMM component VDAC3 (Porin 3) was below the cutoff threshold of −twofold. Similarly, histone

Table 6.2 Prominent deacylation of mitochondrial proteins, particularly of respiratory complex V in PD model brain

Gene (localization)	Normalized fold change All DM: WT	DM1: WT	DM2: WT	DM3: WT	Max intensity	Max %CV Biological	Site	Peptide
Mitochondrial proteins								
Atp5j2	−2.0	−2.5	−2.5	−1.4	5.111.154	33.8	§16	LM#EVK*LGELPSWIM#M#R
Atp5j2	−5.7	−5.6	−5.2	−6.1	1.963.103	14.8	48	YYNK*YINVR
Cox5b	−1.8	−1.4	−2.2	−1.8	3.755.537	16.6	§85	EDPNLVPSISNK*R
Dlat	−1.7	−2.0	−1.7	−1.5	294.537	25.5	§461	ELNK*MLEGK
Dlat	−1.8	−2.1	−1.8	−1.4	533.444	21.9	§632	K*YLEKPITM#LL
Dld	−2.1	−2.0	−2.9	−1.7	5.733.193	27.8	§334	RPFTQNLGLEELGIELDPK*GR
Fahd1	−1.9	−1.6	−2.6	−1.6	311.095	24.0	83	NLHHEVELGVLLGK*R
Glud1	−1.8	−2.2	−1.8	−1.4	812.800	23.9	§352	ELEDFK*LQHGSILGFPK
Got2	−2.3	−1.5	−4.1	−2.1	493.912	43.8	§185	YYDPK*TCGFDFSGALEDISK
Got2	−3.3	−2.9	−5.8	−2.5	744.978	37.9	§404	EFSVYM#TK*DGR
Got2	−3.8	−3.3	−3.6	−4.4	712.023	34.9	§430	ISVAGVTSGNVGYLAHAIHQVTK*
Gpd2	−1.4	−1.2	−1.0	−2.6	626.758	39.4	§634	FHK*FDEDEKGFITIVDVQR
Gpd2	−1.4	−1.2	−1.0	−2.6	626.758	39.4	640	FHKFDEDEK*GFITIVDVQR

Continued

Table 6.2 Prominent deacylation of mitochondrial proteins, particularly of respiratory complex V in PD model brain—cont'd

	Normalized fold change					Max %CV		
Gene (localization)	**All DM: WT**	**DM1: WT**	**DM2: WT**	**DM3: WT**	**Max intensity**	**Biological**	**Site**	**Peptide**
Hsd17b10	−2.1	−1.9	−2.4	−1.9	1.214.999	20.7	§28	GLVAVVTGGASGLGLATAK*R
Hsd17b10	−2.3	−1.8	−2.7	−2.3	324.973	22.1	§52	LVGQGATAVLLDVPDSEGEAQAK*K
Hsdl2 (mitochondria or peroxisome?)	−1.5	−1.5	−2.0	−1.1	305.335	29.9	116	FGGIDILVNNASAISLTNTLDTPTK*R
Idh3b	−2.0	−1.5	−2.5	−2.0	2.451.356	18.5	349	HLNLEYHSSM#IADAVK*K
Idh3b	−2.0	−1.5	−2.5	−2.0	2.451.356	18.5	350	HLNLEYHSSM#IADAVKK*
Mdh2	−2.4	−2.2	−3.5	−1.9	10.238.996	25.8	§296	ETECTYFSTPLLLGK*K
Mdh2	−2.4	−2.2	−3.5	−1.9	10.238.996	25.8	§297	ETECTYFSTPLLLGKK*
Mut	−1.8	−2.1	−1.9	−1.5	1.094.159	18.9	§341	NSK*SLLLR
Ndufv1	−1.7	−1.2	−1.7	−2.3	2.508.562	30.5	§64	RGDWYK*TK
Nipsnap1 (IMM or peroxisome?)	−1.6	−2.1	−1.7	−1.2	2.325.669	30.4	§279	IM#IPLK*ISPLQ
Oxct1	−1.9	−1.6	−2.6	−1.6	430.329	27.9	480	GVFDVDK*K
Pccb	−2.3	−2.8	−2.1	−2.1	2.913.805	17.7	§103	NK*FPGDSVVTGR

Pdha1	−2.0	−1.8	−2.7	−1.7	2.767.672	19.1	313, § 321	SK*SDPIM#LLK*DR
Pdpr	−1.6	−1.3	−2.0	−1.6	717.262	18.8	159	LNVVGIPSEIISPK*K
Sucla2	−2.0	−1.9	−1.9	−2.0	1.669.103	16.7	§88	SSDEAYAIAK*K
Uqcrc1	−1.5	−2.1	−1.4	−1.2	2.520.298	28.6	85	YETEK*NNGAGYFLEHLAFK
Uqcrc2	−2.1	−2.8	−2.2	−1.5	787.987	31.0	199	M#GK*ITSEELHYFVQNHFTSAR
ATP8	−5.4	−5.3	−4.5	−6.3	36.014.743	16.8	§54	TPWELK*WTK
Atp1a3	−1.8	−2.4	−2.1	−1.3	290.790	32.7	35	EVAM#TEHK*MSVEEVCR
Atp1a3	−1.8	−1.3	−2.5	−1.8	1.468.962	27.2	434	GGQDNIPVLK*R
Atp5a1	−2.3	−1.8	−2.9	−2.3	1.009.612	30.0	§132	EGDVVK*R
Atp5a1	−2.3	−1.9	−2.7	−2.2	4.466.488	18.2	§161	VVDALGNAIDGK*GPIGSK
Atp5a1	−1.7	−1.2	−2.3	−1.6	3.720.440	26.5	§230	TSIAIDTIINQK*R
Atp5a1	−1.8	−2.5	−1.6	−1.6	1.376.459	68.5	§240	FNDGTDEKK*K
Atp5f1	−9.3	−9.8	−9.8	−7.9	6.220.081	15.5	§162	AQQALVQK*R
Atp5f1	−5.3	−5.3	−6.1	−4.4	619.675	22.6	§238	ETIAK*CIEDLK
Atp5f1	−5.8	−9.6	−6.1	−3.9	456.868	44.4	§244	CIEDLK*LLAK
Atp5f1	−6.2	−6.6	−7.3	−4.8	3.153.151	20.3	53	PLPPLPEYGGK*VR
Atp5h	−8.9	−9.6	−8.8	−8.0	11.193.316	12.4	§149	KK*YPYWPHQPIENL

Continued

Table 6.2 Prominent deacylation of mitochondrial proteins, particularly of respiratory complex V in PD model brain—cont'd

	Normalized fold change					Max %CV		
Gene (localization)	**All DM: WT**	**DM1: WT**	**DM2: WT**	**DM3: WT**	**Max intensity**	**Biological**	**Site**	**Peptide**
Atp5h	−7.7	−7.0	−10.0	−6.4	3.371.654	20.5	§72	PGLVDDFEK*K
Atp5h	−5.7	−7.7	−5.2	−4.8	839.789	27.1	§85	IPVPEDK*YTALVDQEEKEDVK
Atp5j	−2.1	−2.5	−2.1	−1.8	28.922.741	15.5	§41	ELDPVQK*LFVDK
Atp5j	−2.5	−3.0	−2.4	−2.2	72.513.983	16.5	§99	FDDPK*FEVIDKPQS
Atp5j	−3.8	−6.4	−3.2	−3.1	626.170	37.8	§99. §105	FDDPK*FEVIDK*PQS
Atp5l	−2.0	−1.7	−2.1	−2.3	358.171	30.1	61. §66	IIQSAK*TGSFK*HLTVK
Atp5o	−2.5	−2.9	−2.6	−2.0	47.218.266	27.4	§172	SFLSPNQILK*LEIK
Atp5o	−2.5	−2.3	−2.1	−3.0	20.783.292	19.1	§192	IGEK*YVDMSAK
Atp5o	−2.4	−2.4	−1.9	−2.8	4.800.867	21.6	§54	EKK*LDQVEK
Ndufb3	−2.8	−2.2	−4.1	−2.6	603.554	24.9	§40	IEGTPLETVQK*K
Slc25a31/4/5	−1.9	−1.3	−2.1	−2.9	128.201.210	47.5	§105, §92, §92	YFPTQALNFAFK*DK

Slc25a4/5	−2.3	−2.4	−2.3	2.1	3.553.341	24.5	§272	AFFK*GAWSNVLR
Slc25a5	−1.9	−1.5	−2.3	−1.8	5.628.986	17.9	§105	QIFLGGVDK*R
Slc25a5	−4.0	−4.5	−4.8	−3.0	376.299	36.0	§155	EFK*GLGDCLVK
Vdac3 (OMM)	−1.7	−1.5	−2.4	−1.5	787.090	21.4	§109, §110	LTLDTIFVPNTGK*K
Vdac3 (OMM)	−1.7	−2.0	−1.7	−1.4	839.326	23.8	§63, §64	YK*VCNYGLTFTQK
Dbi (Mito or ER or Golgi?)	−2.4	−2.5	−2.1	−2.5	301.369	24.1	§51	QATVGDVNTDRPGLLDLK*GK
Dlst	−2.1	−1.7	−2.3	−2.1	24.228.860	12.9	§273	HKDAFLK*K
Glul, Zfp811	−1.8	−1.3	−2.4	−2.0	2.996.395	26.1	118, 104	HICK*R
Acat1	−2.1	−1.9	−3.2	−1.6	330.681	32.4	§199	IHM#GNCAENTAK*K
Acat1	−2.1	−1.9	−3.2	−1.6	330.681	32.4	200	IHM#GNCAENTAKK*
Ivd	1.8	1.8	2.0	1.7	27.249.169	19.3	§56	HTISK*FLQENLAPK
Zadh2 (mito or peroxisome?)	1.9	1.9	2.1	1.7	845.613	16.4	197	IAK*CHVIGTCSSDEK
Translation factors								
Fau	−2.4	−2.5	−3.2	−1.8	872.431	29.3	§51	FVNVVPTFGK*K

Continued

Table 6.2 Prominent deacylation of mitochondrial proteins, particularly of respiratory complex V in PD model brain—cont'd

	Normalized fold change					Max %CV		
Gene (localization)	All DM: WT	DM1: WT	DM2: WT	DM3: WT	Max intensity	Biological	Site	Peptide
Lactb	2.6	2.6	2.7	2.7	4.677.576	18.8	§346	ASGYK*YLDYM#QK
Carbohydrate metabolism								
Pfkp	−1.7	−1.6	−2.0	−1.4	457.785	26.5	735	DLLFQPVAELK*K
Eno1	1.8	2.0	1.9	1.5	2.446.933	15.5	§406	LAK*YNQILR
Eno1	−2.0	−1.8	−2.6	−1.7	1.058.352	18.1	§80	TIAPALVSK*K
Enzyme. Misc.								
Bcat1	1.9	2.4	2.3	1.1	4.297.216	34.0	360	GQM#LHIPTM#ENGPK*LASR
Cnp	−1.8	−1.4	−2.1	−1.9	6.219.971	19.6	399	AIFTGYYGK*GKPVPIHGSR
Cnp	−1.8	−1.4	−2.1	−1.9	6.219.971	19.6	401	AIFTGYYGKGK*PVPIHGSR
Fbll1	−1.6	−1.1	−1.8	−2.6	406.320	42.0	16	GGWGWGGGK*GGSK
Npl	2.9	2.6	1.9	4.3	490.274	42.9	173	IPTFQGLK*FTDTDLLDFGQCVDQNHQR
Phosphatase								
Prune2	−1.6	−1.5	−2.0	−1.4	290.156	20.1	2503	VIPTK*GPK
Protein kinase. Tyr (non-receptor)								
Pkm2	−1.7	−1.4	−2.2	−1.6	917.455	21.0	§135	GSGTAEVELK*K

RNA processing								
Npm1	−3.5	−2.4	−6.1	−3.1	222.660	42.4	§154	SAPGGGNKVPQK*K
Transcriptional regulator								
Ing4, Ing4	−2.5	−2.8	−3.2	−1.9	392.405	34.6	§127, §129	QIESSDYDSSSSK*GK*K
Ncl	−2.2	−1.8	−3.3	−2.0	2.246.960	28.7	§87	AAVTPGK*K
G protein or regulator								
Gnao1	−1.5	−2.2	−1.4	−1.2	51.984	28.6	101	K*TDSKMVCDVVSR
Gnao1	−1.5	−2.2	−1.4	−1.2	51.984	28.6	105	KTDSK*MVCDVVSR
Secreted protein								
Serpina1a/c/e	2.1	2.0	1.4	2.9	13.745.000	35.6	§292	ELISK*FLLNR
Adaptor/scaffold								
Dmxl2	−1.9	−1.6	−1.6	−2.7	241.852	31.0	1029	FWK*CCM#ETNSLGNTSDESETYHWR
Adhesion or extracellular matrix protein								
Actl7a	−2.2	−2.0	−2.7	−1.8	2.738.689	17.9	§281	IVEDIK*TR
Cfl1	−1.5	−1.1	−2.3	−1.4	420.679	35.6	44	AVLFCLSEDK*K
Cltc	−2.6	−2.2	−3.5	−2.4	1.028.409	24.0	245	LHIIEVGTPPTGNQPFPK*K
Crebbp	−1.8	−1.6	−2.1	−1.8	236.962	13.7	§1593, §1596, §1598	K*TNK*NK*SSISR

Continued

Table 6.2 Prominent deacylation of mitochondrial proteins, particularly of respiratory complex V in PD model brain—cont'd

	Normalized fold change					Max %CV		
Gene (localization)	**All DM: WT**	**DM1: WT**	**DM2: WT**	**DM3: WT**	**Max intensity**	**Biological**	**Site**	**Peptide**
Rhog	1.7	2.5	1.6	1.1	704.534	36.7	96	HK*WHPEVCHHCPDVPILLVGTK
Tcp1	−2.3	−2.3	−2.7	−2.0	741.343	22.7	§400, §351	SLHDALCVVK*R
Tubb3	1.5	1.5	1.0	2.1	490.308	44.7	19	QIGAK*FWEVISDEHGIDPSGNYVGDSDLQLER
Cell cycle regulation								
Smc3	−2.1	−2.5	−2.2	−1.7	799.161	23.0	§105, §106	VIGAK*K*DQYFLDKK
Chromatin, DNA binding, DNA repair, or DNA replication protein								
H2afv, H2afz	−1.7	−1.5	−2.2	−1.5	1.725.234	25.1	§5, §8	AGGK*AGK*DSGK
Hist1h2bc/e/m/n/p/e/b-ps	−1.9	−1.5	−2.4	−1.9	1.139.701	21.8	§21, §24	KAVTK*AQK*K
Gm10257/6749, H3f3a/b/c/1h3a/g/2h3b	−1.7	−1.4	−2.2	−1.5	6.811.779	19.4	§10, §15, §16, §21, §55, §60	K*STGGK*APR

Gm10257/12657/6749, H3f3/b/1h3a/g/2h3b	−1.6	−1.3	−2.1	−1.6	62.283.361	20.4	§19, §24, §25, §30, §64, §69	K*QLATK*AAR
Tox4	−2.0	−1.8	−2.0	−2.3	259.092	19.7	207	TVVVETGK*K
Cytoskeletal protein								
Mbp	−2.6	−2.2	−4.3	−2.0	1.388.398	32.3	§222, §89, §115,	TQDENPVVHFFK*
Mbp	1.9	2.3	2.4	1.1	1.195.009	36.5	83	SRPGLCHMYK*DSHTR
Sept6	−2.3	−1.5	−3.3	−2.5	326.420	50.5	379	KLHQEEK*K
Unknown function								
Btbd3	1.8	1.7	1.2	2.5	1.565.582	35.2	212	K*YIVPH

Shown are unbiased data of AcLys in whole brain tissue from 18-month-old mice as determined by the PTMscan® technology (Cell Signaling), using trypsin digest and acetyl-lysine motif antibody #11D9B8/16E10E7/21F10C9. Red background (dark gray in print version) = downregulation > twofold, green background (light gray in print version) = upregulation > twofold. * = acetylation, # = oxidized methionine, bold intensity (light gray in print version) = manually reviewed values. Red letters (dark gray in print version) = multiple identifications for one *m/z*. Samples: DM (Double Mutant) 1 PINK1$^{-/-}$ plus A53T-SNCA-PrPmtA = CS 19339, 19340; DM2 PINK1$^{-/-}$ plus A53T-SNCA-PrPmtA = CS 19341, 19342; DM3 PINK1$^{-/-}$ plus A53T-SNCA-PrPmtA = CS 19343, 19344; WT (wild-type control) 1 = CS 19345, 19346; WT 2 = CS 19347, 19348; WT 3 = CS 19349, 19350. WT samples were pooled for bioinformatics. The complete dataset as EXCEL file will be made available by E-mail to interested readers.

(e.g., HIST1H2BC) deacetylation was below the cutoff threshold of −twofold. Tubulin (TUBB3) hyperacetylation was observed, but also below the cutoff threshold of +twofold. Lysine deacetylation was not ubiquitous in the mitochondrial matrix, since Isovaleryl-CoA dehydrogenase appeared hyperacetylated with changes of +1.8-fold. Thus, the data very consistently indicate a prominent deacetylation of mitochondrial matrix proteins in the brain of our PD mouse model, and identify the relevant lysine positions (indicated by * within the peptide sequence in Table 6.1). It is impossible to validate these mass spectrometry findings independently, while site-specific Ac-Lys antibodies are not available. Upon isolation of mitochondrial tissue fractions from the brain of these mice, their subsequent semi-quantitative immunoprecipitation by AcLys antibodies, and immunoblot detection of individual mitochondrial matrix candidate proteins, the changes were not evident in individual samples (data not shown). Since many AcLys positions are known for most of these proteins, and only specific site appeared changed in the mass spectrometry data, these findings are not mutually exclusive. In view of the expression regulation of SIRT3 as the main mitochondrial matrix deacetylase during aging and nutrient excesses, we performed quantitative immunoblots in independent brain tissues from this PD mouse model (Fig. 6.2). In the SDS-soluble tissue fraction where mitochondria are enriched a paradox decrease of SIRT3 levels was observed. Clearly an upregulation would be expected, but it is conceivable that a dramatic elevation of $NAD^+/NADH$ ratio in dysfunctional mitochondria might enhance the biochemical activity of SIRT3 to such a degree that a compensatory expression downregulation occurs. It is also conceivable that the observed mitochondrial matrix deacetylation is due to abnormal lysine acetyltransferase activity, with the SIRT3 downregulation representing a compensatory effort. At present we cannot resolve these doubts, but both observations support the notion that poor bioenergetics and altered mitochondrial matrix acetylation occur in parallel to the known mitophagy deficit of PD brains.

8. THE PREDICTION OF PD

Thus, it is desirable to develop site-specific AcLys antibodies, in particular against motifs in respiratory complex V components, in order to test these preliminary observations further. Such new tools would be useful for cancer research and for studies into the basic mechanisms of mitophagy and mitochondrial bioenergetics. In particular for PD diagnostics, they might

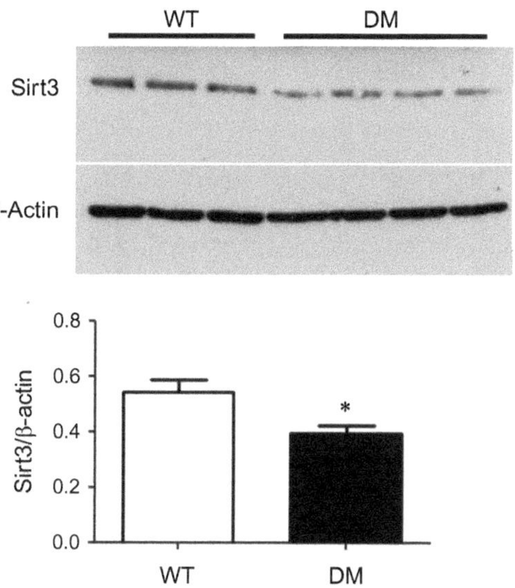

Figure 6.2 SIRT3 protein decrease in PD model mouse brain. Immunoblot of SDS-soluble protein extract from whole brain tissue of 18-month-old mice (3 WT, wild-type, versus 4 DM, double mutants, with PINK1-KO plus A53T-SNCA-PrPmtA), detected by enhanced chemoluminescence with antiSIRT3 (Cell Signaling #5490) and with anti-β-actin (Sigma) antibodies, quantified by densitometry (Pharmacia TotalLab), and statistically evaluated in a bar graph with mean and SEM (GraphPad software). * illustrates p-value < 0.05.

prove useful for the objective documentation of disease progression. PD diagnostics is currently still dependent on the clinical inspection of movement patterns by expert neurologist, on the assessment of therapeutic benefits, and on costly radioactive brain imaging techniques.[131] Objective laboratory procedures would be highly desirable in PD diagnostics.[132] The need is particularly strong for risk detection methods. Most patients who manifested PD at old age are not aware of familial antecedents, so that diagnosis and symptomatic therapy occur at late disease stage when most neurons of the susceptible brain are already irreversibly lost.[133] The development of predictive diagnostics in PD will be a prerequisite for the success of neuroprotective therapies. Given that mitophagy deficits occur long before neuron loss in PD, and that mitochondrial acetylation changes were found to precede mitophagy, the development of AcLys antibodies that target specific sites in mitochondrial matrix proteins might indeed become useful. Similarly, the study of NAD^+-dependent expression changes in tissues with PD vulnerability may provide risk diagnostics and a better molecular definition of early disease stages.

9. CONCLUSIONS

Recent progress has documented the role of sirtuins for longevity, in particular the role of mitochondrial AcLys for mitophagy. Mitophagy deficits are at the core of PD pathogenesis, and emerging data have identified AcLys-modifications in respiratory complex V and SIRT3 level changes in this process, but novel diagnostic tools such as site-specific AcLys-antibodies and an expression signature for atrophic changes remain to be developed, before our knowledge can be implemented in routine PD patient management.

ACKNOWLEDGMENTS

The study was financed by the BMBF through the NGFN2 and NGFNplus NeuroNet Parkinson networks and by the DFG (GI342/1-1 and GI342/3-1). We are grateful for the participation of PD patients and controls, for the technical assistance by Birgitt Meseck-Selchow, and for the critical revision of the manuscript by H. Osiewacz and M. Hirschey.

REFERENCES

1. Vijg J, Campisi J. Puzzles, promises and a cure for ageing. *Nature*. 2008; 454(7208):1065–1071.
2. Kaeberlein M, McVey M, Guarente L. The SIR2/3/4 complex and SIR2 alone promote longevity in Saccharomyces cerevisiae by two different mechanisms. *Genes Dev*. 1999;13(19):2570–2580.
3. Tissenbaum HA, Guarente L. Increased dosage of a sir-2 gene extends lifespan in Caenorhabditis elegans. *Nature*. 2001;410(6825):227–230.
4. Imai S, Armstrong CM, Kaeberlein M, Guarente L. Transcriptional silencing and longevity protein Sir2 is an NAD-dependent histone deacetylase. *Nature*. 2000; 403(6771):795–800.
5. Hirschey MD, Shimazu T, Huang JY, Schwer B, Verdin E. SIRT3 regulates mitochondrial protein acetylation and intermediary metabolism. *Cold Spring Harb Symp Quant Biol*. 2011;76:267–277.
6. Howitz KT, Bitterman KJ, Cohen HY, et al. Small molecule activators of sirtuins extend Saccharomyces cerevisiae lifespan. *Nature*. 2003;425(6954):191–196.
7. Rizki G, Iwata TN, Li J, et al. The evolutionarily conserved longevity determinants HCF-1 and SIR-2.1/SIRT1 collaborate to regulate DAF-16/FOXO. *PLoS Genet*. 2011;7(9):e1002235.
8. Tseng AH, Shieh SS, Wang DL. SIRT3 deacetylates FOXO3 to protect mitochondria against oxidative damage. *Free Radic Biol Med*. 2013;63:222–234.
9. Frescas D, Valenti L, Accili D. Nuclear trapping of the forkhead transcription factor FoxO1 via Sirt-dependent deacetylation promotes expression of glucogenetic genes. *J Biol Chem*. 2005;280(21):20589–20595.
10. van der Horst A, Tertoolen LG, de Vries-Smits LM, Frye RA, Medema RH, Burgering BM. FOXO4 is acetylated upon peroxide stress and deacetylated by the longevity protein hSir2(SIRT1). *J Biol Chem*. 2004;279(28):28873–28879.

11. Motta MC, Divecha N, Lemieux M, et al. Mammalian SIRT1 represses forkhead transcription factors. *Cell*. 2004;116(4):551–563.
12. Schmeisser K, Mansfeld J, Kuhlow D, et al. Role of sirtuins in lifespan regulation is linked to methylation of nicotinamide. *Nat Chem Biol*. 2013;9(11):693–700.
13. Hirschey MD. Old enzymes, new tricks: sirtuins are NAD(+)-dependent de-acylases. *Cell Metab*. 2011;14(6):718–719.
14. Guarente L. Sirtuins, aging, and metabolism. *Cold Spring Harb Symp Quant Biol*. 2011;76:81–90.
15. Lee IH, Cao L, Mostoslavsky R, et al. A role for the NAD-dependent deacetylase Sirt1 in the regulation of autophagy. *Proc Natl Acad Sci U S A*. 2008;105(9):3374–3379.
16. Kawahara TL, Michishita E, Adler AS, et al. SIRT6 links histone H3 lysine 9 deacetylation to NF-kappaB-dependent gene expression and organismal life span. *Cell*. 2009;136(1):62–74.
17. Jiang H, Khan S, Wang Y, et al. SIRT6 regulates TNF-alpha secretion through hydrolysis of long-chain fatty acyl lysine. *Nature*. 2013;496(7443):110–113.
18. Barber MF, Michishita-Kioi E, Xi Y, et al. SIRT7 links H3K18 deacetylation to maintenance of oncogenic transformation. *Nature*. 2012;487(7405):114–118.
19. Ramakrishnan G, Davaakhuu G, Kaplun L, et al. Sirt2 deacetylase is a novel AKT binding partner critical for AKT activation by insulin. *J Biol Chem*. 2014;289(9):6054–6066.
20. North BJ, Marshall BL, Borra MT, Denu JM, Verdin E. The human Sir2 ortholog, SIRT2, is an NAD+-dependent tubulin deacetylase. *Mol Cell*. 2003;11(2):437–444.
21. Maxwell MM, Tomkinson EM, Nobles J, et al. The Sirtuin 2 microtubule deacetylase is an abundant neuronal protein that accumulates in the aging CNS. *Hum Mol Genet*. 2011;20(20):3986–3996.
22. Bell EL, Guarente L. The SirT3 divining rod points to oxidative stress. *Mol Cell*. 2011;42(5):561–568.
23. Outeiro TF, Kontopoulos E, Altmann SM, et al. Sirtuin 2 inhibitors rescue alpha-synuclein-mediated toxicity in models of Parkinson's disease. *Science*. 2007;317(5837):516–519.
24. Wareski P, Vaarmann A, Choubey V, et al. PGC-1{alpha} and PGC-1{beta} regulate mitochondrial density in neurons. *J Biol Chem*. 2009;284(32):21379–21385.
25. Duda JE, Giasson BI, Mabon ME, et al. Concurrence of alpha-synuclein and tau brain pathology in the Contursi kindred. *Acta Neuropathol*. 2002;104(1):7–11.
26. Lill CM, Roehr JT, McQueen MB, et al. Comprehensive research synopsis and systematic meta-analyses in Parkinson's disease genetics: the PDGene database. *PLoS Genet*. 2012;8(3):e1002548.
27. Kang L, Janowska MK, Moriarty GM, Baum J. Mechanistic insight into the relationship between N-terminal acetylation of alpha-synuclein and fibril formation rates by NMR and fluorescence. *PLoS One*. 2013;8(9):e75018.
28. Dikiy I, Eliezer D. N-terminal acetylation stabilizes N-terminal helicity in lipid- and micelle-bound alpha-synuclein and increases its affinity for physiological membranes. *J Biol Chem*. 2014;289(6):3652–3665.
29. Min SW, Cho SH, Zhou Y, et al. Acetylation of tau inhibits its degradation and contributes to tauopathy. *Neuron*. 2010;67(6):953–966.
30. Haigis MC, Mostoslavsky R, Haigis KM, et al. SIRT4 inhibits glutamate dehydrogenase and opposes the effects of calorie restriction in pancreatic beta cells. *Cell*. 2006;126(5):941–954.
31. Du J, Zhou Y, Su X, et al. Sirt5 is a NAD-dependent protein lysine demalonylase and desuccinylase. *Science*. 2011;334(6057):806–809.
32. Newman JC, He W, Verdin E. Mitochondrial protein acylation and intermediary metabolism: regulation by sirtuins and implications for metabolic disease. *J Biol Chem*. 2012;287(51):42436–42443.

33. Shi T, Wang F, Stieren E, Tong Q. SIRT3, a mitochondrial sirtuin deacetylase, regulates mitochondrial function and thermogenesis in brown adipocytes. *J Biol Chem.* 2005;280(14):13560–13567.
34. Lanza IR, Short DK, Short KR, et al. Endurance exercise as a countermeasure for aging. *Diabetes.* 2008;57(11):2933–2942.
35. Palacios OM, Carmona JJ, Michan S, et al. Diet and exercise signals regulate SIRT3 and activate AMPK and PGC-1alpha in skeletal muscle. *Aging (Albany NY).* 2009;1(9):771–783.
36. Hirschey MD, Shimazu T, Goetzman E, et al. SIRT3 regulates mitochondrial fatty-acid oxidation by reversible enzyme deacetylation. *Nature.* 2010;464(7285):121–125.
37. Jing E, Emanuelli B, Hirschey MD, et al. Sirtuin-3 (Sirt3) regulates skeletal muscle metabolism and insulin signaling via altered mitochondrial oxidation and reactive oxygen species production. *Proc Natl Acad Sci U S A.* 2011;108(35):14608–14613.
38. Caton PW, Holness MJ, Bishop-Bailey D, Sugden MC. PPARalpha-LXR as a novel metabolostatic signalling axis in skeletal muscle that acts to optimize substrate selection in response to nutrient status. *Biochem J.* 2011;437(3):521–530.
39. Hirschey MD, Shimazu T, Jing E, et al. SIRT3 deficiency and mitochondrial protein hyperacetylation accelerate the development of the metabolic syndrome. *Mol Cell.* 2011;44(2):177–190.
40. Bao J, Scott I, Lu Z, et al. SIRT3 is regulated by nutrient excess and modulates hepatic susceptibility to lipotoxicity. *Free Radic Biol Med.* 2010;49(7):1230–1237.
41. Kong X, Wang R, Xue Y, et al. Sirtuin 3, a new target of PGC-1alpha, plays an important role in the suppression of ROS and mitochondrial biogenesis. *PLoS One.* 2010;5(7):e11707.
42. Cohen HY, Miller C, Bitterman KJ, et al. Calorie restriction promotes mammalian cell survival by inducing the SIRT1 deacetylase. *Science.* 2004;305(5682):390–392.
43. Chen D, Bruno J, Easlon E, et al. Tissue-specific regulation of SIRT1 by calorie restriction. *Genes Dev.* 2008;22(13):1753–1757.
44. Hayashida S, Arimoto A, Kuramoto Y, et al. Fasting promotes the expression of SIRT1, an NAD+-dependent protein deacetylase, via activation of PPARalpha in mice. *Mol Cell Biochem.* 2010;339(1–2):285–292.
45. Ghosh S, George S, Roy U, Ramachandran D, Kolthur-Seetharam U. NAD: a master regulator of transcription. *Biochim Biophys Acta.* 2010;1799(10–12):681–693.
46. Peek CB, Affinati AH, Ramsey KM, et al. Circadian clock NAD+ cycle drives mitochondrial oxidative metabolism in mice. *Science.* 2013;342(6158):1243417.
47. Koltai E, Szabo Z, Atalay M, et al. Exercise alters SIRT1, SIRT6, NAD and NAMPT levels in skeletal muscle of aged rats. *Mech Ageing Dev.* 2010;131(1):21–28.
48. Yang H, Yang T, Baur JA, et al. Nutrient-sensitive mitochondrial NAD+ levels dictate cell survival. *Cell.* 2007;130(6):1095–1107.
49. Schwer B, Eckersdorff M, Li Y, et al. Calorie restriction alters mitochondrial protein acetylation. *Aging Cell.* 2009;8(5):604–606.
50. Nasrin N, Wu X, Fortier E, et al. SIRT4 regulates fatty acid oxidation and mitochondrial gene expression in liver and muscle cells. *J Biol Chem.* 2010;285(42):31995–32002.
51. Fujino T, Kondo J, Ishikawa M, Morikawa K, Yamamoto TT. Acetyl-CoA synthetase 2, a mitochondrial matrix enzyme involved in the oxidation of acetate. *J Biol Chem.* 2001;276(14):11420–11426.
52. Sone H, Shimano H, Sakakura Y, et al. Acetyl-coenzyme A synthetase is a lipogenic enzyme controlled by SREBP-1 and energy status. *Am J Physiol Endocrinol Metab.* 2002;282(1):E222–E230.
53. Weinert BT, Wagner SA, Horn H, et al. Proteome-wide mapping of the Drosophila acetylome demonstrates a high degree of conservation of lysine acetylation. *Sci Signal.* 2011;4(183):ra48.

54. Scott I, Webster BR, Li JH, Sack MN. Identification of a molecular component of the mitochondrial acetyltransferase programme: a novel role for GCN5L1. *Biochem J.* 2012;443(3):655–661.
55. Baker SP, Grant PA. The SAGA continues: expanding the cellular role of a transcriptional co-activator complex. *Oncogene.* 2007;26(37):5329–5340.
56. Bell EL, Emerling BM, Ricoult SJ, Guarente L. SirT3 suppresses hypoxia inducible factor 1alpha and tumor growth by inhibiting mitochondrial ROS production. *Oncogene.* 2011;30(26):2986–2996.
57. Qiu X, Brown K, Hirschey MD, Verdin E, Chen D. Calorie restriction reduces oxidative stress by SIRT3-mediated SOD2 activation. *Cell Metab.* 2010;12(6):662–667.
58. Tao R, Coleman MC, Pennington JD, et al. Sirt3-mediated deacetylation of evolutionarily conserved lysine 122 regulates MnSOD activity in response to stress. *Mol Cell.* 2010;40(6):893–904.
59. Kim HS, Patel K, Muldoon-Jacobs K, et al. SIRT3 is a mitochondria-localized tumor suppressor required for maintenance of mitochondrial integrity and metabolism during stress. *Cancer Cell.* 2010;17(1):41–52.
60. Finley LW, Carracedo A, Lee J, et al. SIRT3 opposes reprogramming of cancer cell metabolism through HIF1alpha destabilization. *Cancer Cell.* 2011;19(3):416–428.
61. Huang KH, Hsu CC, Fang WL, et al. SIRT3 expression as a biomarker for better prognosis in gastric cancer. *World J Surg.* 2013;.
62. Yang B, Fu X, Shao L, Ding Y, Zeng D. Aberrant expression of SIRT3 is conversely correlated with the progression and prognosis of human gastric cancer. *Biochem Biophys Res Commun.* 2014;443(1):156–160.
63. Li H, Feng Z, Wu W, Li J, Zhang J, Xia T. SIRT3 regulates cell proliferation and apoptosis related to energy metabolism in non-small cell lung cancer cells through deacetylation of NMNAT2. *Int J Oncol.* 2013;43(5):1420–1430.
64. Xiao K, Jiang J, Wang W, et al. Sirt3 is a tumor suppressor in lung adenocarcinoma cells. *Oncol Rep.* 2013;30(3):1323–1328.
65. Zhang B, Qin L, Zhou CJ, Liu YL, Qian HX, He SB. SIRT3 expression in hepatocellular carcinoma and its impact on proliferation and invasion of hepatoma cells. *Asian Pac J Trop Med.* 2013;6(8):649–652.
66. Zhang CZ, Liu L, Cai M, et al. Low SIRT3 expression correlates with poor differentiation and unfavorable prognosis in primary hepatocellular carcinoma. *PLoS One.* 2012;7(12):e51703.
67. Lai CC, Lin PM, Lin SF, et al. Altered expression of SIRT gene family in head and neck squamous cell carcinoma. *Tumour Biol.* 2013;34(3):1847–1854.
68. Shackelford R, Hirsh S, Henry K, Abdel-Mageed A, Kandil E, Coppola D. Nicotinamide phosphoribosyltransferase and SIRT3 expression are increased in well-differentiated thyroid carcinomas. *Anticancer Res.* 2013;33(8):3047–3052.
69. Ashraf N, Zino S, Macintyre A, et al. Altered sirtuin expression is associated with node-positive breast cancer. *Br J Cancer.* 2006;95(8):1056–1061.
70. Alhazzazi TY, Kamarajan P, Verdin E, Kapila YL. SIRT3 and cancer: tumor promoter or suppressor? *Biochim Biophys Acta.* 2011;1816(1):80–88.
71. Webster BR, Scott I, Han K, et al. Restricted mitochondrial protein acetylation initiates mitochondrial autophagy. *J Cell Sci.* 2013;126(Pt 21):4843–4849.
72. Scott I, Webster BR, Chan CK, Okonkwo JU, Han K, Sack MN. GCN5-like protein 1 (GCN5L1) controls mitochondrial content through coordinated regulation of mitochondrial biogenesis and mitophagy. *J Biol Chem.* 2014;289(5):2864–2872.
73. Guedes-Dias P, Oliveira JM. Lysine deacetylases and mitochondrial dynamics in neurodegeneration. *Biochim Biophys Acta.* 2013;1832(8):1345–1359.
74. Deas E, Wood NW, Plun-Favreau H. Mitophagy and Parkinson's disease: the PINK1-Parkin link. *Biochim Biophys Acta.* 2011;1813(4):623–633.

75. Ding WX, Yin XM. Mitophagy: mechanisms, pathophysiological roles, and analysis. *Biol Chem*. 2012;393(7):547–564.
76. Whitworth AJ, Pallanck LJ. The PINK1/Parkin pathway: a mitochondrial quality control system? *J Bioenerg Biomembr*. 2009;41(6):499–503.
77. Tan JM, Wong ES, Kirkpatrick DS, et al. Lysine 63-linked ubiquitination promotes the formation and autophagic clearance of protein inclusions associated with neurodegenerative diseases. *Hum Mol Genet*. 2008;17(3):431–439.
78. Geisler S, Holmstrom KM, Skujat D, et al. PINK1/Parkin-mediated mitophagy is dependent on VDAC1 and p62/SQSTM1. *Nat Cell Biol*. 2010;12(2):119–131.
79. Sarraf SA, Raman M, Guarani-Pereira V, et al. Landscape of the PARKIN-dependent ubiquitylome in response to mitochondrial depolarization. *Nature*. 2013;496(7445):372–376.
80. Lee JY, Koga H, Kawaguchi Y, et al. HDAC6 controls autophagosome maturation essential for ubiquitin-selective quality-control autophagy. *EMBO J*. 2010;29(5):969–980.
81. Lee JY, Nagano Y, Taylor JP, Lim KL, Yao TP. Disease-causing mutations in Parkin impair mitochondrial ubiquitination, aggregation, and HDAC6-dependent mitophagy. *J Cell Biol*. 2010;189(4):671–679.
82. de Vries RL, Przedborski S. Mitophagy and Parkinson's disease: be eaten to stay healthy. *Mol Cell Neurosci*. 2013;55:37–43.
83. Springer W, Kahle PJ. Regulation of PINK1-Parkin-mediated mitophagy. *Autophagy*. 2011;7(3):266–278.
84. Tofaris GK. Lysosome-dependent pathways as a unifying theme in Parkinson's disease. *Mov Disord*. 2012;27(11):1364–1369.
85. Pallanck LJ. Culling sick mitochondria from the herd. *J Cell Biol*. 2010;191(7):1225–1227.
86. Klinkenberg M, Gispert S, Dominguez-Bautista JA, Braun I, Auburger G, Jendrach M. Restriction of trophic factors and nutrients induces PARKIN expression. *Neurogenetics*. 2012;13(1):9–21.
87. Mai S, Klinkenberg M, Auburger G, Bereiter-Hahn J, Jendrach M. Decreased expression of Drp1 and Fis1 mediates mitochondrial elongation in senescent cells and enhances resistance to oxidative stress through PINK1. *J Cell Sci*. 2010;123(Pt 6):917–926.
88. Valente EM, Abou-Sleiman PM, Caputo V, et al. Hereditary early-onset Parkinson's disease caused by mutations in PINK1. *Science*. 2004;304(5674):1158–1160.
89. Silvestri L, Caputo V, Bellacchio E, et al. Mitochondrial import and enzymatic activity of PINK1 mutants associated to recessive Parkinsonism. *Hum Mol Genet*. 2005;14(22):3477–3492.
90. Becker D, Richter J, Tocilescu MA, Przedborski S, Voos W. Pink1 kinase and its membrane potential (Deltapsi)-dependent cleavage product both localize to outer mitochondrial membrane by unique targeting mode. *J Biol Chem*. 2012;287(27):22969–22987.
91. Kitada T, Asakawa S, Hattori N, et al. Mutations in the Parkin gene cause autosomal recessive juvenile Parkinsonism. *Nature*. 1998;392(6676):605–608.
92. Corti O, Lesage S, Brice A. What genetics tells us about the causes and mechanisms of Parkinson's disease. *Physiol Rev*. 2011;91(4):1161–1218.
93. Petit A, Kawarai T, Paitel E, et al. Wild-type PINK1 prevents basal and induced neuronal apoptosis, a protective effect abrogated by Parkinson disease-related mutations. *J Biol Chem*. 2005;280(40):34025–34032.
94. Hoepken HH, Gispert S, Morales B, et al. Mitochondrial dysfunction, peroxidation damage and changes in glutathione metabolism in PARK6. *Neurobiol Dis*. 2007;25(2):401–411.

95. Grunewald A, Gegg ME, Taanman JW, et al. Differential effects of PINK1 nonsense and missense mutations on mitochondrial function and morphology. *Exp Neurol.* 2009;219(1):266–273.
96. Abramov AY, Gegg M, Grunewald A, Wood NW, Klein C, Schapira AH. Bioenergetic consequences of PINK1 mutations in Parkinson disease. *PLoS One.* 2011;6(10): e25622.
97. Exner N, Treske B, Paquet D, et al. Loss-of-function of human PINK1 results in mitochondrial pathology and can be rescued by Parkin. *J Neurosci.* 2007;27(45):12413–12418.
98. Klinkenberg M, Thurow N, Gispert S, et al. Enhanced vulnerability of PARK6 patient skin fibroblasts to apoptosis induced by proteasomal stress. *Neuroscience.* 2010;166(2):422–434.
99. Corti O, Brice A. Mitochondrial quality control turns out to be the principal suspect in Parkin and PINK1-related autosomal recessive Parkinson's disease. *Curr Opin Neurobiol.* 2013;23(1):100–108.
100. Rakovic A, Grunewald A, Kottwitz J, et al. Mutations in PINK1 and Parkin impair ubiquitination of mitofusins in human fibroblasts. *PLoS One.* 2011;6(3):e16746.
101. Rakovic A, Grunewald A, Seibler P, et al. Effect of endogenous mutant and wild-type PINK1 on Parkin in fibroblasts from Parkinson disease patients. *Hum Mol Genet.* 2010;19(16):3124–3137.
102. Seibler P, Graziotto J, Jeong H, Simunovic F, Klein C, Krainc D. Mitochondrial Parkin recruitment is impaired in neurons derived from mutant PINK1 induced pluripotent stem cells. *J Neurosci.* 2011;31(16):5970–5976.
103. Rakovic A, Shurkewitsch K, Seibler P, et al. Phosphatase and tensin homolog (PTEN)-induced putative kinase 1 (PINK1)-dependent ubiquitination of endogenous Parkin attenuates mitophagy: study in human primary fibroblasts and induced pluripotent stem cell-derived neurons. *J Biol Chem.* 2013;288(4):2223–2237.
104. Greene JC, Whitworth AJ, Kuo I, Andrews LA, Feany MB, Pallanck LJ. Mitochondrial pathology and apoptotic muscle degeneration in Drosophila Parkin mutants. *Proc Natl Acad Sci U S A.* 2003;100(7):4078–4083.
105. Clark IE, Dodson MW, Jiang C, et al. Drosophila pink1 is required for mitochondrial function and interacts genetically with Parkin. *Nature.* 2006;441(7097):1162–1166.
106. Park J, Lee SB, Lee S, et al. Mitochondrial dysfunction in Drosophila PINK1 mutants is complemented by Parkin. *Nature.* 2006;441(7097):1157–1161.
107. Goldberg MS, Fleming SM, Palacino JJ, et al. Parkin-deficient mice exhibit nigrostriatal deficits but not loss of dopaminergic neurons. *J Biol Chem.* 2003;278(44):43628–43635.
108. Perez FA, Palmiter RD. Parkin-deficient mice are not a robust model of Parkinsonism. *Proc Natl Acad Sci U S A.* 2005;102(6):2174–2179.
109. Itier JM, Ibanez P, Mena MA, et al. Parkin gene inactivation alters behaviour and dopamine neurotransmission in the mouse. *Hum Mol Genet.* 2003;12(18):2277–2291.
110. Von Coelln R, Thomas B, Savitt JM, et al. Loss of locus coeruleus neurons and reduced startle in Parkin null mice. *Proc Natl Acad Sci U S A.* 2004;101(29):10744–10749.
111. Kitada T, Pisani A, Porter DR, et al. Impaired dopamine release and synaptic plasticity in the striatum of PINK1-deficient mice. *Proc Natl Acad Sci U S A.* 2007;104(27):11441–11446.
112. Gandhi S, Vaarmann A, Yao Z, Duchen MR, Wood NW, Abramov AY. Dopamine induced neurodegeneration in a PINK1 model of Parkinson's disease. *PLoS One.* 2012;7(5):e37564.
113. Gispert S, Ricciardi F, Kurz A, et al. Parkinson phenotype in aged PINK1-deficient mice is accompanied by progressive mitochondrial dysfunction in absence of neurodegeneration. *PLoS One.* 2009;4(6):e5777.

114. Dehorter N, Lozovaya N, Mdzomba BJ, et al. Subthalamic lesion or levodopa treatment rescues giant GABAergic currents of PINK1-deficient striatum. *J Neurosci.* 2012;32(50):18047–18053.
115. Zhu XR, Maskri L, Herold C, et al. Non-motor behavioural impairments in Parkin-deficient mice. *Eur J Neurosci.* 2007;26(7):1902–1911.
116. Stichel CC, Zhu XR, Bader V, Linnartz B, Schmidt S, Lubbert H. Mono- and double-mutant mouse models of Parkinson's disease display severe mitochondrial damage. *Hum Mol Genet.* 2007;16(20):2377–2393.
117. Schmidt S, Linnartz B, Mendritzki S, et al. Genetic mouse models for Parkinson's disease display severe pathology in glial cell mitochondria. *Hum Mol Genet.* 2011;20(6):1197–1211.
118. Polymeropoulos MH, Lavedan C, Leroy E, et al. Mutation in the alpha-synuclein gene identified in families with Parkinson's disease. *Science.* 1997;276(5321):2045–2047.
119. Singleton AB, Farrer M, Johnson J, et al. Apha-synuclein locus triplication causes Parkinson's disease. *Science.* 2003;302(5646):841.
120. Kamp F, Exner N, Lutz AK, et al. Inhibition of mitochondrial fusion by alpha-synuclein is rescued by PINK1, Parkin and DJ-1. *EMBO J.* 2010;29(20):3571–3589.
121. Todd AM, Staveley BE. Expression of Pink1 with alpha-synuclein in the dopaminergic neurons of Drosophila leads to increases in both lifespan and healthspan. *Genet Mol Res.* 2012;11(2):1497–1502.
122. Hoepken HH, Gispert S, Azizov M, et al. Parkinson patient fibroblasts show increased alpha-synuclein expression. *Exp Neurol.* 2008;212(2):307–313.
123. Samaranch L, Lorenzo-Betancor O, Arbelo JM, et al. PINK1-linked Parkinsonism is associated with Lewy body pathology. *Brain.* 2010;133(Pt 4):1128–1142.
124. Liu W, Vives-Bauza C, Acin-Perez R, et al. PINK1 defect causes mitochondrial dysfunction, proteasomal deficit and alpha-synuclein aggregation in cell culture models of Parkinson's disease. *PLoS One.* 2009;4(2):e4597.
125. Burre J, Sharma M, Tsetsenis T, Buchman V, Etherton MR, Sudhof TC. Alpha-synuclein promotes SNARE-complex assembly in vivo and in vitro. *Science.* 2010;329(5999):1663–1667.
126. Gispert S, Del Turco D, Garrett L, et al. Transgenic mice expressing mutant A53T human alpha-synuclein show neuronal dysfunction in the absence of aggregate formation. *Mol Cell Neurosci.* 2003;24(2):419–429.
127. Kurz A, Double KL, Lastres-Becker I, et al. A53T-alpha-synuclein overexpression impairs dopamine signaling and striatal synaptic plasticity in old mice. *PLoS One.* 2010;5(7):e11464.
128. Tozzi A, Costa C, Siliquini S, et al. Mechanisms underlying altered striatal synaptic plasticity in old A53T-alpha synuclein overexpressing mice. *Neurobiol Aging.* 2012; 33(8):1792–1799.
129. Kurz A, May C, Schmidt O, et al. A53T-alpha-synuclein-overexpression in the mouse nigrostriatal pathway leads to early increase of 14-3-3 epsilon and late increase of GFAP. *J Neural Transm.* 2012;119(3):297–312.
130. Platt NJ, Gispert S, Auburger G, Cragg SJ. Striatal dopamine transmission is subtly modified in human A53Talpha-synuclein overexpressing mice. *PLoS One.* 2012; 7(5):e36397.
131. Agarwal PA, Stoessl AJ. Biomarkers for trials of neuroprotection in Parkinson's disease. *Mov Disord.* 2013;28(1):71–85.
132. Parnetti L, Castrioto A, Chiasserini D, et al. Cerebrospinal fluid biomarkers in Parkinson disease. *Nat Rev Neurol.* 2013;9(3):131–140.
133. Ferrer I, Lopez-Gonzalez I, Carmona M, Dalfo E, Pujol A, Martinez A. Neurochemistry and the non-motor aspects of PD. *Neurobiol Dis.* 2012;46(3):508–526.

CHAPTER SEVEN

Mitochondrial Dysfunction: Cause and Consequence of Alzheimer's Disease

Kristina Friedland-Leuner*, Carola Stockburger†, Isabel Denzer*,‡, Gunter P. Eckert†, Walter E. Müller†

*Molecular and Clinical Pharmacy, Department of Chemistry and Pharmacy, Friedrich-Alexander-Universität Erlangen-Nürnberg, Erlangen, Germany
†Department of Pharmacology, Biocenter, University of Frankfurt, Frankfurt, Germany
‡Henriette Schmidt-Burkhardt Chair of Food Chemistry, Department of Chemistry and Pharmacy, Friedrich-Alexander-Universität Erlangen-Nürnberg, Erlangen, Germany

Contents

Abstract

The etiology of common, nonfamiliar late-onset Alzheimer's disease (LOAD) is only partly understood and seems to be extremely complex including many genetic and environmental factors. The most important environmental risk factor to develop LOAD is aging itself. Aging and LOAD are considered to be strongly linked to mitochondrial dysfunction and enhanced oxidative stress. In this review, we focus on the interaction between mitochondrial dysfunction in aging especially on defects of the respiratory chain of the oxidative phosphorylation system resulting in enhanced oxidative stress and the interplay between aging-associated mitochondrial defects and LOAD-associated mitochondrial failure. The deleterious effects of the two hallmarks of LOAD, amyloid beta, and hyperphosphorylated tau, on mitochondrial function, movement, and morphology are described as well as the toxic effects of the most relevant genetic

Progress in Molecular Biology and Translational Science, Volume 127
ISSN 1877-1173
http://dx.doi.org/10.1016/B978-0-12-394625-6.00007-6

risk factor of LOAD, the apolipoprotein E4 allele. Finally, the review provides an overview about drugs and nutritional ingredients which improve mitochondrial function or/and act as antioxidants and discusses their potential role in the treatment of LOAD.

1. BRAIN AGING: THE ROLE OF OXPHOS AND ROS

Cells in the central nervous system are affected by aging and react to aging by increasing amounts of reactive oxygen species (ROS), perturbed energy homeostasis, altered calcium signaling, accumulation of damaged proteins, lesions of their DNA on the molecular level, impaired function of signaling mechanisms, and altered gene expression at the cellular level.[1] The cell organelles playing the major role in the aging process are the mitochondria, due to their central role in producing ATP as the major source of cellular energy, but also as major source of ROS and as critical regulators of apoptosis during aging.[1–3] With advancing age, neurons may suffer from reduced production of ATP and NAD^+ associated with impaired mitochondrial function. Alterations of mitochondrial efficiency and function are mostly related to alterations in expression and efficiency of the complexes of the respiratory chain but also to reduced plasma membrane redox system.[4] The respiratory chain or the oxidative phosphorylation system (OXPHOS) is located in the inner mitochondrial membrane and comprises of five membrane spanning multiprotein complexes and two mobile electron carriers (ubiquinone and cytochrome *c*), which transfer electrons through a series of oxidation and reduction reactions, culminating into the reduction of oxygen.[5] This energy of the electrochemical gradient is in turn used to phosphorylate ADP via complex V (F1–F0 ATPase). Importantly, complexes I, III, and IV form higher order assemblies called mitochondrial respirasomes. Recent data suggest that defects affecting a single complex may lead to combined respirasome defects. For example, mutations in a subunit of complex IV may also induce complex I defects because the insertion of this subunit into the supercomplex occurs before or in parallel with the incorporation of complex I subunits (NDUFV1 and NDUFS4) leading to altered assembly or stability of these subunits in the supercomplexes. Recently, a study suggested that complex I is scarcely detectable as an individual entity in rat cortex.[6] In cortices isolated from aged rats (30 months), the content of complexes I and IV in the respective supercomplexes decreased significantly during aging. These findings are in line with data suggesting that complexes I and IV show significantly decreased enzymatic activities in mitochondria isolated from aged rat and mouse brain.[7–9] Moreover, the production of ROS seems to be strongly increased when the association of the supercomplexes, e.g., the prevention of the association between complexes I and III is experimentally altered. Complexes II and III of the respiratory chain are almost unaffected by normal brain aging.[6–8] In our own experiments, we could confirm a decrease of complex I activity as a rather early event within the scenario of the aging mouse brain.[10] Complex I activity declines substantially during normal brain suggesting complex I as the major player of the brain aging scenario and ROS generation.[6,11–13] In addition, complex I function and functional expression seem to be extremely sensitive to oxidative stress because of two different factors. First, complex I is in the firsthand exposition to oxidative modification. The high sensitivity of complex I results from containing iron–sulfur clusters which can be a site of direct ROS and reactive nitrogen species (RNS) attack. These oxidative modifications can manifest in the impairment of enzymatic activity and dysfunction. Interestingly, for RNS-mediated inhibition of complex I three mechanisms were proposed: *S*-nitrosylation, tyrosine nitration, and damage to iron centers.[14] Second, seven subunits of its catalytic core are encoded in the mitochondrial DNA (mtDNA) which makes this subunit enzyme complex tremendously susceptible to ROS-mediated mtDNA mutations which in turn results in reduced complex I activity and enhanced ROS production. Third, a progressive alteration of mitochondrial gene expression in aging was observed in rats and humans, which is involved in electron transfer and the assembly of the respirasomes.[6,11]

Besides complex I, each complex consists of multiple polypeptide subunits encoded by either nuclear or mtDNA.[5] Only complex II is exclusively encoded by the nuclear genome. Aging-dependent increase in the level of damaged DNA can be detected through biomarkers, e.g., the formation of 8-oxo-2′-deoxyguanosine (oxo^{8}dG). The levels of oxo^{8}dG were found to be significantly higher in mtDNA compared to nuclear DNA.[11] These differences can be explained by the proximity of mtDNA to ROS generated by the mitochondrial respiratory chain itself, the lack of any protective histone covering and a deficient repair mechanism compared to nuclear DNA.[15] Therefore, mitochondria themselves are extremely sensitive to oxidative stress. ROS are generated in multiple compartments and by multiple enzymes within the cell but mainly as overflow by products of complexes I, II, and III.[3,11,16] Approximately, 90% of cellular ROS can be traced back to mitochondria. Electrons from complexes I, II, and III can spin off prematurely, univalently reducing O_2 to produce the superoxide anion radical which is dismutated to H_2O_2.[17] Mitochondrial membrane lipids are highly susceptible to ROS, especially the long-chain polyunsaturated fatty acid components. Furthermore, the inner mitochondrial membrane proteins such as complexes I, II, and III are themselves directly susceptible to effects of oxidative stress leading to membrane depolarization and subsequently impaired mitochondrial function including increased oxidized proteins and mtDNA damage with age which in turn may cause further dysfunction.[1,4,18] mtDNA damage comprises reduction of mtDNA deletion and an increase in mtDNA mutations, e.g., point mutations in the mtDNA region of cytochrome oxidase subunit I in humans.[19] Under healthy conditions, mitochondria themselves have several powerful mechanisms which remove the consequences of ROS activity such as intramitochondrial proteases and molecular chaperones maintain mitochondrial proteasis as well as mitochondrial dynamics comprising mitochondrial fission and fusion which isolate damaged components of mitochondria and thereby target them for mitophagy.[20–22] During the aging process, these mitochondrial defense mechanisms as well as the cellular defense mechanisms including antioxidant enzymes partly fail and lead to enhanced ROS levels.[23]

2. MITOCHONDRIAL DYSFUNCTION IN ALZHEIMER'S DISEASE

Increasing evidence suggests an important role of mitochondrial dysfunction and oxidative stress in the pathogenesis of many aging-related neurodegenerative diseases, especially Alzheimer's disease (AD).[4,18,24–30] Late-onset Alzheimer's disease (LOAD) is a progressive disorder that leads to dementia and affects approximately 10% of the population older than 65 years of age. In contrast to familiar AD (FAD; only affecting 1% of all AD cases worldwide) which is caused by mutations in the amyloid precursor protein (APP) or the Presenilin 1 and 2 (PS1 and 2), the most important risk factor to develop LOAD is aging itself.[31] The etiology of common, nonfamiliar LOAD is only partly understood and seems to be extremely complex including many genetic and environmental factors.[31,32] The most relevant genetic risk factor is the APOE4 allele with a frequency of more than 50% in patients developing LOAD.[33] The clinical symptoms of both FAD and LOAD include a progressive loss of memory and impairment of cognitive ability. Even if the molecular underlying mechanisms probably differ between FAD and LOAD, brains of both patients suffering from FAD and LOAD are marked by severe neurodegenerative alterations, such as the loss of synapses and neurons, atrophy, and the selective depletion of neurotransmitter systems (e.g., acetylcholine) in the hippocampus and cerebral cortex.[34,35] Such defects are mainly observed in the late stage of the disease.

Mitochondrial dysfunction is observed in FAD and LOAD brain and has been proposed as an underlying mechanism of disease pathogenesis since defective energy metabolism is a fundamental component of AD.[32] Furthermore, early defects in glucose utilization in the brain of AD patients suggest possible abnormalities in mitochondrial function.[2] Interestingly, the activities of mitochondrial enzymes, which are reduced in the brains of AD patients, such as α-ketoglutarate dehydrogenase and pyruvate dehydrogenase, were inhibited by amyloid beta (Aβ).[36] The most consistent defect in mitochondrial electron transport enzymes in AD is a deficiency in cytochrome *c* oxidase (COX), which was reported in both AD platelets and postmortem brain samples.[37–39] In platelets obtained from daughters of mothers suffering from LOAD, complex IV activity was significantly reduced suggesting that this defect might be exclusively maternally inherited.[40] Besides reduction of mitochondrial function, mitochondrial biogenesis also seems to be attenuated in AD. Peroxisome proliferator-activated receptor γ coactivator alpha (PGC-1α) mediates changes in mitochondrial mass and coupling efficiency, antioxidant defenses, and cellular respiration by activating downstream target genes including the nuclear factor erythroid 2-related factor 2(NRF-2), the transcription factor A (TFAM), and antioxidant enzyme genes.[41] PGC-1α mRNA expression and protein levels are reduced in AD brain or in AD and mild cognitive impairment (MCI) cybrid cell lines.[42,43]

Besides mitochondrial dysfunction, postmortem tissue provides strong evidence for increased levels of cellular oxidative stress in vulnerable regions of AD brains compared to aged controls.[1,4,18,29] Increased protein oxidation, protein nitration, and lipid peroxidation were detected in brain areas showing neurofibrillary tangles and amyloid plaques.[44,45] Further evidence comes from studies investigating peroxidation products such as 4-hydroxynonenal (HNE) in cerebrospinal fluid and plasma of AD patients.[46,47] Elevated HNE levels were detected which are particular devastating for neuronal function. They impair the function of membrane ion motive ATPases and glucose and glutamate transporters. These changes consecutively lead to a disruption of cellular calcium homeostasis. Additionally, alteration in levels of antioxidants such as gluthathione or antioxidant enzymes such as catalase, Cu/Zn–SOD, and Mn–SOD support the evidence for increased oxidative stress in Alzheimer postmortem tissue and AD animal models.[48–50] In other studies, oxidative modified brain proteins were detected in AD patients with redox proteomics.[45] Importantly, individuals with MCI or very mild AD showed elevated levels of lipid peroxidation and nucleic acid oxidation in postmortem brain tissue and increased levels of lipid peroxidation and nucleic acid oxidation compared to patients suffering from severe AD.[51] Furthermore, decreased levels of plasma antioxidants and total plasma antioxidant activity were observed, as well as enhanced ROS levels in lymphocytes isolated from AD patients.[52–54] These results are supported by studies conducted in AD animal models[18] demonstrating that 3-months-old APP transgenic mice showed increased levels of HNE before Aβ plaques could be detected.[55] These increased HNE levels were accompanied by reduced activity of Cu/Zn-superoxide dismutase.

3. Aβ AND TAU—A DELETERIOUS DUO FOR MITOCHONDRIAL FUNCTION

The toxic effect of Aβ on mitochondrial function could be confirmed by several groups in different neuronal cell models as well as in different AD animal models.[29,35] Mainly oligomeric, intracellular Aβ seems to be the toxic species which impairs mitochondrial membrane potential (MMP) and reduces ATP levels.[10,56] When tracing mitochondrial dysfunction at the level of the respiratory chain, complex IV dysfunction seems to be specifically mediated by Aβ which progressively accumulates in mitochondria and its colocalization is associated with altered mitochondrial morphology and diminished complex IV activity[37,57,58] (see Fig. 7.1). Aβ is imported into the mitochondria through the translocase of the outer membrane (TOMM) and is mostly localized to mitochondrial cristae and the inner mitochondrial membrane.[59] Importantly, a TOMM40 variable length polymorphism was detected in LOAD patients, which increased the risk to develop LOAD in an age over 60 years. Patients carrying this mutation develop on an average of 7 years earlier LOAD.[60] However, these findings could not be reproduced by other groups.[61] Besides Aβ, APP itself accumulates in mitochondrial membranes and forms a stable complex with TOMM and links TOMM and the translocase of the inner membrane together as already mentioned.[22,62] Importantly, in a neuronal cell model mitochondrial localized APP was processed by mitochondrial γ-secretase resulting in Aβ production inside mitochondria.[63] Mitochondrial Aβ seems to directly impair complex IV activity. In neuronal as well as rat liver mitochondria, treatment with

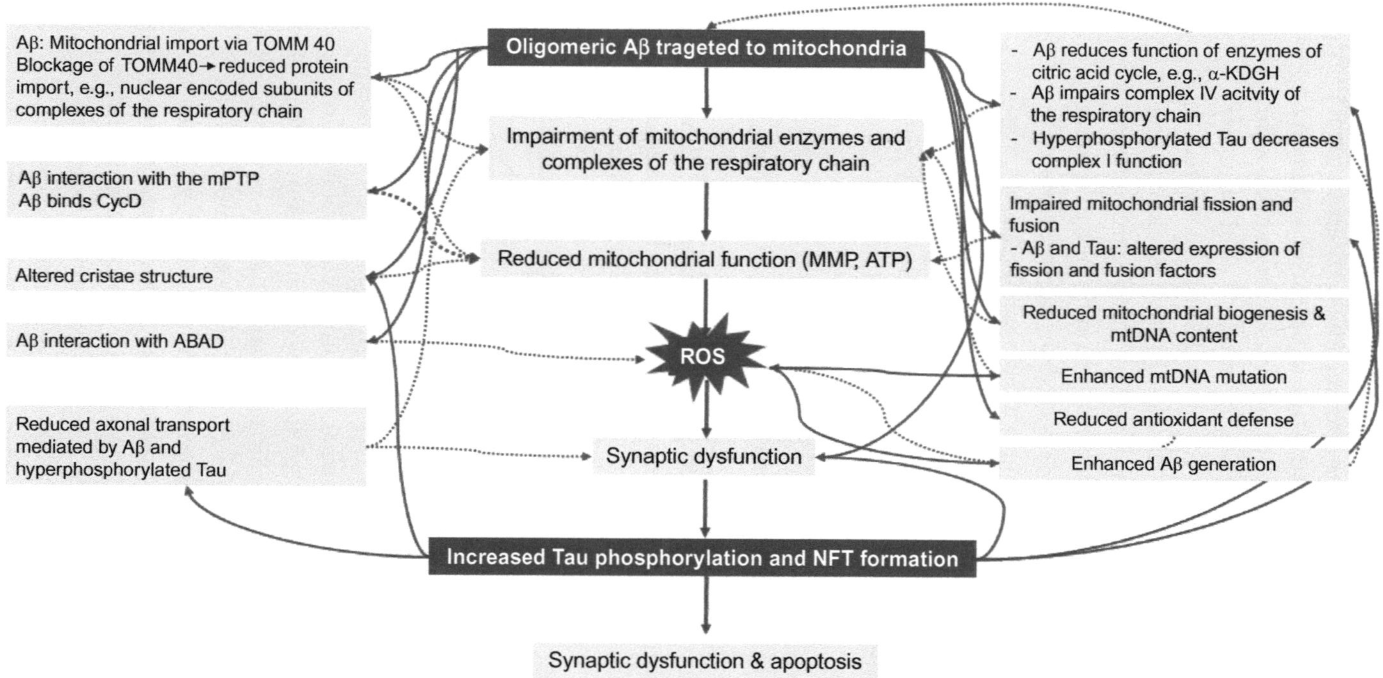

Figure 7.1 Effects of hyperphosphorylated Tau and Aβ on mitochondrial function finally leading to synaptic dysfunction and apoptosis. Aβ and Tau seem to impair mitochondrial function comprising reduced mitochondrial membrane potential (MMP) and ATP levels via different mechanisms. Aβ directly interacts with several different proteins such as ABAD or CycD, a component of the mitochondrial permeability transistion pore (mPTP) probably enhancing mPTP opening. Furthermore, Aβ interacts with the TOMM40 which regulates protein export into mitochondria of nuclear encoded proteins. In addition, Aβ might be present in the inner mitochondrial membrane and the mitochondrial matrix and might thereby impair the function of complex IV of the respiratory chain and enzymes of the citric acid cycle such as the α-ketoglutarate dehydrogenase (α-KGDH) or the pyruvate dehydrogenase. Reduced mitochondrial function leads to enhanced ROS production resulting in enhanced mtDNA mutations, reduced mitochondrial biogenesis, and mtDNA content as well as enhanced Aβ generation. Mitochondrial-derived ROS are not efficiently degraded because the antioxidative defense mechanisms are reduced in AD. The molecular mechanism how hyperphosphorylated Tau alters mitochondrial function including MMP, ATP, impaired complex I function as well as altered mitochondrial fission and fusion is not understood yet. Both, Aβ and hyperphosphorylated Tau, derogate axonal transport, and thereby further impair synaptic dysfunction. (See the color plate.)

truncated Aβ or full-length Aβ lead to decreased complex IV activity.[64] The other complexes remained unaffected. Furthermore, Aβ also leads to ultrastructural changes of mitochondrial cristae structure as described in different cell and animal models.[27,65] Mitochondrial Aβ accumulation might also be in part responsible for impaired mitochondrial dynamics.[27,66–69] The tightly regulated balance between mitochondrial fusion and fission is shifted to enhanced fission which is accompanied by increased protein levels of fission proteins such as Fis-1 and reduced protein levels of fusion proteins like OPA-1 and Drp-1. Importantly, posttranslational modification including phosphorylation and nitrosylation of Drp-1 were detected in postmortem brain tissue from AD patients probably strongly affecting mitochondrial fission.[70] Enhanced nitrosative stress by Aβ is also discussed to be involved in altered mitochondrial dynamics by altering the expression of fusion and fission factors. However, these findings are controversially discussed.[71] But not only Aβ itself but also oxidative stress leads to mitochondrial fragmentation and enhanced mitochondrial fission. Impairment of complex I or complex III, H_2O_2 itself and nitrosative stress induce mitochondrial fission.[27] These defects are associated with increased mRNA expression of FIS-1, Drp-1, and MFN1.[72]

Furthermore, Aβ affects fast axonal transport of mitochondria. Three mechanisms are discussed to be involved in this defect: first by increased phosphorylation of kinesin-1 light chains and subsequent release of kinesin from its cargoes. Second, some recent studies provide evidence that extracellular Aβ binds to a so far unknown receptor causing an increase of Ca^{2+} levels, activation of several kinases including MARK, cdk5, and JNK, and Tau phosphorylation, leading to traffic defects and mitochondria depletion as a consequence of microtubule destabilization. Third, oligomeric Aβ is discussed to induce an activation of GSK3β which cause dissociation of kinesin-1 from its cargo.[73]

Recent evidence supports the involvement of the mitochondrial permeability transition pore (mPTP) in the proteolytic cleavage of OPA-1. After mPTP opening and collapse of MMP, specific proteases which are sensitive to MMP changes are activated and cleave OPA-1 leading to impaired mitochondrial fusion.[74–76] In the last few years the formation of mPTP, an unselective mitochondrial pore, leading to mitochondrial swelling, severe mitochondrial dysfunction, the release of pro-apoptotic factors and in the end eventually cell death, is gaining more attention regarding the pathophysiology of neurodegenerative diseases. Although the complete components (the involvement of cyclophilin D, voltage-dependent anion channel, adenine nucleotide translocase, and mitochondrial phosphate carrier is highly probable) and the exact mechanism how Aβ is involved in this process, are not yet fully elucidated, recent findings argue for a definitely strong impact of Aβ on mPTP formation.[74,77] For example, the inhibition of mPTP and selective reduction in mitochondrial superoxide flashes significantly ameliorated the negative effects of Aβ on neural progenitor cell proliferation and survival.[78] More specifically Aβ is supposed to interact with cyclophin-D and thereby potentiates mitochondrial, neuronal, and synaptic stress, giving rise to the suspicion that blocking cyclophilin D might be a promising therapeutic target for AD.[79] In addition, Lustbader *et al.* demonstrated that Aβ-binding alcohol dehydrogenase (ABAD) is a direct molecular link between Aβ and mitochondrial toxicity.[80,81] ABAD catalyses with the help of NAD^+/NADH the reduction of aldehydes and ketons and oxidation of alcohols for energy production utilizing different substrates.[82] In a simplified view, Aβ binds to Aβ and impairs the function of ABAD and finally results in neuronal apoptosis.

The other hallmark of AD, the appearance of neurofibrillary tangles, primarily composed of aggregated hyperphosphorylated tau also seem to be involved in mitochondrial dysfunction[83,84] (see Fig. 7.1). Truncated tau alone induces mitochondrial fragmentation whereas full-length tau enhances mitochondrial size.[85] Using transgenic mice overexpressing the disease-associated mutant P301L, we could demonstrate mitochondrial dysfunction by proteomic and functional analyses in these mice.[86]

Our functional analysis demonstrated reduced NADH-ubiquinone oxidoreductase (complex I) activity and, with age, impaired mitochondrial respiration and ATP synthesis in P301L tau mice. In particular, the reduction in state-3 respiration reflects a reduced capacity of mitochondria to metabolize oxygen and the complex I substrates in the presence of a limited quantity of ADP. Accordingly, higher levels of reactive oxygen species in aged transgenic mice were detected and P301L tau mitochondria displayed increased vulnerability towards Aβ peptide insult, suggesting a synergistic action of tau and Aβ pathology on the mitochondria. The synergistic interaction of aging with AD specific alterations like Tau pathology, APP as well as PS1 mutations in leading to mitochondrial dysfunction has recently been confirmed using double and triple transgenic mice.[57,87] Yao *et al.* showed that even in embryonic hippocampal neurons mitochondrial respiration was significantly decreased and increased glycolysis.[87] These defects are markedly exacerbated during aging. Using a different triple transgenic mouse model, Rhein *et al.* detected a massive deregulation of mitochondrial proteins mainly related to complexes I and IV. Importantly, deregulation of complex I was tau dependent, whereas complex IV dysfunction is mediated by Aβ.[88] In contrast to Aβ, where different molecular mechanisms are proposed regarding its effects on complex IV, the molecular mechanism how tau alters complex I activity is not yet understood. Besides the described alteration of OXPHOS activities, tau also interferes with mitochondrial movement, fission, and fusion.[89,90] Recently, a mechanism why mitochondrial fission is altered by tau was proposed. DRP1 localization to mitochondria depends on the interaction of DRP1 and mitochondria with filamentous (F)-actin prior to their colocalization.[85] Tau induces stabilization and bundling of F-actin. In Tau-overexpressing models such as Tau transgenic mice or flies, increased F-actin disrupts the physical interaction between DRP1 and mitochondria leading to mitochondrial elongation and reduced fission. The dysbalance between fission and fusion results in neurotoxicity, which can be rescued by reducing mitochondrial fusion.[89]

Several groups showed that phosphorylated Tau impairs mitochondria anterograde axonal transport via different mechanisms including trapping the kinesin adaptor molecule JIP1 in the soma thereby preventing the loading of cargoes onto the kinesin machinery for transport down the axon or by activating the axonal protein phosphatase 1 and the glycogen synthase kinase 3.[22,91,92] Other experiments suggest that Tau itself is a cargo of kinesin displacing other cargoes from kinesin such as mitochondria.[22] Our group showed a significant decrease in mitochondrial movement in neuronal cells overexpressing the tau mutation P301L. In addition, both fission and fusion factors mRNA levels were decreased.[90] The mutation P301L is known to destabilize the microtubule network, which leads to a perinuclear localization of the mitochondria—as seen in our cell model—and to decreased dynamics of microtubules affecting the transport of vesicles and organelles.[89] Furthermore, cristae morphology was also significantly altered in this cell model suggesting that mitochondrial respirasomes might be altered due to the impairment of complex I activity and reduced axonal transport which could also affect the transport of nuclear coded complex subunits.

4. MITOCHONDRIAL-DERIVED ROS INDUCE Aβ GENERATION—FOCUS ON COMPLEXES I AND III

Several lines of evidence suggest that elevated ROS production might initiate toxic APP processing and thereby trigger toxic Aβ generation. The group of Tabaton showed during the last years that oxidative stress in the form of HNE or H_2O_2 leads to enhanced Aβ production in cell models.[93,94] In addition, recent data suggest that complex I-derived ROS contribute to amyloidogenic APP processing.[27] In cell and animal models, complex I dysfunction induced by functional or genetically mediated inhibition leads to enhanced Aβ levels. These results are in line with recently published findings showing that in APP transgenic mice (Tg2576) the pesticide paraquat which enhances ROS levels also increases Aβ levels and leads to cognitive impairment.[95] Further support comes from a completely different aging model, the SAMP8 mouse. This premature aging mouse model is characterized by a unique missence A11181G mutation in the *mt-ND4* gene which encodes one of the subunits of complex I, NADH dehydrogenase subunit 4 (ND4). Interestingly, these mice exhibit a 50% Aβ increase over age.[96]

Are only complex I-derived ROS able to induce amyloidogenic APP processing or are these common features of mitochondrial-derived ROS? We propose that this is a shared property of all mitochondrial-derived ROS. This hypothesis is supported by findings that the AD prevalence is increased in patients with a stroke history.[97] Interestingly, hypoxia increases ROS via reduced complex III activity.[16] In animal and cell models, several groups showed that hypoxia increases Aβ production *in vitro* and *in vivo*.[98,99] Our group also showed that antimycin, a selective complex III inhibitor, also leads to increased Aβ levels in HEK293 cells.[27] Recent data from different groups provide evidence that mitochondrial-derived ROS lead to an increase of APP processing via enhanced BACE1 and γ-secretase activity.[93] Importantly, several lines of evidence suggest γ-secretase regulates BACE1 expression.[93] Several pathways are discussed to be involved in this phenomenon such as HIF-1α, a molecule that regulates oxygen homeostasis, NF-κB and JNK/cjun.

5. INTERPLAY BETWEEN AGING AND AD: THE BALANCE BETWEEN SYNERGISTIC DYSFUNCTION AND FUNCTIONAL COMPENSATION

In LOAD, the slow process of brain degeneration begins decades before clinical symptoms appear around midlife.[29] Several factors synergistically lead to reduced energy metabolism and enhanced reactive oxygen species (ROS) production. Enhanced oxidative stress in "normal brain aging" is an important factor, due to age-related alterations in ROS producing (for example, complex I defects) or ROS detoxifyng (antioxidant enzymes) mechanisms. Moreover, around midlife, normal "low-level" Aβ production is slowly elevated, likely triggered by physiological, environmental, and genetic risk factors, e.g., APOE4.[32,100,101] This initiates a vicious cycle comprising mitochondrial dysfunction, enhanced ROS production, and increase in Aβ levels in its center. Mitochondrial dysfunction results in enhanced ROS production, ROS trigger Aβ generation, and Aβ itself impairs mitochondrial function thereby closing this deleterious and self-promoting pathophysiological cycle. Low-level elevation of Aβ over baseline, many decades before clinical symptoms appear, seems to be highly important in the initiation of mitochondrial dysfunction.[24,28]

As explained above, Aβ itself elevates ROS production, which in turn is able to accelerate Aβ production, leading to a vicious cycle and initiating a process which leading to increasingly severe brain damage and finally clinical manifestation of AD.[24,28] In humans, very little data are available about the increase of Aβ levels associated with normal aging or the early presymptomatic stage. In several postmortem studies, only slightly elevated Aβ levels were observed until midlife, afterwards, around the age of 60, Aβ levels start to increase several times in different brain regions.[100,102–105] Similarly, in cognitively intact subjects, Aβ oligomer levels increased very little for the first few decades of life and became only marginally elevated (less than five times up to an age of 50 years[106]). Importantly, while data regarding normal aging in humans are confounded by presymptomatic AD, data about soluble Aβ levels in rat brains also suggest small increases up to 9–12 months, where after Aβ levels show a substantial increase.[107,108]

To investigate the effects of low-level Aβ elevation on mitochondrial function, we used neuronal cells overexpressing APP_{wt} producing approximately twice as much Aβ compared to control cells. This increase is very moderate with respect to other cells treated with high-exogenous Aβ concentrations or cell lines transfected with human AβPP mutants relevant for early onset AD[109] and considerably less than those seen in AD mouse models.[110] Interestingly neuronal cells suffering from this very mild Aβ elevation exhibit a reduced MMP, decreased mitochondrial respiration, and decreased ATP levels. Mitochondrial size and shape was also significantly different between APP_{wt} and SH-control cells, but surprisingly the changes of markers for mitochondrial dynamics (Mfn1, Fis1) were rather small. This suggests that there are additional factors determining the extent of fragmentation, for example, properties of the mitochondrial membranes.[111,112] In contrast to previous studies of AD mice, cell line models, and patient samples whereby a large degree of cristae swelling and disruption[113–115] could be demonstrated, investigation of the three-dimensional structure of APP_{wt} neuronal cells by cryo-ET revealed little change with respect to mitochondrial ultrastructure and cristae morphology, suggesting that at a low-Aβ concentrations mitochondria are still able to maintain their characteristic cristae membrane morphology.

Our hypothesis of mitochondrial dysfunction in AD suggests that a small elevation of Aβ occurring in the normal brain until midlife may not be sufficient to initiate major mitochondrial damage alone. But indeed low-level Aβ could enhance vulnerability to additional mitochondrial defects, e.g., those initiated by enhanced oxidative stress during aging in combination with the APOE4 phenotype. The APOE4 allele is the most common genetic risk factor for LOAD. Heterozygosity of APOE4 is associated with a threefold higher LOAD risk compared to the common APOE3 allele. Homozygosity increases the risk even 10-fold. The APOE4 protein differs in its tertiary protein structure and seems to be involved in reduced degradation of extracellular Aβ, in lipid and cholesterol trafficking, APP processing, increased Tau-hyperphosphorylation, and reduced synaptic plasticity.[116,117] APOE4 also affects mitochondrial function. APOE4 expression was shown to reduce expression of several complexes of the respiratory chain including complexes I and IV.[118] Furthermore, complex IV enzymatic activity was also reduced. The lipid-binding region of APOE4 seems to be critically for the interaction with mitochondria. It was demonstrated in a neuronal cell model that APOE4 (residues 272–299) binds the subunits ubiquinol cytochrome *c* reductase core protein 2 and cytochrome CI of complex III and COX subunit 4 isoform 1 of complex IV and significantly reduces respiratory function of complexes III and IV.[119]

6. PHARMACOLOGICAL STRATEGIES TO IMPROVE MITOCHONDRIAL FUNCTION

While the concept of Aβ- and tau-induced mitochondrial dysfunction in AD has received substantial support over the last decade, improving mitochondrial function as a target for new drug development has not. Until recently, scientific interest was mostly focused on drugs leading to reduced Aβ load. However, as several disease-modifying compounds failed to show clinical effectiveness in AD trials.[90] Therefore, improving mitochondrial dysfunction might be a promising concept to treat LOAD[24,26] (also see Table 7.1).

Table 7.1 Overview about interesting compounds targeting mitochondria which were tested AD models

Drugs	Treatment	Effects	References	Clinical evidence
MitoQ	Treatment with MitoQ in transgenic AD mice, transgenic *Caenorhabditis elegans* model of AD, and AD cell models	Reduced Aβ-induced pathology, reduced cognitive decline, reduced Aβ accumulation, synaptic loss, and protection of complexes I and IV function	114,120–122	None in AD
CoQ10	Treatment with CoQ10 in transgenic AD mice and AD neuronal stem cell model	Reduced amyloid generation, improved cognition, and improved proliferation of neuronal stem cells	123,124	No alteration of CSF biomarkes in AD patients[125]
Ginkgo biloba	Treatment with EGb 761 in old mice and transgenic AD mice, transgenic *Caenorhabditis elegans* model of AD, and AD cell models	Improved mitochondrial function (improved MMP, ATP, and function of complexes I, III, and IV), reduced Aβ generation, and reduced ROS levels	126–132	Different clinical trials and a recent meta analysis[133]
Piracetam	Treatment with piracetam in old mice and transgenic AD mice and AD cell models	Improved mitochondrial function (improved MMP, ATP, and function of the complexes of the respiratory chain)	112,134,135	No clinical trials with ADAS-COG scale. Several older clinical trials with elderly patients suffering from cognitive

Table 7.1 Overview about interesting compounds targeting mitochondria which were tested AD models—cont'd

Drugs	Treatment	Effects	References	Clinical evidence
		and reduced Aβ generation		impairment or dementia[136]
Dimebon	Treatment with dimebon in transgenic AD mice and AD cell models	Improved mitochondrial function (enhanced MMP, ATP, function of the complexes of the respiratory chain), reduced Aβ pathology, and improved cognitive performance	137–141	Controversial data[142,143]

7. ANTIOXIDANTS, FLAVONOIDS, POLYPHENOLS, AND GINKGO

Several antioxidants have a long history as possible treatments for AD and even have been and are used in this context. Initially mainly vitamin E or vitamin C or the combination of both has been investigated. While both at high concentrations definitively show antioxidant properties *in vitro* and *in vivo*, their therapeutical benefit to improve or even prevent age-related cognitive impairment in AD is still under discussion.[23,144]

Another important class of naturally accruing antioxidants are flavonoids or other polyphenols, which also are fairly good antioxidants reducing oxidative stress *in vitro* and *in vivo*.[145–147] Flavonoids also improve mitochondrial dysfunction and seem to have therapeutical benefit for long-term treatment of age-related cognitive impairment in animals and men. The significant reduction of the risk in getting AD by Mediterranean diet is very likely explained by an impotent part by the high daily intake of flavonoids.[148] In general, even if the effectiveness of those natural occurring antioxidants to protect against AD seems to be limited, they seem to be the major players of diet in reducing oxidative stress and acting as a second however weaker defense system.[145] As the tissue level or even brain levels of those dietary antioxidants are rather low, it has been suggested that their protective properties might be associated with the phenomenon of neurohormesis, sensitization of the whole antioxidant system by low-levels exposure, probably by inactivation of free radicals relevant for cellular signaling.[149] Polyphenols, e.g., resveratrol increases PGC-1α activation via sirtuins. Sirtuins are members of the NAD^+-dependent histone deacetylase family which regulates cellular metabolism, energy metabolism, cell survival, and aging.[147,150] Another polyphenol with promising antioxidant- and mitochondrial-protecting effects is curcumin.[147,151,152] Other dietary antioxidants are NRF-2 activators such as sulforaphane isolated from broccoli. Sulforaphane was investigated in an AD animal model[153] and showed protective effects in memory tasks. However, additional experiments are needed to further strengthen its antioxidant effects and protective effects on mitochondrial function in AD.

Another case of an herbal drug is the standardized Ginkgo biloba extract (EGb 761), which has been used for many years as a prescription or OTC drug in many countries to treat aging-related cognitive disorders including AD. EGb 761 contains 24% of flavonoids and 6% of terpenens. The terpene lactones are represented by the Ginkgolides A, B, C, J, and M and bilobalide. The flavonoid fraction is composed of quercetin, kaempferol, and isorhamnetin, which are mainly responsible for the free radical scavenging properties. However, the activity of EGb 761 also includes substantial mitochondria-protecting properties which have been described in several publications. Our group showed protection of mitochondrial function in a neuronal like cell line and in dissociated brain cells and isolated mitochondria of EGb 761-treated animals.[126–128,154,155] In accordance with our data, treatment with EGb 761 prevented age-associated changes in mitochondrial morphology, mitochondrial glutathione levels, and respiratory function of rat brain mitochondria.[128] Several studies provide evidence for the antioxidant properties of EGb 761, e.g., lymphocytes and *Caenorhabditis elegans* worms.[129,155] Smith and Luo showed a direct attenuation of ROS by the flavonoid fraction of EGb 761.[130]

Other possible treatment options are mitochondrial targeted antioxidants such as CoQ10, the CoQ10 analog idebenone or MitoQ. CoQ10 is an endogenous biological substrate for the electron transport chain and an important antioxidant in mitochondrial membranes. MitoQ is a form of coenzyme Q ubiquinone linked to triphosphonium ions through covalent attachment. Both antioxidants showed protective effects in AD models.[26,120,121]

8. METABOLIC ENHANCER

While in many cases there is a substantial overlap between antioxidant and mitochondria-protecting properties, it is important to note that both do not share the same mechanism and several compounds show significant mitochondrial protection without having antioxidant or radical scavenging properties. A drug which has been extensively characterized in this respect is the metabolic enhancer piracetam,[112,134,135,156] which shows no antioxidant properties but exhibits pronounced mitochondrial protection *ex vivo* as well as *in vitro*. Evidences that piracetam might improve disturbed mitochondrial function originate from observations that piracetam improves glucose uptake and utilization as well as ATP production.[156] In line with these data are our previous observations of significant mitochondrial protection and enhanced ATP synthesis by piracetam against experimentally induced oxidative as well as nitrosative stress *in vitro* and after *ex vivo* treatment, where again aged animals with well-characterized mitochondrial dysfunction benefited most.[134,135] In agreement with the assumption of mitochondrial membranes as primary target, the beneficial effect of piracetam was similar after experimental impairment of each of the five respiratory chain complexes. Piracetam also improves mitochondrial function in AD cell and animal models. Piracetam shows protective effects on MMP, OXPHOS activity, ATP levels, altered mitochondrial morphology as well as it reduced intracellular Aβ levels.[112,134]

9. DIMEBON

For the old Russian antiallergic drug dimebon several findings suggested positive effects of dimebon on impaired cognitive functions in AD.[142] In most AD animal models, dimebon improved Aβ pathology and improved cognition.[137,157] Only in five FAD mice, dimebon showed no effect on Aβ pathology and behavior.[138] However, its efficacy to treat AD is highly discussed due to negative clinical trials. Its molecular mechanism of action is still a matter of debate.[143] Recent publications reported evidence for mitochondrial protection by dimebon. In micromolar concentrations, dimebon inhibits calcium-induced swelling of rat brain mitochondria, without affecting cytochrome *c* release or calcium retention.[139,158,159] The concept of mitochondrial protection by dimebon is further supported and substantially extended by our recent indicating that dimebon at nanomolar concentrations shows distinct effects on mitochondrial dynamics and OXPHOS activity.[140]

10. CONCLUSION AND FURTHER PERSPECTIVE

Impaired mitochondrial metabolism associated with respiratory chain dysfunction and the oxidative stress is considered to be a major pathological mechanism in a number of neurodegenerative diseases including AD. In contrast to Aβ plaques and tau tangles seen in the late stage of AD, mitochondrial dysfunction, and oxidative stress are two early events in the pathology of AD cumulating with aging-associated changes in mitochondrial function, morphology, dynamics, and oxidative stress. Aβ-mediated complex IV impairment together with complexes I, III, and IV reduced activities in aging as well as hyperphosphorylated tau-mediated complex I impairment lead to severe defects in mitochondrial energy supply. Furthermore, the most important genetic risk factor for LOAD, APOE4, also impairs mitochondrial function.[117,119,160] These changes result in a vicious cycle inducing electron leakage from the ETC leading to enhanced production of ROS damaging ETC components and mtDNA and further accelerating mitochondrial dysfunction and oxidative stress. When phenotypic threshold and severe energy deprivation is reached, neuronal, and synaptic dysfunction appears which is enhanced by depletion of mitochondria from axons and dendrites. Thus, both mitochondrial dysfunction and oxidative stress clearly play an important role in the pathogenesis of AD and might be promising targets for the development of new antidementive drugs which are used early in the disease process or even for the prevention of AD. However, recent antioxidants and drugs which improve mitochondrial function only show moderate effects in clinical trials with LOAD patients suggesting that early treatment of patients before clinical symptoms arise might be necessary to improve clinical efficacy of these drugs. To achieve this goal, reliable biomarkers detecting LOAD decades before clinical symptoms appear in patients are essential.

REFERENCES

1. Stranahan AM, Mattson MP. Recruiting adaptive cellular stress responses for successful brain ageing. *Nat Rev Neurosci.* 2012;13(3):209–216. http://dx.doi.org/10.1038/nrn3151.
2. Kapogiannis D, Mattson MP. Disrupted energy metabolism and neuronal circuit dysfunction in cognitive impairment and Alzheimer's disease. *Lancet Neurol.* 2011;10(2):187–198. http://dx.doi.org/10.1016/S1474-4422(10)70277-5.
3. Brand MD, Orr aL, Perevoshchikova IV, Quinlan CL. The role of mitochondrial function and cellular bioenergetics in ageing and disease. *Br J Dermatol.* 2013;169-(Suppl):1–8. http://dx.doi.org/10.1111/bjd.12208.
4. Gibson GE, Starkov A, Blass JP, Ratan RR, Beal MF. Cause and consequence: mitochondrial dysfunction initiates and propagates neuronal dysfunction, neuronal death and behavioral abnormalities in age-associated neurodegenerative diseases. *Biochim Biophys Acta.* 2010;1802(1):122–134. http://dx.doi.org/10.1016/j.bbadis.2009.08.010.
5. Morân M, Moreno-Lastres D, Marín-Buera L, Arenas J, Martin Ma, Ugalde C. Mitochondrial respiratory chain dysfunction: implications in neurodegeneration. *Free Radic Biol Med.* 2012;53(3):595–609. http://dx.doi.org/10.1016/j.freeradbiomed.2012.05.009.
6. Frenzel M, Rommelspacher H, Sugawa MD, Dencher NA. Ageing alters the supramolecular architecture of OxPhos complexes in rat brain cortex. *Exp Gerontol.* 2010;45(7–8):563–572. http://dx.doi.org/10.1016/j.exger.2010.02.003.
7. Navarro A, López-Cepero JM, Bândez MJ, et al. Hippocampal mitochondrial dysfunction in rat aging. *Am J Physiol Regul Integr Comp Physiol.* 2008;294(2):R501–R509. http://dx.doi.org/10.1152/ajpregu.00492.2007.
8. Navarro A, Boveris A. Rat brain and liver mitochondria develop oxidative stress and lose enzymatic activities on aging. *Am J Physiol Regul Integr Comp Physiol.* 2004;287(5): R1244–R1249. http://dx.doi.org/10.1152/ajpregu.00226.2004.
9. Tatarková Z, Kuka S, Račay P, et al. Effects of aging on activities of mitochondrial electron transport chain complexes and oxidative damage in rat heart. *Physiol Res.* 2011;60(2):281–289. Available at, http://www.ncbi.nlm.nih.gov/pubmed/21114360, Accessed 24.03.14.
10. Hauptmann S, Scherping I, Dröse S, et al. Mitochondrial dysfunction: an early event in Alzheimer pathology accumulates with age in AD transgenic mice. *Neurobiol Aging.* 2009;30(10):1574–1586. http://dx.doi.org/10.1016/j.neurobiolaging.2007.12.005.
11. Li D, Chen H, Florentino A, et al. Enzymatic dysfunction of mitochondrial complex I of the *Candida albicans* goa1 mutant is associated with increased reactive oxidants and cell death. *Eukaryot Cell.* 2011;10(5):672–682. http://dx.doi.org/10.1128/EC.00303-10.
12. Maranzana E, Barbero G, Falasca AI, Lenaz G, Genova ML. Mitochondrial respiratory supercomplex association limits production of reactive oxygen species from complex I. *Antioxid Redox Signal.* 2013;19(13):1469–1480. http://dx.doi.org/10.1089/ars.2012.4845.
13. Jones TT, Brewer GJ. Age-related deficiencies in complex I endogenous substrate availability and reserve capacity of complex IV in cortical neuron electron transport. *Biochim Biophys Acta.* 2010;1797(2):167–176. http://dx.doi.org/10.1016/j.bbabio.2009.09.009.
14. Papa S, De Rasmo D. Complex I deficiencies in neurological disorders. *Trends Mol Med.* 2013;19(1):61–69. http://dx.doi.org/10.1016/j.molmed.2012.11.005.
15. Bereiter-Hahn J. Do we age because we have mitochondria? *Protoplasma.* 2014;251(1):3–23. http://dx.doi.org/10.1007/s00709-013-0515-x.

16. Bleier L, Dröse S. Superoxide generation by complex III: from mechanistic rationales to functional consequences. *Biochim Biophys Acta.* 1827;11–12:1320–1331. http://dx.doi.org/10.1016/j.bbabio.2012.12.002.
17. Mailloux RJ, Jin X, Willmore WG. Redox regulation of mitochondrial function with emphasis on cysteine oxidation reactions. *Redox Biol.* 2013;2:123–139. http://dx.doi.org/10.1016/j.redox.2013.12.011.
18. Leuner K, Hauptmann S, Abdel-Kader R, et al. Mitochondrial dysfunction: the first domino in brain aging and Alzheimer's disease? *Antioxid Redox Signal.* 2007;9(10): 1659–1675. http://dx.doi.org/10.1089/ars.2007.1763.
19. Kazachkova N, Ramos A, Santos C, Lima M. Mitochondrial DNA damage patterns and aging: revising the evidences for humans and mice. *Aging Dis.* 2013;4(6): 337–350. http://dx.doi.org/10.14336/AD.2013.0400337.
20. Frank M, Duvezin-Caubet S, Koob S, et al. Mitophagy is triggered by mild oxidative stress in a mitochondrial fission dependent manner. *Biochim Biophys Acta.* 2012;1823(12):2297–2310. http://dx.doi.org/10.1016/j.bbamcr.2012.08.007.
21. Barnett A, Brewer GJ. Autophagy in aging and Alzheimer's disease: pathologic or protective? *J Alzheimers Dis.* 2011;25(3):385–394. http://dx.doi.org/10.3233/JAD-2011-101989.
22. Eckert A, Nisbet R, Grimm A, Götz J. March separate, strike together—role of phosphorylated TAU in mitochondrial dysfunction in Alzheimer's disease. *Biochim Biophys Acta.* 2013;1842(8):1258–1266. http://dx.doi.org/10.1016/j.bbadis.2013.08.013.
23. Heo J-H, Hyon-Lee, Lee KM. The possible role of antioxidant vitamin C in Alzheimer's disease treatment and prevention. *Am J Alzheimers Dis Other Demen.* 2013;28(2):120–125. http://dx.doi.org/10.1177/1533317512473193.
24. Müller WE, Eckert A, Kurz C, Eckert GP, Leuner K. Mitochondrial dysfunction: common final pathway in brain aging and Alzheimer's disease-therapeutic aspects. *Mol Neurobiol.* 2010;41(2–3):159–171. http://dx.doi.org/10.1007/s12035-010-8141-5.
25. Beal MF. Mitochondrial dysfunction and oxidative damage in Alzheimer's and Parkinson's diseases and coenzyme Q10 as a potential treatment. *J Bioenerg Biomembr* 2004;36(4):381–386. http://dx.doi.org/10.1023/B:JOBB.0000041772.74810.92.
26. Chaturvedi RK, Flint Beal M. Mitochondrial diseases of the brain. *Free Radic Biol Med.* 2013;63:1–29. http://dx.doi.org/10.1016/j.freeradbiomed.2013.03.018.
27. Leuner K, Schütt T, Kurz C, et al. Mitochondrion-derived reactive oxygen species lead to enhanced amyloid beta formation. *Antioxid Redox Signal.* 2012;16(12):1421–1433. http://dx.doi.org/10.1089/ars.2011.4173.
28. Swerdlow RH, Burns JM, Khan SM. The Alzheimer's disease mitochondrial cascade hypothesis. *J Alzheimers Dis.* 2010;20(Suppl 2):S265–S279. http://dx.doi.org/10.3233/JAD-2010-100339.
29. Swerdlow RH, Burns JM, Khan SM. The Alzheimer's disease mitochondrial cascade hypothesis: progress and perspectives. *Biochim Biophys Acta.* 2014;1842(8):1219–1231. http://dx.doi.org/10.1016/j.bbadis.2013.09.010.
30. Swerdlow RH. Alzheimer's disease pathologic cascades: who comes first, what drives what. *Neurotox Res.* 2012;22(3):182–194. http://dx.doi.org/10.1007/s12640-011-9272-9.
31. Jiang T, Yu J-T, Tian Y, Tan L. Epidemiology and etiology of Alzheimer's disease: from genetic to non-genetic factors. *Curr Alzheimer Res.* 2013;10(8):852–867. Available at, http://www.ncbi.nlm.nih.gov/pubmed/23919770, Accessed 16.01.14.
32. Rhinn H, Fujita R, Qiang L, Cheng R, Lee JH, Abeliovich A. Integrative genomics identifies APOE ε4 effectors in Alzheimer's disease. *Nature.* 2013;500(7460):45–50. http://dx.doi.org/10.1038/nature12415.

33. Liraz O, Boehm-Cagan A, Michaelson DM. ApoE4 induces Aβ42, tau, and neuronal pathology in the hippocampus of young targeted replacement apoE4 mice. *Mol Neurodegener.* 2013;8(1):16. http://dx.doi.org/10.1186/1750-1326-8-16.
34. Haass C, Selkoe DJ. Soluble protein oligomers in neurodegeneration: lessons from the Alzheimer's amyloid beta-peptide. *Nat Rev Mol Cell Biol.* 2007;8(2):101–112. http://dx.doi.org/10.1038/nrm2101.
35. Querfurth HW, Laferla FM. Alzheimer's disease. *N Engl J Med.* 2013;362(4):329–344.
36. Gibson GE, Blass JP, Beal MF, Bunik V. The alpha-ketoglutarate-dehydrogenase complex: a mediator between mitochondria and oxidative stress in neurodegeneration. *Mol Neurobiol.* 2005;31(1–3):43–63. http://dx.doi.org/10.1385/MN:31:1-3:043.
37. Kish SJ, Bergeron C, Rajput A, et al. Brain cytochrome oxidase in Alzheimer's disease. *J Neurochem.* 1992;59(2):776–779. Available at, http://www.ncbi.nlm.nih.gov/pubmed/1321237, Accessed 26.03.14.
38. Cardoso SM, Proença MT, Santos S, Santana I, Oliveira CR. Cytochrome *c* oxidase is decreased in Alzheimer's disease platelets. *Neurobiol Aging.* 2004;25(1):105–110. Available at, http://www.ncbi.nlm.nih.gov/pubmed/14675736, Accessed 26.03.14.
39. Krishnan KJ, Ratnaike TE, De Gruyter HLM, Jaros E, Turnbull DM. Mitochondrial DNA deletions cause the biochemical defect observed in Alzheimer's disease. *Neurobiol Aging.* 2012;33(9):2210–2214. http://dx.doi.org/10.1016/j.neurobiolaging.2011.08.009.
40. Mosconi L, de Leon M, Murray J, et al. Reduced mitochondria cytochrome oxidase activity in adult children of mothers with Alzheimer's disease. *J Alzheimers Dis.* 2011;27(3):483–490. http://dx.doi.org/10.3233/JAD-2011-110866.
41. Cheng A, Wan R, Yang J-L, et al. Involvement of PGC-1α in the formation and maintenance of neuronal dendritic spines. *Nat Commun.* 2012;3:1250. http://dx.doi.org/10.1038/ncomms2238.
42. Silva DF, Selfridge JE, Lu J, et al. Bioenergetic flux, mitochondrial mass and mitochondrial morphology dynamics in AD and MCI cybrid cell lines. *Hum Mol Genet.* 2013;22(19):3931–3946. http://dx.doi.org/10.1093/hmg/ddt247.
43. Sheng B, Wang X, Su B, et al. Impaired mitochondrial biogenesis contributes to mitochondrial dysfunction in Alzheimer's disease. *J Neurochem.* 2012;120(3):419–429. http://dx.doi.org/10.1111/j.1471-4159.2011.07581.x.
44. Hardas SS, Sultana R, Clark AM, et al. Oxidative modification of lipoic acid by HNE in Alzheimer disease brain. *Redox Biol.* 2013;1(1):80–85. http://dx.doi.org/10.1016/j.redox.2013.01.002.
45. Sultana R, Perluigi M, Allan Butterfield D. Lipid peroxidation triggers neurodegeneration: a redox proteomics view into the Alzheimer disease brain. *Free Radic Biol Med.* 2013;62:157–169. http://dx.doi.org/10.1016/j.freeradbiomed.2012.09.027.
46. Selley ML, Close DR, Stern SE. The effect of increased concentrations of homocysteine on the concentration of (E)-4-hydroxy-2-nonenal in the plasma and cerebrospinal fluid of patients with Alzheimer's disease. *Neurobiol Aging.* 2002;23(3):383–388. Available at, http://www.ncbi.nlm.nih.gov/pubmed/11959400, Accessed 27.03.14.
47. Völkel W, Sicilia T, Pähler A, et al. Increased brain levels of 4-hydroxy-2-nonenal glutathione conjugates in severe Alzheimer's disease. *Neurochem Int.* 2006;48(8):679–686. http://dx.doi.org/10.1016/j.neuint.2005.12.003.
48. Schuessel K, Frey C, Jourdan C, et al. Aging sensitizes toward ROS formation and lipid peroxidation in PS1M146L transgenic mice. *Free Radic Biol Med.* 2006;40(5):850–862. http://dx.doi.org/10.1016/j.freeradbiomed.2005.10.041.
49. Marcus DL, Thomas C, Rodriguez C, et al. Increased peroxidation and reduced antioxidant enzyme activity in Alzheimer's disease. *Exp Neurol.* 1998;150(1):40–44. http://dx.doi.org/10.1006/exnr.1997.6750.

50. Mandal PK, Tripathi M, Sugunan S. Brain oxidative stress: detection and mapping of anti-oxidant marker "glutathione" in different brain regions of healthy male/female, MCI and Alzheimer patients using non-invasive magnetic resonance spectroscopy. *Biochem Biophys Res Commun.* 2012;417(1):43–48. http://dx.doi.org/10.1016/j.bbrc.2011.11.047.
51. Aluise CD, Robinson RAS, Cai J, Pierce WM, Markesbery WR, Butterfield DA. Redox proteomics analysis of brains from subjects with amnestic mild cognitive impairment compared to brains from subjects with preclinical Alzheimer's disease: insights into memory loss in MCI. *J Alzheimers Dis.* 2011;23(2):257–269. http://dx.doi.org/10.3233/JAD-2010-101083.
52. Torres LL, Quaglio NB, de Souza GT, et al. Peripheral oxidative stress biomarkers in mild cognitive impairment and Alzheimer's disease. *J Alzheimers Dis.* 2011;26(1):59–68. http://dx.doi.org/10.3233/JAD-2011-110284.
53. Leuner K, Müller WE, Reichert AS. From mitochondrial dysfunction to amyloid beta formation: novel insights into the pathogenesis of Alzheimer's disease. *Mol Neurobiol.* 2012;46(1):186–193. http://dx.doi.org/10.1007/s12035-012-8307-4.
54. Leuner K, Schulz K, Schütt T, et al. Peripheral mitochondrial dysfunction in Alzheimer's disease: focus on lymphocytes. *Mol Neurobiol.* 2012;46(1):194–204. http://dx.doi.org/10.1007/s12035-012-8300-y.
55. Schuessel K, Schäfer S, Bayer TA, et al. Impaired Cu/Zn-SOD activity contributes to increased oxidative damage in APP transgenic mice. *Neurobiol Dis.* 2005;18(1):89–99. http://dx.doi.org/10.1016/j.nbd.2004.09.003.
56. Eckert A, Hauptmann S, Scherping I, et al. Oligomeric and fibrillar species of beta-amyloid (A beta 42) both impair mitochondrial function in P301L tau transgenic mice. *J Mol Med (Berl).* 2008;86(11):1255–1267. http://dx.doi.org/10.1007/s00109-008-0391-6.
57. Rhein V, Baysang G, Rao S, et al. Amyloid-beta leads to impaired cellular respiration, energy production and mitochondrial electron chain complex activities in human neuroblastoma cells. *Cell Mol Neurobiol.* 2009;29(6–7):1063–1071. http://dx.doi.org/10.1007/s10571-009-9398-y.
58. Bobba A, Amadoro G, Valenti D, Corsetti V, Lassandro R, Atlante A. Mitochondrial respiratory chain complexes I and IV are impaired by β-amyloid via direct interaction and through complex I-dependent ROS production, respectively. *Mitochondrion.* 2013;13(4):298–311. http://dx.doi.org/10.1016/j.mito.2013.03.008.
59. Hansson Petersen CA, Alikhani N, Behbahani H, et al. The amyloid beta-peptide is imported into mitochondria via the TOM import machinery and localized to mitochondrial cristae. *Proc Natl Acad Sci U S A.* 2008;105(35):13145–13150. http://dx.doi.org/10.1073/pnas.0806192105.
60. Roses AD, Lutz MW, Amrine-Madsen H, et al. A TOMM40 variable-length polymorphism predicts the age of late-onset Alzheimer's disease. *Pharmacogenomics J.* 2010;10(5):375–384. http://dx.doi.org/10.1038/tpj.2009.69.
61. Roses AD, Lutz MW, Crenshaw DG, Grossman I, Saunders AM, Gottschalk WK. TOMM40 and APOE: requirements for replication studies of association with age of disease onset and enrichment of a clinical trial. *Alzheimers Dement.* 2013;9(2):132–136. http://dx.doi.org/10.1016/j.jalz.2012.10.009.
62. Devi L, Prabhu BM, Galati DF, Avadhani NG, Anandatheerthavarada HK. Accumulation of amyloid precursor protein in the mitochondrial import channels of human Alzheimer's disease brain is associated with mitochondrial dysfunction. *J Neurosci.* 2006;26(35):9057–9068. http://dx.doi.org/10.1523/JNEUROSCI.1469-06.2006.
63. Pavlov PF, Wiehager B, Sakai J, et al. Mitochondrial γ-secretase participates in the metabolism of mitochondria-associated amyloid precursor protein. *FASEB J.* 2011;25(1):78–88. http://dx.doi.org/10.1096/fj.10-157230.

64. Casley CS, Canevari L, Land JM, Clark JB, Sharpe MA. Beta-amyloid inhibits integrated mitochondrial respiration and key enzyme activities. *J Neurochem.* 2002;80(1):91–100. Available at, http://www.ncbi.nlm.nih.gov/pubmed/11796747, Accessed 27.03.14.
65. Khalifat N, Puff N, Dliaa M, Angelova MI. Amyloid-β and the failure to form mitochondrial cristae: a biomimetic study involving artificial membranes. *J Alzheimers Dis.* 2012;28(1):33–48. http://dx.doi.org/10.3233/JAD-2011-110389.
66. Santos RX, Correia SC, Wang X, et al. A synergistic dysfunction of mitochondrial fission/fusion dynamics and mitophagy in Alzheimer's disease. *J Alzheimers Dis.* 2010; 20(Suppl 2):S401–S412. http://dx.doi.org/10.3233/JAD-2010-100666.
67. Manczak M, Reddy PH. Abnormal interaction between the mitochondrial fission protein Drp1 and hyperphosphorylated tau in Alzheimer's disease neurons: implications for mitochondrial dysfunction and neuronal damage. *Hum Mol Genet.* 2012; 21(11):2538–2547. http://dx.doi.org/10.1093/hmg/dds072.
68. Cho B, Choi SY, Cho HM, Kim HJ, Sun W. Physiological and pathological significance of mitochondrial fission in the nervous system. *Exp Neurobiol.* 2013;22(3):149–157.
69. Wang X, Su B, Lee H, et al. Impaired balance of mitochondrial fission and fusion in Alzheimer's disease. *J Neurosci.* 2009;29(28):9090–9103. http://dx.doi.org/10.1523/JNEUROSCI.1357-09.2009.
70. Cho D-H, Nakamura T, Fang J, et al. S-nitrosylation of Drp1 mediates beta-amyloid-related mitochondrial fission and neuronal injury. *Science.* 2009;324(5923):102–105. http://dx.doi.org/10.1126/science.1171091.
71. Bossy B, Petrilli A, Klinglmayr E, et al. S-nitrosylation of DRP1 does not affect enzymatic activity and is not specific to Alzheimer's disease. *J Alzheimers Dis.* 2010;20(Suppl 2):S513–S526. http://dx.doi.org/10.3233/JAD-2010-100552.
72. Jendrach M, Mai S, Pohl S, Vöth M, Bereiter-Hahn J. Short- and long-term alterations of mitochondrial morphology, dynamics and mtDNA after transient oxidative stress. *Mitochondrion.* 2008;8(4):293–304. http://dx.doi.org/10.1016/j.mito.2008.06.001.
73. Riemer J, Kins S. Axonal transport and mitochondrial dysfunction in Alzheimer's disease. *Neurodegener Dis.* 2013;12(3):111–124. http://dx.doi.org/10.1159/000342020.
74. Du H, Yan SS. Mitochondrial permeability transition pore in Alzheimer's disease: cyclophilin D and amyloid beta. *Biochim Biophys Acta.* 2010;1802(1):198–204. http://dx.doi.org/10.1016/j.bbadis.2009.07.005.
75. Gazaryan IG, Brown AM. Intersection between mitochondrial permeability pores and mitochondrial fusion/fission. *Neurochem Res.* 2007;32(4–5):917–929. http://dx.doi.org/10.1007/s11064-006-9252-2.
76. Piquereau J, Caffin F, Novotova M, et al. Down-regulation of OPA1 alters mouse mitochondrial morphology, PTP function, and cardiac adaptation to pressure overload. *Cardiovasc Res.* 2012;94(3):408–417. http://dx.doi.org/10.1093/cvr/cvs117.
77. Halestrap AP. What is the mitochondrial permeability transition pore? *J Mol Cell Cardiol* 2009;46(6):821–831. http://dx.doi.org/10.1016/j.yjmcc.2009.02.021.
78. Hou Y, Ghosh P, Wan R, et al. Permeability transition pore-mediated mitochondrial superoxide flashes mediate an early inhibitory effect of amyloid beta1-42 on neural progenitor cell proliferation. *Neurobiol Aging.* 2014;35(5):975–989. http://dx.doi.org/10.1016/j.neurobiolaging.2013.11.002.
79. Du H, Guo L, Fang F, et al. Cyclophilin D deficiency attenuates mitochondrial and neuronal perturbation and ameliorates learning and memory in Alzheimer's disease. *Nat Med.* 2008;14(10):1097–1105. http://dx.doi.org/10.1038/nm.1868.
80. Lustbader JW, Cirilli M, Lin C, et al. ABAD directly links Abeta to mitochondrial toxicity in Alzheimer's disease. *Science.* 2004;304(5669):448–452. http://dx.doi.org/10.1126/science.1091230.

81. Yao J, Du H, Yan S, et al. Inhibition of amyloid-beta (Abeta) peptide-binding alcohol dehydrogenase-Abeta interaction reduces Abeta accumulation and improves mitochondrial function in a mouse model of Alzheimer's disease. *J Neurosci.* 2011;31(6):2313–2320. http://dx.doi.org/10.1523/JNEUROSCI.4717-10.2011.
82. Borger E, Aitken L, Du H, Zhang W, Gunn-Moore FJ, Du Yan SS. Is amyloid binding alcohol dehydrogenase a drug target for treating Alzheimer's disease? *Curr Alzheimer Res.* 2013;10(1):21–29. Available at, http://www.pubmedcentral.nih.gov/articlerender.fcgi?artid=3674883&tool=pmcentrez&rendertype=abstract, Accessed 13.05.14.
83. Selkoe D, Mandelkow E, Holtzman D. Deciphering Alzheimer disease. *Cold Spring Harb Perspect Med.* 2012;2(1):a011460. http://dx.doi.org/10.1101/cshperspect.a011460.
84. Mandelkow E-M, Mandelkow E. Biochemistry and cell biology of tau protein in neurofibrillary degeneration. *Cold Spring Harb Perspect Med.* 2012;2(7):a006247. http://dx.doi.org/10.1101/cshperspect.a006247.
85. DuBoff B, Feany M, Götz J. Why size matters—balancing mitochondrial dynamics in Alzheimer's disease. *Trends Neurosci.* 2013;36(6):325–335. http://dx.doi.org/10.1016/j.tins.2013.03.002.
86. David DC, Hauptmann S, Scherping I, et al. Proteomic and functional analyses reveal a mitochondrial dysfunction in P301L tau transgenic mice. *J Biol Chem.* 2005;280(25):23802–23814. http://dx.doi.org/10.1074/jbc.M500356200.
87. Yao J, Irwin RW, Zhao L, Nilsen J, Hamilton RT, Brinton RD. Mitochondrial bioenergetic deficit precedes Alzheimer's pathology in female mouse model of Alzheimer's disease. *Proc Natl Acad Sci U S A.* 2009;106(34):14670–14675. http://dx.doi.org/10.1073/pnas.0903563106.
88. Rhein V, Song X, Wiesner A, et al. Amyloid-beta and tau synergistically impair the oxidative phosphorylation system in triple transgenic Alzheimer's disease mice. *Proc Natl Acad Sci U S A.* 2009;106(47):20057–20062. http://dx.doi.org/10.1073/pnas.0905529106.
89. DuBoff B, Götz J, Feany MB. Tau promotes neurodegeneration via DRP1 mislocalization in vivo. *Neuron.* 2012;75(4):618–632. http://dx.doi.org/10.1016/j.neuron.2012.06.026.
90. Schulz KL, Eckert A, Rhein V, et al. A new link to mitochondrial impairment in tauopathies. *Mol Neurobiol.* 2012;46(1):205–216. http://dx.doi.org/10.1007/s12035-012-8308-3.
91. Iijima-Ando K, Sekiya M, Maruko-Otake A, et al. Loss of axonal mitochondria promotes tau-mediated neurodegeneration and Alzheimer's disease-related tau phosphorylation via PAR-1. *PLoS Genet.* 2012;8(8):e1002918. http://dx.doi.org/10.1371/journal.pgen.1002918.
92. Reddy PH. Abnormal tau, mitochondrial dysfunction, impaired axonal transport of mitochondria, and synaptic deprivation in Alzheimer's disease. *Brain Res.* 2011;1415:136–148. http://dx.doi.org/10.1016/j.brainres.2011.07.052.
93. Tamagno E, Guglielmotto M, Monteleone D, Tabaton M. Amyloid-β production: major link between oxidative stress and BACE1. *Neurotox Res.* 2012;22(3):208–219. http://dx.doi.org/10.1007/s12640-011-9283-6.
94. Guglielmotto M, Aragno M, Autelli R, et al. The up-regulation of BACE1 mediated by hypoxia and ischemic injury: role of oxidative stress and HIF1alpha. *J Neurochem.* 2009;108(4):1045–1056. http://dx.doi.org/10.1111/j.1471-4159.2008.05858.x.
95. Chen L, Yoo S-E, Na R, Liu Y, Ran Q. Cognitive impairment and increased Aβ levels induced by paraquat exposure are attenuated by enhanced removal of mitochondrial H(2)O(2). *Neurobiol Aging.* 2012;33(2):432.e15–432.e26. http://dx.doi.org/10.1016/j.neurobiolaging.2011.01.008.

96. Morley JE, Armbrecht HJ, Farr SA, Kumar VB. The senescence accelerated mouse (SAMP8) as a model for oxidative stress and Alzheimer's disease. *Biochim Biophys Acta* 2012;1822(5):650–656. http://dx.doi.org/10.1016/j.bbadis.2011.11.015.
97. Orsucci D, Mancuso M, Ienco EC, Simoncini C, Siciliano G, Bonuccelli U. Vascular factors and mitochondrial dysfunction: a central role in the pathogenesis of Alzheimer's disease. *Curr Neurovasc Res.* 2013;10(1):76–80. Available at, http://www.ncbi.nlm.nih.gov/pubmed/23151073, Accessed 21.03.14.
98. Shiota S, Takekawa H, Matsumoto S-E, et al. Chronic intermittent hypoxia/reoxygenation facilitate amyloid-β generation in mice. *J Alzheimers Dis.* 2013;37(2):325–333. http://dx.doi.org/10.3233/JAD-130419.
99. Gao L, Tian S, Gao H, Xu Y. Hypoxia increases Aβ-induced tau phosphorylation by calpain and promotes behavioral consequences in AD transgenic mice. *J Mol Neurosci.* 2013;51(1):138–147. http://dx.doi.org/10.1007/s12031-013-9966-y.
100. Jack CR, Knopman DS, Jagust WJ, et al. Hypothetical model of dynamic biomarkers of the Alzheimer's pathological cascade. *Lancet Neurol.* 2010;9(1):119–128. http://dx.doi.org/10.1016/S1474-4422(09)70299-6.
101. Reitz C, Brayne C, Mayeux R. Epidemiology of Alzheimer disease. *Nat Rev Neurol.* 2011;7(3):137–152. http://dx.doi.org/10.1038/nrneurol.2011.2.
102. Hellström-Lindahl E, Ravid R, Nordberg A. Age-dependent decline of neprilysin in Alzheimer's disease and normal brain: inverse correlation with A beta levels. *Neurobiol Aging.* 2008;29(2):210–221. http://dx.doi.org/10.1016/j.neurobiolaging.2006.10.010.
103. Hellström-Lindahl E, Viitanen M, Marutle A. Comparison of Abeta levels in the brain of familial and sporadic Alzheimer's disease. *Neurochem Int.* 2009;55(4):243–252. http://dx.doi.org/10.1016/j.neuint.2009.03.007.
104. Morishima-Kawashima M, Oshima N, Ogata H, et al. Effect of apolipoprotein E allele epsilon4 on the initial phase of amyloid beta-protein accumulation in the human brain. *Am J Pathol.* 2000;157(6):2093–2099.
105. Funato H, Yoshimura M, Kusui K, et al. Quantitation of amyloid beta-protein (A beta) in the cortex during aging and in Alzheimer's disease. *Am J Pathol.* 1998;152(6):1633–1640.
106. Lesné SE. Breaking the code of amyloid-β oligomers. *Int J Cell Biol.* 2013;2013:950783. http://dx.doi.org/10.1155/2013/950783.
107. Silverberg GD, Miller MC, Messier AA, et al. Amyloid deposition and influx transporter expression at the blood–brain barrier increase in normal aging. *J Neuropathol Exp Neurol.* 2010;69(1):98–108. http://dx.doi.org/10.1097/NEN.0b013e3181c8ad2f.
108. Chiu C, Miller MC, Caralopoulos IN, et al. Temporal course of cerebrospinal fluid dynamics and amyloid accumulation in the aging rat brain from three to thirty months. *Fluids Barriers CNS.* 2012;9(1):3. http://dx.doi.org/10.1186/2045-8118-9-3.
109. Keil U, Bonert A, Marques CA, et al. Amyloid beta-induced changes in nitric oxide production and mitochondrial activity lead to apoptosis. *J Biol Chem.* 2004;279(48):50310–50320. http://dx.doi.org/10.1074/jbc.M405600200.
110. Schmitt K, Grimm A, Kazmierczak A, Strosznajder JB, Götz J, Eckert A. Insights into mitochondrial dysfunction: aging, amyloid-β, and tau-A deleterious trio. *Antioxid Redox Signal.* 2012;16(12):1456–1466. http://dx.doi.org/10.1089/ars.2011.4400.
111. Eckmann J, Eckert SH, Leuner K, Muller WE, Eckert GP. Mitochondria: mitochondrial membranes in brain ageing and neurodegeneration. *Int J Biochem Cell Biol.* 2013;45(1):76–80. http://dx.doi.org/10.1016/j.biocel.2012.06.009.
112. Stockburger C, Kurz C, Koch KA, Eckert SH, Leuner K, Müller WE. Improvement of mitochondrial function and dynamics by the metabolic enhancer piracetam. *Biochem Soc Trans.* 2013;41(5):1331–1334. http://dx.doi.org/10.1042/BST20130054.

113. Cha M-Y, Han S-H, Son SM, et al. Mitochondria-specific accumulation of amyloid β induces mitochondrial dysfunction leading to apoptotic cell death. *PLoS One*. 2012;7(4):e34929. http://dx.doi.org/10.1371/journal.pone.0034929.
114. Manczak M, Mao P, Calkins MJ, et al. Mitochondria-targeted antioxidants protect against amyloid-beta toxicity in Alzheimer's disease neurons. *J Alzheimers Dis*. 2010;-20(Suppl. 2):S609–S631. http://dx.doi.org/10.3233/JAD-2010-100564.
115. Baloyannis SJ, Costa V, Michmizos D. Mitochondrial alterations in Alzheimer's disease. *Am J Alzheimers Dis Other Demen*. 2004;19(2):89–93.
116. Bu G. Apolipoprotein E, and its receptors in Alzheimer's disease: pathways, pathogenesis and therapy. *Nat Rev Neurosci*. 2009;10(5):333–344. http://dx.doi.org/10.1038/nrn2620.
117. Liu C-C, Liu C-C, Kanekiyo T, Xu H, Bu G. Apolipoprotein E and Alzheimer disease: risk, mechanisms and therapy. *Nat Rev Neurol*. 2013;9(2):106–118. http://dx.doi.org/10.1038/nrneurol.2012.263.
118. Chen H-K, Ji Z-S, Dodson SE, et al. Apolipoprotein E4 domain interaction mediates detrimental effects on mitochondria and is a potential therapeutic target for Alzheimer disease. *J Biol Chem*. 2011;286(7):5215–5221. http://dx.doi.org/10.1074/jbc.M110.151084.
119. Nakamura T, Watanabe A, Fujino T, Hosono T, Michikawa M. Apolipoprotein E4 (1-272) fragment is associated with mitochondrial proteins and affects mitochondrial function in neuronal cells. *Mol Neurodegener*. 2009;4:35. http://dx.doi.org/10.1186/1750-1326-4-35.
120. Fang Ng L, Gruber J, Cheah I, et al. The mitochondria-targeted antioxidant MitoQ extends life span and improves health span of a transgenic *Caenorhabditis elegans* model of Alzheimer disease. *Free Radic Biol Med*. 2014;71C:390–401. http://dx.doi.org/10.1016/j.freeradbiomed.2014.03.003.
121. Mao P, Manczak M, Shirendeb UP, Reddy PH. MitoQ, a mitochondria-targeted antioxidant, delays disease progression and alleviates pathogenesis in an experimental autoimmune encephalomyelitis mouse model of multiple sclerosis. *Biochim Biophys Acta*. 2013;1832(12):2322–2331. http://dx.doi.org/10.1016/j.bbadis.2013.09.005.
122. McManus MJ, Murphy MP, Franklin JL. The mitochondria-targeted antioxidant MitoQ prevents loss of spatial memory retention and early neuropathology in a transgenic mouse model of Alzheimer's disease. *J Neurosci*. 2011;31(44):15703–15715. http://dx.doi.org/10.1523/JNEUROSCI.0552-11.2011.
123. Dumont M, Kipiani K, Yu F, et al. Coenzyme Q10 decreases amyloid pathology and improves behavior in a transgenic mouse model of Alzheimer's disease. *J Alzheimers Dis*. 2011;27(1):211–223. http://dx.doi.org/10.3233/JAD-2011-110209.
124. Choi H, Park H-H, Lee K-Y, et al. Coenzyme Q10 restores amyloid beta-inhibited proliferation of neural stem cells by activating the PI3K pathway. *Stem Cells Dev*. 2013;22(15):2112–2120. http://dx.doi.org/10.1089/scd.2012.0604.
125. Galasko DR, Peskind E, Clark CM, et al. Antioxidants for Alzheimer disease: a randomized clinical trial with cerebrospinal fluid biomarker measures. *Arch Neurol*. 2012;69(7):836–841. http://dx.doi.org/10.1001/archneurol.2012.85.
126. Sastre J, Lloret A, Borrás C, et al. Ginkgo biloba extract EGb 761 protects against mitochondrial aging in the brain and in the liver. *Cell Mol Biol (Noisy-le-Grand)*. 2002;48(6):685–692. Available at, http://www.ncbi.nlm.nih.gov/pubmed/12396080, Accessed 21.03.14.
127. Abdel-Kader R, Hauptmann S, Keil U, et al. Stabilization of mitochondrial function by Ginkgo biloba extract (EGb 761). *Pharmacol Res*. 2007;56(6):493–502. http://dx.doi.org/10.1016/j.phrs.2007.09.011.
128. Rhein V, Giese M, Baysang G, et al. Ginkgo biloba extract ameliorates oxidative phosphorylation performance and rescues abeta-induced failure. *PLoS One*. 2010;5(8):e12359. http://dx.doi.org/10.1371/journal.pone.0012359.

129. Luo Y. Alzheimer's disease, the nematode Caenorhabditis elegans, and ginkgo biloba leaf extract. *Life Sci.* 2006;78(18):2066–2072. http://dx.doi.org/10.1016/j.lfs.2005.12.004.
130. Smith JV, Luo Y. Elevation of oxidative free radicals in Alzheimer's disease models can be attenuated by Ginkgo biloba extract EGb 761. *J Alzheimers Dis.* 2003;5(4):287–300. Available at, http://www.ncbi.nlm.nih.gov/pubmed/14624024, Accessed 28.03.14.
131. Kellermann AJ, Kloft C. Is there a risk of bleeding associated with standardized Ginkgo biloba extract therapy? A systematic review and meta-analysis. *Pharmacotherapy.* 2011;31(5):490–502. http://dx.doi.org/10.1592/phco.31.5.490.
132. Wu Y, Wu Z, Butko P, et al. Amyloid-beta-induced pathological behaviors are suppressed by Ginkgo biloba extract EGb 761 and ginkgolides in transgenic *Caenorhabditis elegans*. *J Neurosci.* 2006;26(50):13102–13113. http://dx.doi.org/10.1523/JNEUROSCI.3448-06.2006.
133. Weinmann S, Roll S, Schwarzbach C, Vauth C, Willich SN. Effects of Ginkgo biloba in dementia: systematic review and meta-analysis. *BMC Geriatr.* 2010;10:14. http://dx.doi.org/10.1186/1471-2318-10-14.
134. Kurz C, Ungerer I, Lipka U, et al. The metabolic enhancer piracetam ameliorates the impairment of mitochondrial function and neurite outgrowth induced by beta-amyloid peptide. *Br J Pharmacol.* 2010;160(2):246–257. http://dx.doi.org/10.1111/j.1476-5381.2010.00656.x.
135. Keil U, Scherping I, Hauptmann S, Schuessel K, Eckert A, Müller WE. Piracetam improves mitochondrial dysfunction following oxidative stress. *Br J Pharmacol.* 2006;147(2):199–208. http://dx.doi.org/10.1038/sj.bjp.0706459.
136. Waegemans T, Wilsher CR, Danniau A, Ferris SH, Kurz A, Winblad B. Clinical efficacy of piracetam in cognitive impairment: a meta-analysis. *Dement Geriatr Cogn Disord.* 2002;13(4):217–224. http://dx.doi.org/10.1159/000057700.
137. Steele JW, Lachenmayer ML, Ju S, et al. Latrepirdine improves cognition and arrests progression of neuropathology in an Alzheimer's mouse model. *Mol Psychiatry.* 2013;18(8):889–897. http://dx.doi.org/10.1038/mp.2012.106.
138. Peters OM, Shelkovnikova T, Tarasova T, et al. Chronic administration of Dimebon does not ameliorate amyloid-β pathology in 5xFAD transgenic mice. *J Alzheimers Dis.* 2013;36(3):589–596. http://dx.doi.org/10.3233/JAD-130071.
139. Cano-Cuenca N, Solís-García del Pozo JE, Jordán J. Evidence for the efficacy of latrepirdine (Dimebon) treatment for improvement of cognitive function: a meta-analysis. *J Alzheimers Dis.* 2014;38(1):155–164.
140. Eckert SH, Eckmann J, Renner K, Eckert GP, Leuner K, Muller WE. Dimebon ameliorates amyloid-β induced impairments of mitochondrial form and function. *J Alzheimers Dis.* 2012;31(1):21–32. http://dx.doi.org/10.3233/JAD-2012-120310.
141. Peters OM, Connor-Robson N, Sokolov VB, et al. Chronic administration of dimebon ameliorates pathology in TauP301S transgenic mice. *J Alzheimers Dis.* 2013;33(4):1041–1049. http://dx.doi.org/10.3233/JAD-2012-121732.
142. Doody RS, Gavrilova SI, Sano M, et al. Effect of dimebon on cognition, activities of daily living, behaviour, and global function in patients with mild-to-moderate Alzheimer's disease: a randomised, double-blind, placebo-controlled study. *Lancet.* 2008;372(9634):207–215. http://dx.doi.org/10.1016/S0140-6736(08)61074-0.
143. Bezprozvanny I. The rise and fall of Dimebon. *Drug News Perspect.* 2010;23(8):518–523. http://dx.doi.org/10.1358/dnp.2010.23.8.1500435.
144. Farina N, Isaac MGEKN, Clark AR, Rusted J, Tabet N. Vitamin E for Alzheimer's dementia and mild cognitive impairment. *Cochrane Database Syst Rev.* 2012;11: CD002854. http://dx.doi.org/10.1002/14651858.CD002854.pub3.
145. Williams RJ, Spencer JPE. Flavonoids, cognition, and dementia: actions, mechanisms, and potential therapeutic utility for Alzheimer disease. *Free Radic Biol Med.* 2012;52(1):35–45. http://dx.doi.org/10.1016/j.freeradbiomed.2011.09.010.

146. Choi D-Y, Lee Y-J, Hong JT, Lee H-J. Antioxidant properties of natural polyphenols and their therapeutic potentials for Alzheimer's disease. *Brain Res Bull.* 2012;87(2–3): 144–153. http://dx.doi.org/10.1016/j.brainresbull.2011.11.014.
147. Darvesh AS, Carroll RT, Bishayee A, Geldenhuys WJ, Van der Schyf CJ. Oxidative stress and Alzheimer's disease: dietary polyphenols as potential therapeutic agents. *Expert Rev Neurother.* 2010;10(5):729–745. http://dx.doi.org/10.1586/ern.10.42.
148. Scarmeas N. Mediterranean food for thought? *J Neurol Neurosurg Psychiatry.* 2013;84(12):1297. http://dx.doi.org/10.1136/jnnp-2013-305153.
149. Eckert GP, Renner K, Eckert SH, et al. Mitochondrial dysfunction—a pharmacological target in Alzheimer's disease. *Mol Neurobiol.* 2012;46(1):136–150. http://dx.doi.org/10.1007/s12035-012-8271-z.
150. Hofer A, Noe N, Tischner C, et al. Defining the action spectrum of potential PGC-1α activators on a mitochondrial and cellular level in vivo. *Hum Mol Genet.* 2014;23(9): 2400–2415. http://dx.doi.org/10.1093/hmg/ddt631.
151. Jat D, Parihar P, Kothari SC, Parihar MS. Curcumin reduces oxidative damage by increasing reduced glutathione and preventing membrane permeability transition in isolated brain mitochondria. *Cell Mol Biol (Noisy-le-Grand).* 2013;59(Suppl): OL1899–OL1905. Available at, http://www.ncbi.nlm.nih.gov/pubmed/24461330, Accessed 13.05.14.
152. Lee W-H, Loo C-Y, Bebawy M, Luk F, Mason RS, Rohanizadeh R. Curcumin and its derivatives: their application in neuropharmacology and neuroscience in the 21st century. *Curr Neuropharmacol.* 2013;11(4):338–378. http://dx.doi.org/10.2174/1570159X11311040002.
153. Kim HV, Kim HY, Ehrlich HY, Choi SY, Kim DJ, Kim Y. Amelioration of Alzheimer's disease by neuroprotective effect of sulforaphane in animal model. *Amyloid.* 2013;20(1):7–12. http://dx.doi.org/10.3109/13506129.2012.751367.
154. Eckert A, Keil U, Scherping I, Hauptmann S, Müller WE. Stabilization of mitochondrial membrane potential and improvement of neuronal energy metabolism by Ginkgo biloba extract EGb 761. *Ann N Y Acad Sci.* 2005;1056:474–485. http://dx.doi.org/10.1196/annals.1352.023.
155. Schindowski K, Leutner S, Kressmann S, Eckert A, Müller WE. Age-related increase of oxidative stress-induced apoptosis in mice prevention by Ginkgo biloba extract (EGb761). *J Neural Transm.* 2001;108(8–9):969–978. Available at, http://www.ncbi.nlm.nih.gov/pubmed/11716149, Accessed 21.03.14.
156. Leuner K, Kurz C, Guidetti G, Orgogozo JM, Müller WE. Improved mitochondrial function in brain aging and Alzheimer disease—the new mechanism of action of the old metabolic enhancer piracetam. *Front Neurosci.* 2010;4(44):1–11. http://dx.doi.org/10.3389/fnins.2010.00044.
157. Perez SE, Nadeem M, Sadleir KR, et al. Dimebon alters hippocampal amyloid pathology in 3xTg-AD mice. *Int J Physiol Pathophysiol Pharmacol.* 2012;4(3):115–127. Available at, http://www.pubmedcentral.nih.gov/articlerender.fcgi?artid=3466490&tool=pmcentrez&rendertype=abstract, Accessed 31.03.14.
158. Naga KK, Geddes JW. Dimebon inhibits calcium-induced swelling of rat brain mitochondria but does not alter calcium retention or cytochrome *c* release. *Neuromolecular Med.* 2011;13(1):31–36. http://dx.doi.org/10.1007/s12017-010-8130-x.
159. Shevtsova EF, Vinogradova DV, Kireeva EG, Reddy VP, Aliev G, Bachurin SO. Dimebon attenuates the Aβ-induced mitochondrial permeabilization. *Curr Alzheimer Res.* 2014; Available at, http://www.ncbi.nlm.nih.gov/pubmed/24801220, Accessed 12.05.14.
160. James R, Searcy JL, Le Bihan T, et al. Proteomic analysis of mitochondria in APOE transgenic mice and in response to an ischemic challenge. *J Cereb Blood Flow Metab.* 2012;32(1):164–176. http://dx.doi.org/10.1038/jcbfm.2011.120.

CHAPTER EIGHT

Mitochondria in Cancer: Why Mitochondria Are a Good Target for Cancer Therapy

Lan-Feng Dong*, Jiri Neuzil*,†
*School of Medical Science, Griffith University, Southport, Queensland, Australia
†Institute of Biotechnology, Academy of Sciences of the Czech Republic, Prague, Czech Republic

Contents

Abstract

Cancer can be characterized as a state of multifaceted cellular deregulation including control of proliferation and bioenergetics. The latter involves in particular mitochondria, the site of the generation of ATP, essential for the proper cellular function (including proliferation). Mitochondria also contain a variety of proteins that are necessary for the induction/promotion, as well as for the prevention of cell death. Therefore, mitochondria are pivotal in deciding the fate of a cell. In cancer, mitochondria are dysfunctional, which was observed as early as in the 1930s by Otto Warburg. Due to the central role of mitochondria, these organelles, endowed with its own DNA, are a focus of research as possible "culprits" for the malignancy of cancer cells (or at least contributing to this phenotype) and, importantly, as emerging targets for anticancer therapy.

1. MITOCHONDRIA IN MALIGNANT CELLS—CULPRITS OR VICTIMS?

In the 1930s, the visionary discovery by Otto Warburg brought about the notion that malignant cells alter their metabolic profile compared to nonmalignant cells toward increased glycolysis and glucose consumption, which was described in more detail in two landmark *Science* papers[1,2];

Progress in Molecular Biology and Translational Science, Volume 127
ISSN 1877-1173
http://dx.doi.org/10.1016/B978-0-12-394625-6.00008-8

this paradigm is now known after its discoverer as the Warburg effect.[3,4] The importance of this discovery was of such as caliber that Warburg was awarded for it the Nobel Prize. Owing to the fact that mitochondria considerably differ in malignant cells from those in their nonmalignant counterparts, there has been intensive research attempting to clearly delineate the role of mitochondria in cancer: how these organelles contribute to the malignant conversion and, as well, how they amplify the malignant status governed, primarily, by mutations in the nuclear genes.

Mitochondria are organelles that are unique in their structure and function. They are defined by the mitochondrial outer membrane (MOM) with the underlying intermembrane space limited on the other side by the mitochondrial inner membrane (MIM) that is continuous with the cristae. Contained by the MIM is the mitochondrial matrix. Each of the mitochondrial components has its distinct and complementary role(s). A unique characteristics of the mitochondrion is that it, besides the nucleus as the only organelle in a mammalian cell, carries its genetic information (at least a fragment of it), comprised in mitochondrial DNA (mtDNA). Interestingly, every cell, depending on its type, contains a number of mitochondria (some 100 or more), each with multiple copies of mtDNA.

Mammalian mitochondrial genome is circular, with slightly over 16.5 kb in size. It codes for 13 mRNAs (all subunits of oxidative phosphorylation, OXPHOS), 2 rRNAs, and 22 tRNAs.[5] The mitochondrial proteome comprises some 1500 proteins, of which vast majority is coded for by the nucleus, translated on free ribosomes and imported into the proper compartment of the mitochondrion, often in the final orientation.[6] This process requires a precisely regulated cross talk between mitochondria and the nucleus[7–9] as well as coordinated regulation of expression of a whole suite of genes, controlled in a concerted manner by the transcription factors of NRF1 and NRF2, and by coactivators like PGC1α/β.[10] Of particular importance is the mitochondrial transcription factor TFAM encoded by the nucleus. This protein, together with several other mitochondria-regulatory proteins (e.g., DNA polymerase γ, the helicase TWINKEL, etc.), binds to the D-LOOP sequence of mtDNA.[11,12] TFAM not only acts as the main transcription factor for mtDNA but also regulates its replication, and it is essential for packaging of mtDNA into spheroid nucleoids tethered to the MIM.[12] TFAM has been also found to coat mtDNA, offering it a certain level of protection (possibly similar as histones do in the case of the nuclear genome), and to regulate its unwinding.[13–16]

Since mitochondria contain products of DNA-encoded genes that are essential for the proper function of OXPHOS, it is essential that the mitochondrial genome is "intact," i.e., mitochondria can produce ATP via OXPHOS to meet the bioenergetic demands of the cancer cells. Although it is now recognized that mtDNA is not "naked" being coated by TFAM, it is still prone to mutations, perhaps due to the fact that it does not contain noncoding regions and because mitochondria are a source of reactive oxygen species (ROS).[17] ROS are formed in particular at the level of the mitochondrial complexes CI and CIII[17] as well as, as documented recently, CII.[18,19] Certain regions of mtDNA appear to be more prone to mutations, as shown in particular for neurodegenerative diseases and as frequently seen also in cancers, all tissues with high demands for energy.[20–23] The D-LOOP of mtDNA has been referred to as the "hot spot" for mutations.[21,24] It has been shown for breast cancer that majority of mutations occur in the D-LOOP region with additional mutations found in the 16S rRNA, ND2, and ATP6 loci.[25–27] These findings indicate that mutations in the mitochondrial genome may seriously affect the mtDNA copy number (which is known to vary in cancers, see, for example, Ref. 28) as well as the mitochondrial function, although implications of these alterations are not well understood. One reason is that mtDNA is heteroplasmic, and for a mutation to manifest itself in terms of an altered function, more than 60% of mtDNA copies ought to carry such a mutation.[20]

To overcome the problem with heteroplasmy, cancer cells devoid of mtDNA (ρ^0 cells) prepared by various means (EtBr treatment, knockout of DNA polymerase γ), and the derived cybrids (cells with defined mtDNA mutations) are often used in studies aimed at understanding the role of mutated mtDNA in cancer.[29,30] It has been documented that ρ^0 tumor cells show lower capacity to generate tumors,[31,32] indicating that at least some level of OXPHOS is important to maintain the malignant status of cancer cells (it is now clear that more or less all cancer cells rely, to some extent, on OXPHOS[33–35]). We have been recently studying the efficacy of syngeneic tumor formation using the murine metastatic breast cancer cell line 4T1 devoid of mtDNA.[36] Our data show that the 4T1ρ^0 cells formed a tumor with a considerable delay and, very intriguingly, it acquired mtDNA from the host, most likely by intercellular mitochondrial transfer. Once the tumor cells completely recovered their respiratory capacity, they were able to form a tumor without a delay, recovering their malignant phenotype. This restoration of malignancy required a full assembly of mitochondrial complexes, including the respirasome (a supercomplex composed of CI, CIII, and CIV, directly channelling electrons from CI to CIV[37,38]) and CII. These data indicate not only the intercellular transfer of mitochondria *in vivo* as a thus far unknown characteristics of cancer cells[39–41] but also the importance of OXPHOS for tumorigenicity.

The above data support a role of mitochondria and mtDNA in the tumorigenic as well as malignant status of cancer cells, a notion which is further fuelled by several recent compelling findings. Thus, Petros and colleagues documented that mutations in mtDNA further enhanced the tumorigenic status of prostate cancer.[42] Even more intriguingly, it was reported that swapping mtDNA of non-metastatic caner cells with mtDNA from their metastatic counterparts resulted in their acquisition of metastatic potential, while the originally metastatic cells lost this phenotype.[43] These and other findings[44,45] link mutations in mtDNA with the malignant properties of the cells, corroborating the role of mtDNA mutations in tumorigenic and metastatic properties of cells. One more interesting aspect of mitochondria in terms of the malignant phenotype should be mentioned: It has been recently reported that a cross talk between mitochondria and the nucleus (referred to as mitochondrial retrograde signaling) promotes the "conversion" of cancer cells (shown for breast cancer cells) into cancer stem cells.[46] This is particularly intriguing since mitochondria can, thus, be involved in the maintenance of the stem-like cell subpopulation within tumors, which is important for tumor initiation and reinitiation and which causes a particular complication for cancer therapy.[47–49]

Collectively, the aforementioned as well as other findings (that cannot be included for lack of space) clearly point to mitochondria as a culprit of tumorigenicity or malignancy.[50] This having said, it is imperative to stress that the major reason for the malignant status stems from mutations in nuclear DNA (nDNA). It is beyond the scope of this chapter to discuss the various reasons for malignant transformation of cells. It will suffice to say that, in relation to mitochondria, mutations in nDNA will be relayed by mitochondria, such as (and in particular so) in the altered mitochondrial bioenergetics that is positively associated with malignancy.[3,51,52] Neat examples of the effect of mutations in genes coding for mitochondrial proteins are mutations in subunits of CII and in isocitrate dehydrogenase (IDH).[53–55] All four subunits of CII are now recognized as tumor suppressors, since mutations in their genes were found in certain tumors (e.g., paraganglioma and pheochromocytoma).[56] It is possible that these mutations, in particular in subunits SDHB, SDHC, and SDHD, contribute to dysfunctional role of the succinate quinone reductase (SQR) activity of CII being at the cross-roads of the tricarboxylic acid (TCA) cycle and OXPHOS. Mutations in the IDH-1- and IDH-2-coding genes result in the conversion of the normal product of IDH, 2-oxoglutarate, to its reduced form, 2-hydroxyglutarate, that acts as an oncometabolite.[57] Recent research has witnessed the emerging knowledge of the role of microRNA in mitochondrial dysfunction and modulation of malignancy.[58] In many cases, microRNAs act as tumor suppressors, often being "lost" in cancer patients. We found that microRNA-126, selectively suppressed in malignant mesothelioma patients, affects the mitochondria function via a rather complex pathway affecting a range of genes by initially binding the insulin receptor substrate-1 protein.[59]

In many (if not most) instances, mitochondria "amplify" the signal(s) coming from oncogenic mutations in the nucleus, usually yielding a considerable alteration of the cellular bioenergetics.[3,60] Often, such nuclear perturbations are relayed to the change in the cellular function by means of mitochondria. For example, Huang's group used a cellular system with inducible oncogenic K-Ras (G12V mutation).[61] They found that induction of K-Ras resulted within 12 h in a considerable alteration of the mitochondrial parameters (decrease in mitochondrial potential and respiration, and increase in ROS generation), followed by a decrease of expression of CI and increase in the expression of CII components; after 3 days, increased glucose uptake and lactate production were detected. While these mitochondria-related changes were very fast, it took more than additional 1 month before a malignant transformation could be observed.[61,62] This clearly documents that mitochondria, while being themselves direct drivers of malignancy ("culprits"), also act by amplifying the mutations in the nucleus. Thus, the two DNA-bearing organelles, the nucleus and the mitochondrion, are formidable partners who, together, perhaps by amplifying each other's "oncogenic signals," have a very high capacity to cause and propagate a malignant transformation of a cell and maintain its tumorigenic/metastatic status.

2. MITOCHONDRIA AS TARGETS FOR ANTICANCER THERAPY

Cancer is a pathology of unprecedented heterogeneity. It has been shown that tumors of the same types in different patients vary considerably in their genetic profile and mutational signature.[63–65] Even more intriguingly and problematically from the point of view of cancer treatment, the same tumor features different mutational profiles in its individual parts and in the derived metastases, as shown for kidney cancer.[66] This suggests that treatment of tumors may be harder than expected (and there is ample evidence for this) and also provides some explanation as to why many tumor types are refractory to therapy and, for others, a considerable percentage of patients are not treatable and/or, the tumor recurs. Therefore, we have to search for targets that are common to many different types of tumors across the landscape of the neoplastic pathologies and targeting of which will preferentially kill cancer cells.[67] Such a "general" target may be presented by mitochondria, organelles that only recently started to emerge as a versatile anticancer target.[68–70] Recent research has documented that agents targeting mitochondria have a very good potential to be utilized in cancer therapy, although this area of research is only now starting to emerge and is still awaiting its full exploitation.[71–75]

We have recently coined the term "mitocans" standing for *mito*chondria and *can*cer, i.e., defining a vast group of compounds that induce apoptosis by way of destabilizing mitochondria, which translates into anticancer effects.[76,77] To better understand the molecular mechanism of the function of mitocans, we categorized them into eight classes according to their molecular target.[78] This classification is documented in the cartoon in Fig. 8.1. Class 1 comprises agents that act at the surface of the MOM, inhibiting the very important enzyme hexokinase. Class 2 includes small molecules that act as BH3 mimetics jeopardizing the anticancer Bcl-2 family proteins. Classes 3 and 4 comprise agents with redox-inhibitory function and inhibitors of the VDAC and ANT proteins. Of particular interest and a focus of our research are class 5 compounds targeting the electron transport chain (ETC). Class 6 contains lipophilic compounds that target the MIM, class 7 agents acting on the TCA cycle and its "auxiliary" pathways. Finally, class 8 agents act on mtDNA. Examples of members of individual classes are indicated in Fig. 8.1.

As mentioned, our major interest has been in class 5 mitocans, i.e., agents that affect the ETC, the major component of OXPHOS, comprising complexes I–IV.[79] To date, most studies have been undertaken with a focus on agents targeting CII. One of the attractions of CII is that it is at the crossroads of the TCA cycle and the ETC. Figure 8.2 documents the basic components of CII and its position at the intersection of the TCA cycle and OXPHOS, linking the two essential mitochondrial pathways.[80–82] The complex is composed of four subunits, SDHA (F_P), SDHB (I_P), SDHC (Cyb_L), and SDHD (Cyb_S). Its assembly requires two additional assembly

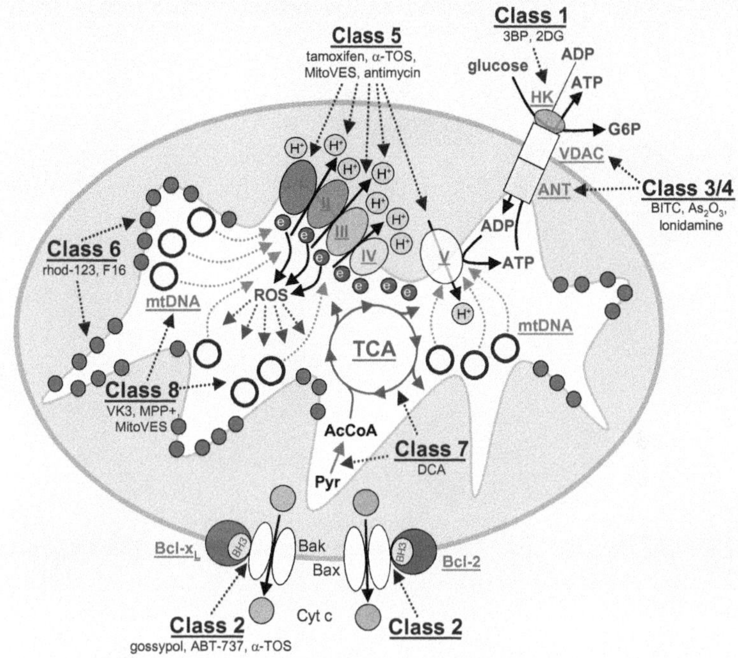

Figure 8.1 Classification of mitocans, agents targeting mitochondria. Individual classes of mitocans are shown, comprising eight groups of compounds with targets in individual compartments of mitochondria. The individual classes are numbered starting from the mitochondrial surface to the mitochondrial matrix. *Adapted from Ref. 78.* (See the color plate.)

factors, SDHAF2 (SDH5) that flavinates SDHA and SDHAF1 involved in the insertion of the [Fe–S] clusters into SDHB. Complete assembly of CII requires functional assembly complexes. For example, it has been shown that a mutation in SDHAF2 results in the lack of flavination of SDHA followed by the degradation of the CII subunit.[83] SDHAF2 was also found mutated in human paraganglioma.[84] The function of CII may go beyond its role in the TCA cycle and OXPHOS. This notion is supported by SDHC that seems to be involved in the formation of the TIM complex and translocation of proteins across the MIM.[85]

Albayrak *et al.* have shown that CII may be involved in apoptosis induced by a variety of agents, including small molecules as well as immunological apoptogens exemplified by the Fas ligand (FasL).[86] The role of CII in apoptosis triggered by the inducers used in this report is, however, unclear. Recently, Sun *et al.* published the crystal structure of a mammalian CII.[87]

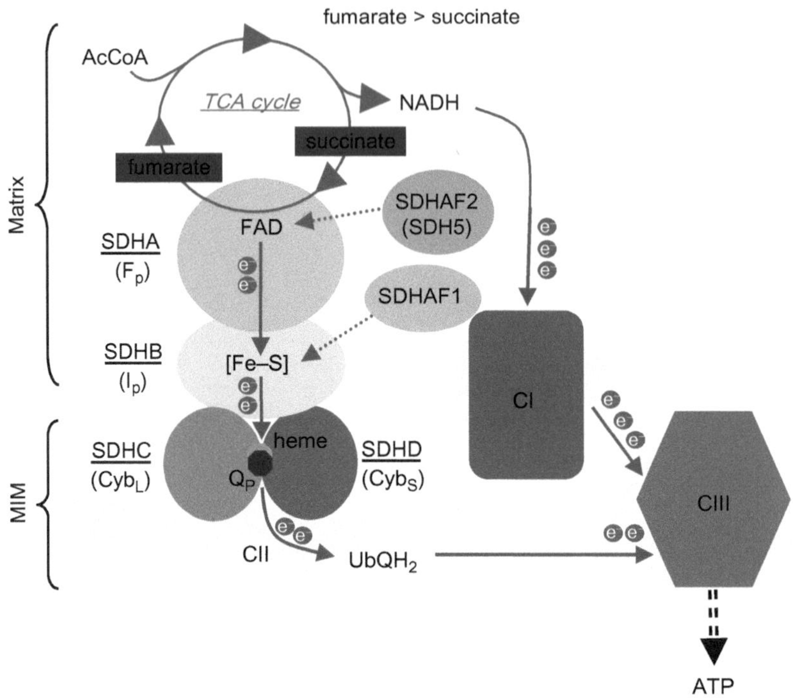

Figure 8.2 Mitochondrial complex II and its targeting with vitamin E analogs. Complex II is composed of four subunits, SDHA (F_p), SDHB (I_p), SDHC (Cyb_L), and SDHD (Cyb_S). All of them are now classified as tumor suppressors. The assembly factors SDHAF2 (flavinates SDHA) and SDHAF1 (inserts [Fe–S] clusters into SDHB) have also been found mutated in certain tumors. The SDH activity of CII converts succinate to fumarate, a constituent of the TCA cycle. The electrons generated during the course of the reaction move toward the MIM portion of CII containing SDHC and SDHD. Both subunits contribute to the ubiquinone (UbQ) binding site and the electron-stabilizing heme site. Two electron reduction of UbQ to UbQH2 results in the departure of the cofactor from CII to donate the two electrons it carries to CIII. Upon losing these electrons, UbQH2 reoxidizes to UbQ that moves back to its site in CII, acting as an electron shuttle between the two complexes. Reduction of UbQ to UbQH2 constitutes the SQR activity of CII. *Adapted from Ref. 78.* (See the color plate.)

This revealed the precise mechanism of the movement of electrons form the SDHA's FAD group via the SDHB's [Fe–S] clusters to the heme and UbQ coordinated by SDHC and SDHD. It also indicated a presence of two UbQ-binding sites, the proximal site (Q_P) and the less well-characterized distal site (Q_D) (proximal and distal with relation to the matrix face of the MIM).

We have been interested for some times in vitamin E (VE) analogs as anticancer agents, epitomized by the ester α-tocopheryl succinate (α-TOS) that selectively kills cancer cells while leaving the normal cells unharmed.[88–90] α-TOS acts by early generation of ROS that then destabilize mitochondria and trigger apoptosis that involves activation of the Mst1 kinase, FoxO1 translocation, and upregulation of the BH3-only protein Noxa that allows for the formation of Bak-dependent pores in the MOM.[91–93] When looking for the molecular target of α-TOS, we tested the mitochondrial complexes with a focus on CII. Using inhibitors (3-bromopyruvate, thenoyltrifluoroacetone), cells with mutated CII and molecular modeling, we found that α-TOS binds to CII, more specifically to the Q_P's S68 (SDHC) and the Q_D's Lys128/135 (SDHD).[94] Importantly, CII was found to be essential for the anticancer activity of α-TOS in a mouse model of cancer.[95] We also documented that α-TOS suppressed angiogenesis *in vitro* and in a tumor model, also acting via CII.[96] The working model of the effect of α-TOS was proposed such that the agent displaces UbQ from its binding sites in CII. Since the agent does not substantially suppress the SDH activity of CII (while strongly inhibiting its SQR activity), the conversion of succinate to fumarate still occurs. However, the electrons released during this reaction cannot be intercepted by UbQ and give rise to superoxide that triggers the activation of the Mst1 kinase activating the apoptotic cascade of reactions.

With the notion of the target for α-TOS being in the MIM, we decided to utilize the principle of mitochondrial targeting published by Murphy and Smith for antioxidants like UbQ.[97] According to this, attachment of a delocalized cation to an aliphatic chain of a compound will direct it to the MIM due to the low potential on its matrix face. Thus, to target UbQ to the MIM, the team of Murphy and Smith tagged the parental compound with a triphenylphosphonium (TPP^+) and showed that, indeed, it exerted its activity "within mitochondria."[98] We adapted this approach to target α-TOS to the MIM: to do this, the agent with a straight aliphatic chain was tagged with TPP^+ at the end of the chain, i.e., containing an 11-C aliphatic chain linking the tocopheryl succinyl moiety and the TPP^+ group.[99,100] It was found that the mitochondrially targeted vitamin E succinate (MitoVES) was some 20–50-fold more apoptogenic than the untargeted α-TOS, maintaining the selectivity for cancer cells.[99,100] Intriguingly, MitoVES also, similarly as α-TOS, suppressed angiogenesis by directly killing proliferating but not growth-arrested endothelial cells, again at a rate much higher than that found for α-TOS.[101] To unequivocally document mitochondrial localization of MitoVES, we tagged it with a fluorophore and, using confocal microscopy, found that overwhelming majority of the agent localized to mitochondria (Fig. 8.3A). The molecular target for MitoVES is the Q_P site of CII within the MIM. This follows from molecular modeling and the finding that shortening the aliphatic chain of MitoVES causes gradual loss of its activity.[99,100] In fact, the Q_P site is a very

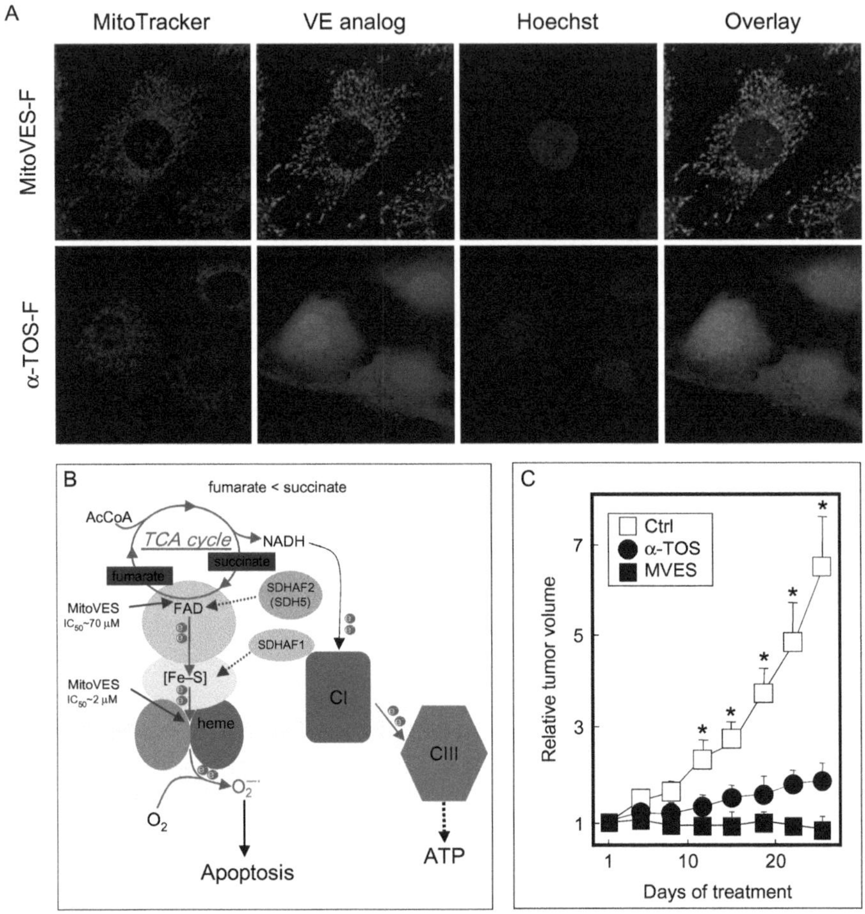

Figure 8.3 MitoVES accumulates in mitochondria and acts via CII. (A) NeuTL cells were incubated with fluorescent probe-tagged α-TOS or MitoVES and co-stained with MitoTracker Red and DAPI. (B) Cartoon of CII indicating the sites of action of MitoVES with the respective approximate IC_{50} values. (C). FVB/N *c-neu* mice with spontaneous breast carcinomas were treated as soon as the tumors appeared with intraperitoneal injection of α-TOS (15 μmol per animal per dose) or MitoVES (1 μmol per animal per dose) twice per week and the tumor volume evaluated using ultrasound imaging. The symbol '*' indicates statistically significant difference with $p<0.05$. *Adapted from Refs. 99 and 100.* (See the color plate.)

advantageous target. It was found that MitoVES suppresses the SDH activity of CII with the IC_{50} of about 70 μM, while the SQR activity was inhibited with the IC_{50} of about 2 μM. While we propose that SQR is inhibited due to direct binding of the oxo groups of the succinyl moiety of MitoVES (most likely resulting in displacing UbQ). Thus, in the presence of MitoVES, the SDH activity is still on, albeit at a somewhat lower rate, generating electrons that cannot be intercepted by UbQ, giving rise to superoxide (Fig. 8.3B). The superior apoptogenic activity of MitoVES compared to α-TOS translates to its higher anticancer activity, as documented using several experimental cancer models (Fig. 8.3C).[99,100]

The abovementioned data and considerations point to the SQR activity of CII as a very good target for anticancer agents. With that in mind, we need to mention atpenins, natural products[102,103] that were found to induce apoptosis in tumor cells.[104,105] These interesting agents were documented to target the SQR site of CII, although the exact mechanism of the interaction of the agents with the site is not clear. Importantly, while killing cancerous cells, atpenins protect nonmalignant tissues from various insults, as shown for cardiac tissue,[106,107] often being damaged during chemotherapy. These findings point to the mitochondrial CII as an intriguing target for anticancer drugs that may act as a protector for nontumor tissues.[108]

3. CONCLUSIONS AND PERSPECTIVES

There is little doubt that mitochondria play a very important role in the malignant phenotype of cancer cells and in their propensity to efficiently form tumors and metastasize. For these reasons, mitochondria may represent the Achilles' heel of cancer, whose targeting may be a very good strategy how to deal with the burden of neoplastic diseases that is on the rise and that needs to be fully exploited for more efficient treatment of tumors. Of the targets that have recently emerged, there is hope in the more wide application of class 5 mitocans, i.e., agents that target the ETC pathway of mitochondria. The individual constituent complexes of ETC present targets with various degree of plausibility to be utilized for clinical applications. A particularly attractive target for anticancer agent is complex II, being at the cross-roads of the TCA cycle and OXPHOS. Its potential use for anticancer therapy stems from the notion that it only rarely mutates. It is often associated with largely nonmalignant neoplasias like pheochromocytomas and paragangliomas, while it has been found mutated in one out of one million breast cancer patients.[109,110] Thus, complex II is a neat example of a potentially very good target and epitomizes the thus far unharnessed use of mitochondria for a potential breakthrough in cancer management.

ACKNOWLEDGMENTS

This project was supported in part by grants from Cancer Council Queensland and Czech Science Foundation P301/10/1937 to J. N. and from BIOCEV European Regional Development Fund CZ.1.05/1.1.00/02.0109.

REFERENCES

1. Warburg O. On the origin of cancer cells. *Science*. 1956;123:309–314.
2. Warburg O. On respiratory impairment in cancer cells. *Science*. 1956;124:269–270.
3. Vander Heiden MG, Cantley LC, Thompson CB. Understanding the Warburg effect: the metabolic requirements of cell proliferation. *Science*. 2009;324:1029–1033.
4. Hanahan D, Weinberg RA. Hallmarks of cancer: the next generation. *Cell*. 2011; 144:646–674.
5. Wallace DC. Mitochondria and cancer. *Nat Rev Cancer*. 2012;12:685–698.
6. Neupert W, Herrmann JM. Translocation of proteins into mitochondria. *Annu Rev Biochem*. 2007;76:723–749.
7. Ryan MT, Hoogenraad NJ. Mitochondrial-nuclear communications. *Annu Rev Biochem*. 2007;76:701–722.
8. Liu Z, Butow RA. Mitochondrial retrograde signaling. *Annu Rev Genet*. 2006; 40:159–185.
9. Guha M, Avadhani NG. Mitochondrial retrograde signaling at the crossroads of tumor bioenergetics, genetics and epigenetics. *Mitochondrion*. 2013;13:577–591.
10. Scarpulla RC. Nuclear activators and coactivators in mammalian mitochondrial biogenesis. *Biochim Biophys Acta*. 2002;1576:1–14.
11. Kukat C, Larsson NG. mtDNA makes a U-turn for the mitochondrial nucleoid. *Trends Cell Biol*. 2013;23:457–463.
12. Falkenberg M, Larsson NG, Gustafsson CM. DNA replication and transcription in mammalian mitochondria. *Annu Rev Biochem*. 2007;76:679–699.
13. Wang Y, Bogenhagen DF. Human mitochondrial DNA nucleoids are linked to protein folding machinery and metabolic enzymes at the mitochondrial inner membrane. *J Biol Chem*. 2006;281:25791–25802.
14. Wang YE, Marinov GK, Wold BJ, Chan DC. Genome-wide analysis reveals coating of the mitochondrial genome by TFAM. *PLoS One*. 2013;8:e74513.
15. Ngo HB, Kaiser JT, Chan DC. The mitochondrial transcription and packaging factor Tfam imposes a U-turn on mitochondrial DNA. *Nat Struct Mol Biol*. 2011; 18:1290–1296.
16. Kukat C, Wurm CA, Spåhr H, Falkenberg M, Larsson NG, Jakobs S. Super-resolution microscopy reveals that mammalian mitochondrial nucleoids have a uniform size and frequently contain a single copy of mtDNA. *Proc Natl Acad Sci USA*. 2011; 108:13534–13539.
17. Murphy MP. How mitochondria produce reactive oxygen species. *Biochem J*. 2009;417:1–13.
18. Quinlan CL, Orr AL, Perevoshchikova IV, Treberg JR, Ackrell BA, Brand MD. Mitochondrial complex II can generate reactive oxygen species at high rates in both the forward and reverse reactions. *J Biol Chem*. 2012;287:27255–27264.

19. Siebels I, Dröse S. Q-site inhibitor induced ROS production of mitochondrial complex II is attenuated by TCA cycle dicarboxylates. *Biochim Biophys Acta*. 1827; 2013:1156–1164.
20. Krishnan KJ, Reeve AK, Samuels DC, et al. What causes mitochondrial DNA deletions in human cells? *Nat Genet*. 2008;40:275–279.
21. Lu J, Sharma LK, Bai Y. Implications of mitochondrial DNA mutations and mitochondrial dysfunction in tumorigenesis. *Cell Res*. 2009;19:802–815.
22. Schon EA, DiMauro S, Hirano M. Human mitochondrial DNA: roles of inherited and somatic mutations. *Nat Rev Genet*. 2012;1:878–890.
23. Brandon M, Baldi P, Wallace DC. Mitochondrial mutations in cancer. *Oncogene*. 2006;25:4647–4662.
24. Lee HC, Wei YH. Mitochondrial DNA instability and metabolic shift in human cancers. *Int J Mol Sci*. 2009;10:674–701.
25. Tan DJ, Bai RK, Wong LJ. Comprehensive scanning of somatic mitochondrial DNA mutations in breast cancer. *Cancer Res*. 2002;62:972–976.
26. Chatterjee A, Mambo E, Sidransky D. Mitochondrial DNA mutations in human cancer. *Oncogene*. 2006;25:4663–4674.
27. Tseng LM, Yin PH, Chi CW, et al. Mitochondrial DNA mutations and mitochondrial DNA depletion in breast cancer. *Genes Chromosomes Cancer*. 2006;45:629–638.
28. Dickinson A, Yeung KY, Donoghue J, et al. The regulation of mitochondrial DNA copy number in glioblastoma cells. *Cell Death Differ*. 2013;20:1644–1653.
29. King MP, Attardi G. Injection of mitochondria into human cells leads to a rapid replacement of the endogenous mitochondrial DNA. *Cell*. 1989;52:811–819.
30. King MP, Attardi G. Human cells lacking mtDNA: repopulation with exogenous mitochondria by complementation. *Science*. 1989;246:500–503.
31. Cavalli LR, Varella-Garcia M, Liang BC. Diminished tumorigenic phenotype after depletion of mitochondrial DNA. *Cell Growth Differ*. 1997;8:1189–1198.
32. Morais R, Zinkewich-Peotti K, Parent M, Wang H, Babai F, Zollinger M. Tumor-forming ability in athymic nude mice of human cell lines devoid of mitochondrial DNA. *Cancer Res*. 1994;54:3889–3896.
33. Rodríguez-Enríquez S, Gallardo-Pérez JC, Marín-Hernández A, et al. Oxidative phosphorylation as a target to arrest malignant neoplasias. *Curr Med Chem*. 2011;18:3156–3167.
34. Jose C, Bellance N, Rossignol R. Choosing between glycolysis and oxidative phosphorylation: a tumor's dilemma? *Biochim Biophys Acta*. 1807;2011:552–561.
35. Smolkova K, Plecita-Hlavata L, Bellance N, Benard G, Rossignol R, Jezek P. Waves of gene regulation suppress and then restore oxidative phosphorylation in cancer cells. *Int J Biochem Cell Biol*. 2011;43:950–968.
36. Tan AS, Baty JW, Dong LF, *et al*. Mitochondrial genome acquisition restores respiratory function and tumorigenic and metastatic potential in cancer cells depleted of their mitochondrial DNA. *Cell Metab*. [under revision].
37. Acin-Perez R, Fernandez-Silva R, Peleato ML, Perez-Martos A, Enriquez JA. Respiratory active mitochondrial supercomplexes. *Mol Cell*. 2008;32:529–539.
38. Lapuente-Brun E, Moreno-Loshuertos R, Acín-Pérez R, et al. Supercomplex assembly determines electron flux in the mitochondrial electron transport chain. *Science*. 2014;340:1567–1570.
39. Rustom A, Saffrich R, Markovic I, Walther P, Gerdes HH. Nanotubular highways for intercellular organelle transport. *Science*. 2004;303:1007–1010.
40. Islam MN, Das SR, Emin MT, et al. Mitochondrial transfer from bone-marrow-derived stromal cells to pulmonary alveoli protects against acute lung injury. *Nat Med*. 2012;18:759–765.
41. Rogers RS, Bhattacharya J. When cells become organelle donors. *Physiology*. 2013;28:414–422.

42. Petros JA, Baumann AK, Ruiz-Pesini E, et al. mtDNA mutations increase tumorigenicity in prostate cancer. *Proc Natl Acad Sci USA*. 2006;102:719–724.
43. Ishikawa K, Takenaga K, Akimoto M, et al. ROS-generating mitochondrial DNA mutations can regulate tumor cell metastasis. *Science*. 2008;320:661–664.
44. Hayashi J, Takemitsu M, Nonaka I. Recovery of the missing tumorigenicity in mitochondrial DNA-less HeLa cells by introduction of mitochondrial DNA from normal human cells. *Somat Cell Mol Genet*. 1992;18:123–129.
45. Imanishi H, Hattori K, Wada R, et al. Mitochondrial DNA mutations regulate metastasis of human breast cancer cells. *PLoS One*. 2011;6:e23401.
46. Guha M, Srinivasan S, Ruthel G, *et al.* Mitochondrial retrograde signaling induces epithelial-mesenchymal transition and generates breast cancer stem cells. *Oncogene*. [in press].
47. Visvader JE, Lindeman GJ. Cancer stem cells: current status and evolving complexities. *Cell Stem Cell*. 2012;10:717–728.
48. Giancotti FG. Mechanisms governing metastatic dormancy and reactivation. *Cell*. 2013;155:750–764.
49. Meacham CE, Morrison SJ. Tumour heterogeneity and cancer cell plasticity. *Nature*. 2013;501:328–337.
50. Frezza C, Gottlieb E. Mitochondria in cancer: not just innocent bystanders. *Semin Cancer Biol*. 2009;19:4–11.
51. Kroemer G, Pouyssegur J. Tumor cell metabolism: cancer's Achilles' heel. *Cancer Cell*. 2008;13:472–482.
52. Bellance N, Lestienne P, Rossignol R. Mitochondria: from bioenergetics to the metabolic regulation of carcinogenesis. *Front Biosci*. 2009;14:4015–4034.
53. Gottlieb E, Tomlinson IP. Mitochondrial tumour suppressors: a genetic and biochemical update. *Nat Rev Cancer*. 2005;5:857–866.
54. Raimundo N, Baysal BE, Shadel GS. Revisiting the TCA cycle: signaling to tumor formation. *Trends Mol Med*. 2011;17:641–649.
55. Cairns RA, Mak TW. Oncogenic isocitrate dehydrogenase mutations: mechanisms, models, and clinical opportunities. *Cancer Discov*. 2013;3:730–741.
56. Baysal BE. Mitochondrial complex II and genomic imprinting in inheritance of paraganglioma tumors. *Biochim Biophys Acta*. 1827;2013:573–577.
57. Losman JA, Kaelin Jr WG. What a difference a hydroxyl makes: mutant IDH, (R)-2-hydroxyglutarate, and cancer. *Genes Dev*. 2013;27:836–852.
58. Tomasetti M, Neuzil J, Dong LF. MicroRNAs as regulators of mitochondrial function: role in cancer suppression. *Biochim Biophys Acta*. 1840;2014:1441–1453.
59. Tomasetti M, Nocchi L, Staffolani S, *et al.* MicroRNA-126 suppresses mesothelioma malignancy by targeting IRS1 and interfering with the mitochondrial function. *Antioxid Redox Signal*. [in press].
60. DeBerardinis RJ, Mancuso A, Daikhin E, et al. Beyond aerobic glycolysis: transformed cells can engage in glutamine metabolism that exceeds the requirement for protein and nucleotide synthesis. *Proc Natl Acad Sci USA*. 2007;104:19345–19350.
61. Hu Y, Lu W, Chen G, et al. K-ras (G12V) transformation leads to mitochondrial dysfunction and a metabolic switch from oxidative phosphorylation to glycolysis. *Cell Res*. 2012;22:399–412.
62. Neuzil J, Rohlena J, Dong LF. K-Ras and mitochondria: dangerous liaisons. *Cell Res*. 2012;22:285–287.
63. Jones S, Zhang X, Parsons DW, et al. Core signaling pathways in human pancreatic cancers revealed by global genomic analyses. *Science*. 2008;321:1801–1806.
64. Parsons DW, Jones S, Zhang X, et al. An integrated genomic analysis of human glioblastoma multiforme. *Science*. 2008;321:1807–1812.

65. Parsons DW, Li M, Zhang X, et al. The genetic landscape of the childhood cancer medulloblastoma. *Science*. 2011;331:435–439.
66. Gerlinger M, Rowan AJ, Horswell S, et al. Intratumor heterogeneity and branched evolution revealed by multiregion sequencing. *N Engl J Med*. 2012; 366:883–892.
67. Hayden E. Cancer complexity quells quest for cure. *Nature*. 2008;455:148.
68. Murphy MP. Mitochondria-a neglected drug target. *Curr Opin Investig Drugs*. 2009;10:1022–1024.
69. Ralph SJ, Rodríguez-Enríquez S, Neuzil J, Moreno-Sanchez R. Bioenergetic pathways in tumor mitochondria as targets for cancer therapy and the importance of the ROS-induced apoptotic trigger. *Mol Asp Med*. 2010;31:29–59.
70. Ralph SJ, Rodríguez-Enríquez S, Neuzil J, Saavedra E, Moreno-Sánchez R. The causes of cancer revisited: "mitochondrial malignancy" and ROS-induced oncogenic transformation: why mitochondria are targets for cancer therapy. *Mol Asp Med*. 2010;31:145–170.
71. Gogvadze V, Orrenius S, Zhivotovsky B. Mitochondria in cancer cells: what is so special about them? *Trends Cell Biol*. 2008;18:165–173.
72. Fulda S, Galluzzi L, Kroemer G. Targeting mitochondria for cancer therapy. *Nat Rev Drug Discov*. 2010;9:447–464.
73. Wallace DC, Fan W, Procaccio V. Mitochondrial energetics and therapeutics. *Annu Rev Pathol*. 2010;5:297–348.
74. Wang F, Ogasawara MA, Huang P. Small mitochondria-targeting molecules as anticancer agents. *Mol Asp Med*. 2010;31:75–92.
75. Fulda S, Kroemer G. Mitochondria as therapeutic targets for the treatment of malignant disease. *Antioxid Redox Signal*. 2011;15:2937–2949.
76. Neuzil J, Wang XF, Dong LF, Low P, Ralph SJ. Molecular mechanism of 'mitocan'-induced apoptosis in cancer cells epitomizes the multiple roles of reactive oxygen species and Bcl-2 family proteins. *FEBS Lett*. 2006;580:5125–5129.
77. Neuzil J, Dong LF, Rohlena J. Mitocans, mitochondria-targeting anti-cancer drugs: focus on mitochondrial complexes. In: Cadenas E, Packer E, Orrenius S, eds. *Oxidative Stress and Disease*. Boca Raton, FL: CRC Press; 2012:55–92.
78. Neuzil J, Dong LF, Rohlena J, Truksa J, Ralph SJ. Classification of mitocans, anticancer drugs acting on mitochondria. *Mitochondrion*. 2013;13:199–208.
79. Kluckova K, Bezework-Geleta A, Rohlena J, Dong LF, Neuzil J. Mitochondrial complex II, a novel intriguing target for anti-cancer agents. *Biochim Biophys Acta*. 1827;2013:552–564.
80. Saraste M. Oxidative phosphorylation at the fin de siecle. *Science*. 1999;283:1488–1493.
81. Cecchini G. Function and structure of complex II of the respiratory chain. *Annu Rev Biochem*. 2003;72:77–109.
82. Iverson TM. Catalytic mechanisms of complex II enzymes: a structural perspective. *Biochim Biophys Acta*. 1827;2013:648–657.
83. Bezawork-Geleta A, Saiyed T, Dougan DA, Truscott KN. Mitochondrial matrix proteostasis is linked to hereditary paraganglioma: LON-mediated turnover of the human flavinylation factor SDH5 is regulated by its interaction with SDHA. *FASEB J*. 2014;28:1794–1804.
84. Hao HX, Khalimonchuk O, Schraders M, et al. SDH5, a gene required for flavination of succinate dehydrogenase, is mutated in paraganglioma. *Science*. 2009;325: 1139–1142.
85. Gebert N, Gebert M, Oeljeklaus S, et al. Dual function of Sdh3 in the respiratory chain and TIM22 protein translocase of the mitochondrial inner membrane. *Mol Cell*. 2011;44:811–818.

86. Albayrak T, Scherhammer V, Schoenfeld N, et al. The tumor suppressor cybL, a component of the respiratory chain, mediates apoptosis induction. *Mol Biol Cell.* 2003;14:3082–3096.
87. Sun F, Huo X, Zhai Y, et al. Crystal structure of mitochondrial respiratory membrane protein complex II. *Cell.* 2005;121:1043–1057.
88. Neuzil J, Weber T, Schröder A, et al. Induction of cancer cell apoptosis by α-tocopheryl succinate: molecular pathways and structural requirements. *FASEB J.* 2001;15:403–415.
89. Neuzil J, Weber T, Gellert N, Weber C. Selective cancer cell killing by α-tocopheryl succinate. *Br J Cancer.* 2001;84:87–89.
90. Weber T, Lu M, Andera L, et al. Vitamin E succinate is a potent novel anti-neoplastic agent with high tumor selectivity and cooperativity with tumor necrosis factor-related apoptosis-inducing ligand (TRAIL, Apo2L) in vivo. *Clin Cancer Res.* 2002;8:863–869.
91. Weber T, Dalen H, Andera L, et al. Mitochondria play a central role in apoptosis induced by α-tocopheryl succinate, an agent with anticancer activity. Comparison with receptor-mediated pro-apoptotic signaling. *Biochemistry.* 2003;42:4277–4291.
92. Prochazka L, Dong LF, Valis K, et al. α-Tocopheryl succinate causes mitochondrial permeabilization by preferential formation of Bak channel. *Apoptosis.* 2010;15: 782–794.
93. Valis K, Prochazka L, Boura E, et al. Hippo/Mst1 stimulates transcription of *NOXA* in a FoxO1-dependent manner. *Cancer Res.* 2011;71:946–954.
94. Dong LF, Low P, Dyason J, et al. α-Tocopheryl succinate induces apoptosis by targeting ubiquinone-binding sites in mitochondrial respiratory complex II. *Oncogene.* 2008;27:4324–4335.
95. Dong LF, Freeman R, Liu J, et al. Suppression of tumour growth *in vivo* by the mitocan α-tocopheryl succinate requires respiratory complex II. *Clin Cancer Res.* 2009; 15:1593–1600.
96. Dong LF, Swettenham E, Eliasson J, et al. Vitamin E analogs inhibit angiogenesis by selective apoptosis induction in proliferating endothelial cells: the role of oxidative stress. *Cancer Res.* 2007;67:11906–11913.
97. Murphy MP, Smith RA. Targeting antioxidants to mitochondria by conjugation to lipophilic cations. *Annu Rev Pharmacol Toxicol.* 2007;47:629–656.
98. Kelso GF, Porteous CM, Coulter CV, et al. Selective targeting of a redox-active ubiquinone to mitochondria within cells: antioxidant and antiapoptotic properties. *J Biol Chem.* 2001;276:4588–4596.
99. Dong LF, Jameson VJA, Tilly D, et al. Mitochondrial targeting of vitamin E succinate enhances its pro-apoptotic and anti-cancer activity via mitochondrial complex II. *J Biol Chem.* 2011;286:3717–3728.
100. Dong LF, Jameson VJA, Tilly D, et al. Mitochondrial targeting of α-tocopheryl succinate enhances its pro-apoptotic efficacy: a new paradigm of efficient anti-cancer therapy. *Free Radic Biol Med.* 2011;50:1546–1555.
101. Rohlena J, Dong LF, Kluckova K, et al. Mitochondrially targeted α-tocopheryl succinate is antiangiogenic. *Antioxid Redox Signal.* 2011;15:2923–2935.
102. Omura S, Tomoda H, Kimura K, et al. Atpenins, new antifungal antibiotics produced by Penicillium sp. Production, isolation, physico-chemical and biological properties. *J Antibiot.* 1998;41:1769–1773.
103. Kawada M, Momose I, Someno T, Tsujiuchi G, Ikeda D. New atpenins, NBRI23477 A and B, inhibit the growth of human prostate cancer cells. *J Antibiot.* 2009;62: 243–246.
104. Oshino K, Kumagai H, Tomoda H, Omura S. Mechanism of action of atpenin B on Raji cells. *J Antibiot.* 1990;43:1064–1068.

105. Miyadera H, Shiomi K, Ui H, et al. Atpenins, potent and specific inhibitors of mitochondrial complex II (succinate-ubiquinone oxidoreductase). *Proc Natl Acad Sci USA*. 2003;100:473–477.
106. Wojtovich AP, Brookes PS. The complex II inhibitor atpenin A5 protects against cardiac ischemia-reperfusion injury via activation of mitochondrial K_{ATP} channels. *Basic Res Cardiol*. 2009;104:121–129.
107. Drose S, Bleier L, Brandt U. A common mechanism links differently acting complex II inhibitors to cardioprotection: modulation of mitochondrial reactive oxygen species production. *Mol Pharmacol*. 2011;79:814–822.
108. Ralph SJ, Moreno-Sanchez R, Neuzil J, Rodriguez-Enriquez S. Inhibitors of the succinate:quinine reductase/complex II regulate production of mitochondrial reactive oxygen species and protect normal cells from ischemic damage but induce specific cancer cell death. *Pharm Res*. 2011;28:2695–2730.
109. Peczkowska M, Cascon A, Prejbisz A, et al. Extra-adrenal and adrenal pheochromocytomas associated with a germline SDHC mutation. *Nat Clin Pract Endocrinol Metab*. 2008;4:111–115.
110. Schiavi F, Boedeker CC, Bausch B, et al. Predictors and prevalence of paraganglioma syndrome associated with mutations of the SDHC gene. *JAMA*. 2005;294:2057–2063.

CHAPTER NINE

Estrogen and Mitochondria Function in Cardiorenal Metabolic Syndrome

Guanghong Jia*,†, Annayya R. Aroor*,†, James R. Sowers*,†,‡
*Division of Endocrinology, Diabetes, and Metabolism, Diabetes Cardiovascular Center, Columbia, Missouri, USA
†Harry S. Truman Memorial Veterans Hospital, Columbia, Missouri, USA
‡Department of Medical Pharmacology and Physiology, Columbia, Missouri, USA

Contents

Abstract

The cardiorenal metabolic syndrome (CRS) consists of a constellation of cardiac, renal, and metabolic disorders including insulin resistance (IR), obesity, metabolic dyslipidemia, high-blood pressure, and evidence of early cardiac and renal disease. Mitochondria dysfunction often occurs in the CRS, and this dysfunction is promoted by excess reactive oxygen species, genetic factors, IR, aging, and altered mitochondrial biogenesis. Recently, it has been shown that there are important sex-related differences in mitochondria function and metabolic, cardiovascular, and renal components. Sex differences in the CRS have mainly been attributed to the estrogen's effects that are mainly

Progress in Molecular Biology and Translational Science, Volume 127
ISSN 1877-1173
http://dx.doi.org/10.1016/B978-0-12-394625-6.00009-X

mediated by estrogen receptor (ER) α, ERβ, and G-protein coupled receptor 30. In this review, we discuss the effects of estrogen on the mitochondrial function, insulin metabolic signaling, glucose transport, lipid metabolism, and inflammatory responses from liver, pancreatic β cells, adipocytes, skeletal muscle, and cardiovascular tissue.

1. INTRODUCTION

The cardiorenal metabolic syndrome (CRS) consists of risk factors that include asatherogenic dyslipidemia, elevated blood pressure, high plasma glucose, prothrombotic disorder, and proinflammatory state (metabolic syndrome).[1] There are a number of important bidirectional interactions among heart, kidney, brain, liver, muscle, and fat tissues, which cause a constellation of cardiac, renal, and metabolic disorders including insulin resistance (IR), obesity, metabolic dyslipidemia, high-blood pressure, and evidence of early cardiac and renal disease (Fig. 9.1). Epidemiological studies have shown that 36.1% of adult men and 32.4% of women had metabolic syndrome in the USA in 2010. This was a considerable increase from 21.8% to 23.7%, respectively, in 2002.[2] The total number of adults with metabolic syndrome ranges from almost 77 million to almost 86 million in America.[2] The first formal definition of the metabolic syndrome was proposed by the World Health Organization in 1998.[3] International Diabetes Federation, American Heart Association, National Heart, Lung, and Blood Institute, World Heart Federation, International Atherosclerosis Society, and International Association for the Study of Obesity in 2009 defined the metabolic syndrome as three of the following elements: enlarged waist circumference with different population- and country-specific criteria; elevated triglycerides (TG) of $\geq$1.7 mmol/L (150 mg/dL); reduced high-density lipoprotein (HDL) cholesterol of $<$1.03 mmol/L (40 mg/dL) in men and $<$1.29 mmol/L (50 mg/dL) in women; elevated blood pressure, with a systolic blood pressure $\geq$130 mm Hg or a diastolic blood pressure $\geq$85 mm Hg; and elevated fasting glucose $\geq$5.6 mmol/L (100 mg/dL), with the inclusion of those individuals using medication to treat hypertriglyceridemia, decrease HDL, hypertension, or hyperglycemia.[4]

Many factors contribute to the genesis of metabolic and cardiovascular and renal abnormalities that characterize the CRS, including genetic predisposition, decreased physical activity, chronic inflammation, oxidative stress, elevated free fatty acids (FFA), hyperglycemia, aldosterone, angiotensin II, western diet as well as mitochondrial dysfunction (Fig. 9.2).[5] Mitochondria

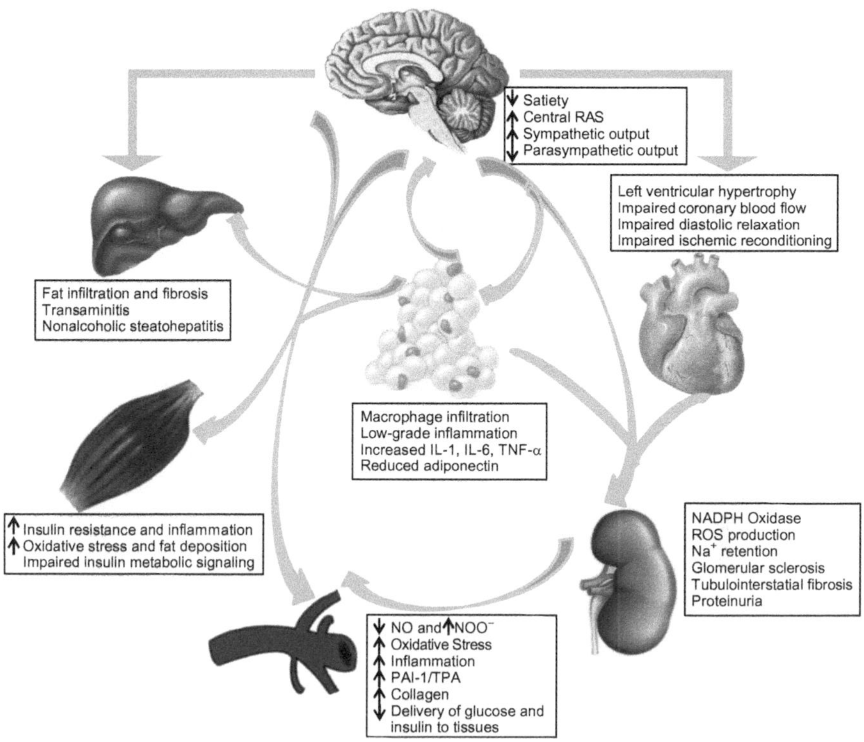

Figure 9.1 Pathophysiological interactions between adiposity and maladaptive changes in the heart and kidney in CRS. The interactions among the environmental, behavioral, hormonal, and genes help to modify an individual's body weight, and further show the effects of the resultant dysfunctional adipose tissue and low-grade inflammation on the structure and function of different organ systems. Abbreviations: RAS, renin–angiotensin–aldosterone system; IL, interleukin; TNF, tumor necrosis factor; NO, nitric oxide; NOO^-, peroxynitrite; ROS, reactive oxygen species; PAI, plasminogen activator inhibitor; TPA, tissue plasminogen activator. (See the color plate.)

function can benefit from the roles of estrogen through binding to estrogen receptor (ER) α/β, such as antioxidant ability and inhibiting renin–angiotensin system. Thus, estrogen reduces IR, diabetes, cardiovascular diseases (CVDs), and CRS (Fig. 9.2). Recently, our investigative group reported that there is a gender difference in mitochondrial function and that this difference contributes to development of the CRS in mice fed a western diet high in fat and fructose.[6] Indeed, the rate of development of CVD related to CRS, such as diastolic dysfunction, coronary vascular stiffness, and impaired vasorelaxation are different in males and females. Normally women develop CVD approximately 10 years later than men, but women

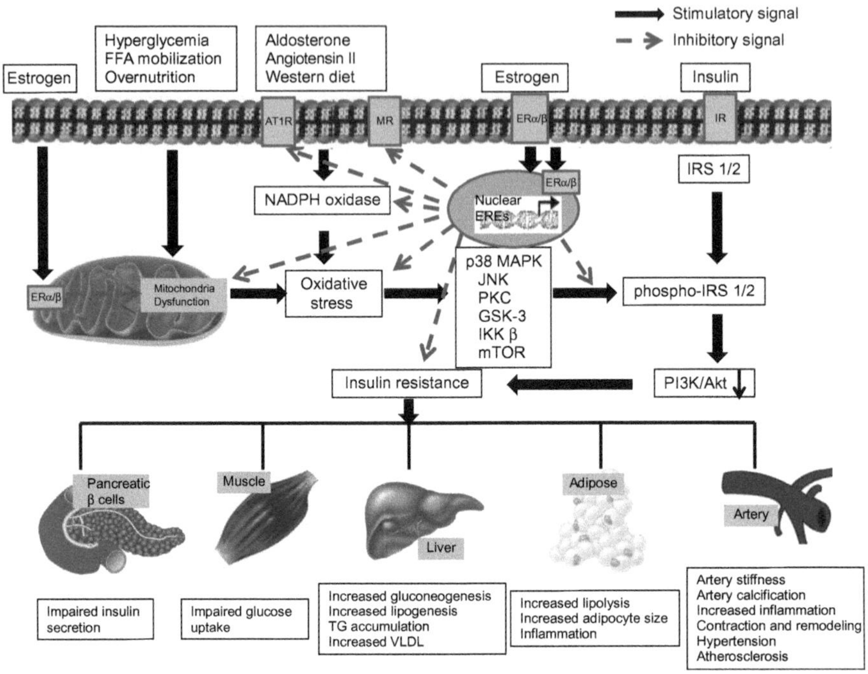

Figure 9.2 Proposed roles of estrogen and its receptors in the maintenance of mitochondria function and reduction of insulin resistance in CRS. Overnutrients, FFA mobilization, hyperglycemia, aldosterone, angiotensin II, and western diet cause mitochondrial dysfunction and increase ROS production, resulting in insulin resistance in CRS. Estrogen binding to ERα/β protects mitochondria function, increases antioxidant ability, inhibits renin–angiotensin system, and thus reduces insulin resistance, diabetes, and cardiovascular diseases. Abbreviations: FFA, free fat acids; AT1R, angiotensin II receptor 1; MR, mineralocorticoid receptor; IR, insulin receptors; NADPH oxidase, nicotinamide adenine dinucleotide phosphate-oxidase; IRS 1/2, insulin receptor substrate 1/2; MAPK, mitogen-activated protein kinase; JNK, c-Jun N-terminal kinase; PKC, protein kinase C; GSK-3, glycogen synthase kinase 3; IKK β, IκB kinase; mTOR, mammalian target of rapamycin; PI3K/Akt, phosphatidylinositide 3-kinases/protein kinase B; VLDL, very low-density lipoprotein cholesterol. (See the color plate.)

show a marked increase in CVD during the postmenopausal period.[7] The increased risk of CVD in postmenopausal women has been linked to the decrease in plasma estrogen levels.[8] However, our understanding of the biological relationship between estrogen signaling, mitochondria function, and the development CRS is still in its infancy. The objectives of this review are to provide a basic overview of the role of estrogen in regulating mitochondrial function and how abnormalities of this regulation and function contribute to the development of the CRS.

2. MITOCHONDRIA IN CRS

Mitochondria are essential for intermediary metabolism as well as adenosine triphosphate (ATP) production, and normally provide more than 90% of cellular energy.[9] It has been established that mitochondrial respiratory chain function is responsible for energy metabolism and ATP production through the tricarboxylic acid (TCA) cycle, coupling of oxidative phosphorylation (OXPHOS), and electron transfer.[10] In the past decade, research has demonstrated that mitochondria play a key role in modulating cell cycle progression, cell survival, apoptosis, as well as reactive oxygen species (ROS) generation which is a "side product" in metabolism.[9] Mitochondria dysfunction is recognized as playing a central role in the development of various abnormalities, including disturbed glucose homeostasis, IR, abdominal fat accumulation, dyslipidemia, hypertension and associated cardiac and renal pathology.

2.1. Mitochondria structure and function

The TCA cycle and electron transport chain (ETC) are two important mitochondrial processes in cellular ATP production. The overall pathway for TCA cycle is as follows: catabolism of glucose in the cytosol produces two molecules of pyruvate, which pass through the mitochondrion's double membrane and enter the TCA cycle.[11] When there are relatively high concentrations of ATP, pyruvate carboxylase is activated and shuttles pyruvate in the direction of gluconeogenesis. When energy demands and the concentration of ATP are relatively low, two pyruvate molecules pass through the pyruvate dehydrogenase (PDH) complex to produce two molecules of acetyl-coenzyme A (acetyl CoA), which enter the TCA cycle. Subsequently, each acetyl CoA produces three molecules of nicotinamide adenine dinuceotide (NADH) and two molecules of flavin adenine dinucleotide (FADH), for a total of six NADH and four FADH per one molecule of pyruvate.[12] Additionally, acetyl CoA can be produced by oxidation of FFA, which then requires the nutrient L-carnitine to shuttle the acetyl CoA into the mitochondria to enter the TCA cycle.[13] The ETC consists of five enzymatic complexes (I–V) of integral membrane proteins. They are NADH-CoQ reductase (complex I), succinate-CoQ reductase (complex II), CoQ-cytochrome c reductase (complex III), cytochrome c oxidase (complex IV), and ATP synthase (complex V).[9] Electrons from NADH or FADH2 are transferred through a series of respiratory chain complexes to

O_2, which finally generates H_2O. A proton gradient across the membrane is the driving force of F0F1-ATPase to produce ATP from adenosine diphosphate (ADP). ATP is transported to the cytosol by exchanging ADP through an adenine nucleotide translocator and is important for various biological events.[14] Additionally, mitochondria generate heat by a mechanism called the "proton leak." Proton leak from the intermembrane space to the matrix reduces proton-motive force and generates heat instead of ATP. Much of the proton leak is a catalytic property of specific molecules termed uncoupling proteins (UCPs), such as UCP1, UCP2, and UCP3. UCPs play an important role in reducing the proton gradient and mitochondrial function by regulating both heat and ROS generation.[14]

Oxygen consumed in mitochondrial is associated with production of oxygen free radicals (ROS) in most cells and tissues. Mitochondrial electron transport generates superoxide radical ($O_2^{\bullet -}$) as an inevitable by-product in complex I and complex III.[15] Biologically, ROS include $O_2^{\bullet -}$, hydrogen peroxide (H_2O_2), and the hydroxyl radical ($OH^{\bullet}$). It has been estimated that about 0.2–2% of oxygen consumed are converted into $O_2^{\bullet -}$ by ETC in human cells.[16] However, the mitochondria of cardiomyocytes have high respiratory rates, and approximately 90% of basal cellular ROS are from mitochondria.[17] Moreover, seven other sites were also suggested to generate $O_2^{\bullet -}$ in the mitochondria. They are cytochrome b five reductase, monoamine oxidase, dihydroorotate dehydrogenase, dehydrogenase of α-glycerophosphate, succinate dehydrogenase, aconitase, and α-ketoglutarate dehydrogenase complex.[9] Although this review is concerned with mitochondrial generation of ROS, it should be recognized that ROS are derived from outside of mitochondria, such as oxygen radicals from peroxisomal β-oxidation of fatty acids, NAD(P)H oxidase, xanthine oxidase, arachidonic acid metabolism, microsomal P-450 enzymes, and prooxidant heme molecule.[18] Several enzymatic means may protect against ROS in mitochondria. These include conversion of superoxide to H_2O_2 by superoxide dismutase (SOD), scavenging of H_2O_2 by catalase, glutathione peroxidase, and peroxiredoxin III.[18] Ca^{2+} transport can cross the mitochondrial inner membrane and is taken up by mitochondria both through a uniporter and a pulsed or rapid mode.[19] The most studied Ca^{2+} influx mechanism is the uniporter, which transports Ca^{2+} down its electrochemical gradient.[20] Several important functions have been proposed for these processes, including control of the metabolic rate for cellular energy production, modulation of cytosolic Ca^{2+} transients, and induction of apoptosis through release of cytochrome c.[21]

2.2. Risk factors in mitochondria function

Mitochondrial dysfunction, such as mitochondrial loss and overproduction of oxidants, has been suggested to be involved in the development of impaired insulin metabolic signaling/IR in various tissues of skeletal muscle, liver, fat, heart, and the kidney. Sedentary lifestyle, genetic factors, oxidative stress, maladaptive immunomodulation, and aging may adversely affect mitochondrial function, leading to impaired insulin metabolic signaling and progression of the CRS (Fig. 9.2). Physical activity plays an important role in the regulation of muscle mitochondrial function because aerobic exercise potently activates mitochondrial function. On the contrary, sedentary behavior is linked with reduced activity, with obesity and other metabolic disorders.[22] Thus, it is likely that some mitochondrial defects in overweight or obese insulin-resistant subjects can be explained by low levels of physical activity. Mitochondrial proteins are encoded by both nuclear and mitochondrial genomes, and there is some evidence to suggest that mitochondria DNA (mtDNA) deletions or mutations in nuclear-encoded genes are linked with various abnormalities characterizing the CRS. It has been hypothesized that the mitochondrial genome is more susceptible to various mutagenic stressors because mitochondrial genes are more proximal to the ROS source.[23] Our previous research showed that a naturally occurring thymidine-to-cytidine mutation in the mitochondrial $tRNA^{ILE}$ gene is associated with phenotypes of hypertension, hypercholesterolemia, and hypomagnesemia.[14] *A3243G*, another mutation gene on mtDNA that encodes tRNA and causes impaired insulin secretion. In addition, patients with defects in acyl CoA dehydrogenase have phenotypes of cardiomyopathy, liver dysfunction, and neurological disorders.[24] Furthermore, polymorphisms in the promoter of UCP2 are associated with decreased incidence of obesity, reduced insulin secretion, and a high prevalence of type 2 diabetes mellitus (T2DM).[14] Thus, genetic factors that are inherited through nuclear or mitochondrial genes may influence the pathogenesis of the CRS and associated CVD through functional impairment of mitochondria.

ROS production occurs mainly at complex I and complex III in mitochondria.[25] Under conditions of glucose and fatty acid overnutrition, nutrient overflow into cells prompts electrons transferring to oxygen without ATP production and further favors a state of increased ROS, which potentially leads to oxidative damage within mitochondria.[26] Therefore, ROS generated from mitochondria damages proteins, DNA, and lipid membrane components, which results in mitochondrial dysfunction. In aging, increasing mutations in mitochondrial DNA have been found.[27] Old mitochondria have changes in morphology in addition to increased ROS production and decreased ATP production. Abnormalities in aging include cumulative DNA damage, mitochondrial dysfunction, telomere loss, altered gene expression, and oxidative damages.[14] In addition, mitochondrial biogenesis may be impaired by age-dependent accumulations of point mutations in human.[14] Thus, increased ROS production associated with aging also contribute to mitochondrial dysfunction in the CRS.

2.3. Mitochondria dysfunction in CRS

Mitochondrial abnormalities in diabetes were initially described in the maternally inherited diabetes and deafness syndrome, which is associated with *A3243G* mtDNA mutation. Loss of pancreas β-cells due to increased apoptosis, influx of fatty acids, and activation of stress-related kinases have been reported in T2DM, which is also named secondary mitochondria dysfunction.[28] A decrease in mitochondrial fatty acid oxidation results in increased levels of fatty acyl CoA and diacylglycerol, which in turn activate stress-related serine (ser)/threonine (thr) kinase activity and inhibit glucose transport.[29] Oxidative stress contributes to impaired insulin signaling by increased UCP2 activity. Activated UCP2 induces a proton leak across the inner mitochondrial membrane; thereby resulting in the uncoupling of glucose oxidative metabolism from mitochondrial ATP production.[30] Furthermore, T2DM and IR have a profound effect on lipid profiles. IR results in the increased production of FFA through loss of its inhibition of hormone-sensitive lipase of adipocytes and decreased endothelial lipoprotein lipase function. The increased influx of FFAs to the liver and insulin stimulation of hepatic lipogenesis lead to increased hepatic TG production in the form of very low-density lipoprotein (VLDL) and steatosis as the TG are stored in the liver.[4] Adipocyte production of cholesterol ester transferase protein allows for the transfer of cholesteryl esters from HDL to VLDL. There is increased clearance of HDL by the kidneys, and the liver takes up HDL and produces VLDL, therefore leading to the low HDL and elevated TG levels seen in CRS.[4] Moreover, it is known that an abnormal lipid profile is considered to initiate atherosclerosis by impairing the bioactivity of endothelial nitric oxide (NO) and promoting leukocyte adhesion, inflammation, thrombosis, and vascular smooth muscle cell (VSMC) proliferation.[31] Taken together, these abnormalities in mitochondria are involved in the pathophysiology of lipid accumulation, IR, T2DM, and CVD.

The molecular mechanism of mitochondrial dysfunction in CRS is driven via abnormalities involving the nuclear respiratory factors (NRF) 1 and 2, the cyclic adenosine monophosphate (AMP) response element binding protein (CREB), and transcription factor A.[23] Upstream, the transcriptional coactivator peroxisome proliferators activated receptor gamma coactivator 1 (PGC-1) has been known as an inducible integrator of transcriptional circuits controlling mitochondrial biogenesis and function in a variety of tissue and cell types.[32] PGC-1α is a coactivator of nuclear transcription factors in NRF 1 and peroxisome proliferators activated receptors (PPARs).[33] PGC-1α is also regulated by the endothelial NO synthase (eNOS)/NO/cyclic guanosine monophosphate (cGMP). The activation of eNOS plays an important role in mitochondrial biogenesis.[34] Meanwhile, NRF 1 regulates the expression of many mitochondrial genes including OXPHOS genes and mitochondrial transcription factor A, which are crucial for mitochondrial gene expression and replication (Fig. 9.3).[35] Another important factor related to mitochondrial biogenesis is AMP-activated protein kinase (AMPK).[36] Studies have shown that decreased expression of PGC-1α-associated impairment of mitochondrial biogenesis may be responsible for various metabolic abnormalities in the CRS.[32] Thus, the impairment of any of the complex steps in regulation of mitochondria biogenesis may contribute to the pathogenesis of IR and CVD in CRS.

3. ESTROGEN AND MITOCHONDRIAL FUNCTION

Normal estrogen signaling through its receptors ERα and ERβ regulates mitochondrial structure, promotes mitochondrial ETC efficiency, preserves mitochondrial integrity, and maintains normal physiological function in various tissue types, such as vascular, fat, brain, and heart.

3.1. Estrogen, ERs, and their signaling pathways

In premenopausal women, natural estrogens are synthesized in the ovaries from cholesterol. In postmenopausal women, the major precursors of estrogen from peripheral tissues are from circulating androstenedione, testosterone, and estrogen through aromatase activity.[8] There are four distinct ERs: two ligand-activated transcription factors ERα and ERβ, one G-protein coupled receptor 30 (GPR30), and a less defined ER-X in the brain.[37] It is remarkable that abnormalities of ER signaling in the mitochondria and nucleus can promote the pathophysiological abnormalities characteristic of the CRS (Figs. 9.2 and 9.3). The mitochondrial genome contains DNA

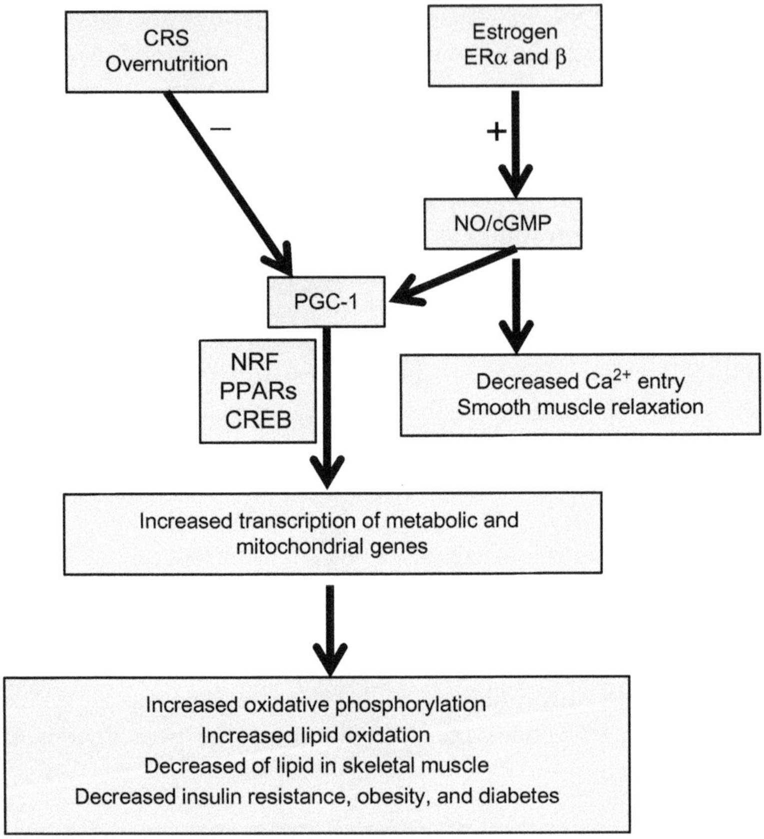

Figure 9.3 Depiction of molecular mechanism and signaling pathways of estrogen in CRS. The molecular mechanism of mitochondrial dysfunction in CRS is involved in PGC-1, NRF 1/2, CREB, and PPARs signaling pathways. Estrogen increases PGC-1 expression and prevents the development of CRS via its receptors and NO/cGMP signaling pathways. Abbreviations: PGC-1, peroxisome proliferators activated receptor gamma coactivator 1; NRF, nuclear respiratory factors, PPARs, peroxisome proliferators activated receptors; CREB, cyclic AMP response element binding protein; NO, nitric oxide; cGMP, cyclic guanosine monophosphate.

sequences that resemble half the palindromic nuclear estrogen response element (ERE) sequence. In an early experiment, ERα and ERβ were shown to directly bind mtDNA *in vitro* through mitochondrial EREs and the binding response was increased with 17β-estradiol treatment.[38] ERs are also widely expressed in the vascular system, specifically in endothelial cells (ECs) and VSMCs. Some studies suggest that ERβ predominates in ECs, whereas ERα expression is greater in VSMCs.[8] GPR30 is widely distributed in the brain and peripheral tissues. GPR30 has also been found in human internal mammary arteries and saphenous veins.[8]

Classical estrogen signaling occurs through a direct binding of ER dimers to EREs in the regulatory regions of estrogen target genes. Estrogen also modulates gene expression by ERs interacting with other transcription factors, such as activating protein-1 and stimulating protein-1.[39] Estrogen exerts rapid nongenomic actions by activating plasma membrane-associated ERs, leading to intracellular protein kinase-mediated phosphorylation signaling cascades, such as protein kinase A (PKA), protein kinase C (PKC), and mitogen-activated protein kinase (MAPK).[40] GPR30 is structurally unrelated to ERα and ERβ. Recently, GPR30 was reported to mediate nongenomic and rapid estrogen signaling, including release of intracellular Ca^{2+}, activation of Ca^{2+}–calmodulin-dependent kinases, activation of MAPK, and phosphoinositide 3-kinase (PI3K) signaling pathways.[41]

3.2. Roles of estrogen in mitochondria function

Estrogen may exert direct or indirect effects on mitochondrial function. It has been shown that estrogen increases levels of several mitochondrial respiratory chain proteins, including cytochrome c and complex IV subunits in ECs.[42] It is noted that estrogen has been recognized as antioxidants in a variety of *in vivo* and *in vitro* models, including ischemic refusion injury, heart failure, brain injury, and neurodegenerative disorders where the delivery of oxygen and glucose are inadequate and mitochondrial OXPHOS is inhibited resulting in more ROS generation.[43] Mitochondria are not only a primary source of ROS, but also a main target of ROS. Mitochondria undergo several prominent alterations in the early phase of OXPHOS that may contribute to eventual cell death, including changes in production of ROS, loss of mitochondrial membrane potential, and disruption of energy metabolism.[42] ERα and ERβ have been identified in mitochondria and to bind to the D-loop of mouse and human mtDNA, which encode three subunits of cytochrome oxidase, seven subunits of NADH-CoQ reductase, one subunit of cytochrome b, two subunits of ATP synthase, rRNAs, and tRNAs.[44] It is not clear whether ERα and ERβ directly regulate mtDNA transcription or whether this effect is mediated through nuclear effects on nuclear-encoded genes that in turn regulate mtDNA transcription.[45] Nonetheless, the presence of ERs in both nucleus and mitochondria provide a possible mechanism for estrogen coordination of the expression of mtDNA and nuclear-encoded mitochondrial respiratory complex genes.[45]

Mitochondria ERα and ERβ participate in the coordination of the cell's energy requirement and in the OXPHOS biosynthesis, affecting ROS generation and induction of apoptosis (Fig. 9.2).[46]

In addition to roles in mitochondria, estrogen also suppresses ROS through other sites. Estrogen treatment reduces angiotensin II (Ang II)— and aldosterone signaling through the Ang II receptor 1 (AT1R) and the mineralocorticoid receptor (MR), respectively (Fig. 9.2).[47] Normally stimulation of the AT1R and the MR induces free radical, superoxide production, NADPH oxidase activity, and intracellular generation of ROS in VSMCs and ECs.[47] Some of our studies confirmed a critical role for increased AT1R and MR in conjunction with an altered redox-mediating impaired endothelial, cardiac, and renal function in the CRS.[48] Furthermore, in VSMCs estrogen treatment increases protein levels of SOD.[37] Although estrogen has only weak radical scavenging activity, estrogen is able to inhibit oxidative stress markers such as lipid peroxidation, protein oxidation, and DNA damage. In cell-free systems, estrogen inhibits iron-induced lipid peroxidation, LDL oxidation, cholesterol oxidation, and conjugated diene formation.[49] These potent antioxidant activities are likely due to a novel redox cycling mechanism. Thus, it could be expected that estrogen plays a crucial role in prevention of mitochondrial dysfunction and the associated development of the CRS.

Estrogen moderates a Ca^{2+}-induced permeability transition during excitoxicity that would clearly be beneficial in mitochondria function. Studies have found that estradiol augments sequestration of cytosolic Ca^{2+} in mitochondria since estrogen promotes influx of Ca^{2+} *via* L-type Ca^{2+} channels, which in turns activates the Src/ERK/CREB signaling pathways through activating the PI3K.[50] Further, estrogen increases mitochondrial sequestration of Ca^{2+}, preventing adverse consequences of excess cytoplasmic Ca^{2+}, and subsequently inhibit the cell apoptosis.[49] The death receptor pathway and mitochondrial pathway have been characterized to mediate apoptosis *via* bcl-2 family protein. Once the mitochondrial apoptotic pathway is activated, the mitochondrial permeability transition pore is open due to elevated mitochondrial calcium levels and excessive ROS. This is followed by cytochrome c release from mitochondria to cytosol with subsequent caspase activation initiating cell self-digestion and nuclear DNA fragmentation.[9,51] Therefore, estrogen may prevent cell apoptosis by suppressing mitochondrial oxidative stress, maintaining Ca^{2+} homeostasis, and increasing bcl-2 family expression in regulation of mitochondria function.

4. ABNORMALITIES IN ESTROGEN SIGNALING PROMOTES DEVELOPMENT OF THE CRS

Our investigation has shown that there are important sex-related differences in the development of diet-induced IR and other key components of the CRS.[6] Normally premenopausal women have a lower prevalence of heart and kidney disease relative to age-matched men. Estrogen participates in the modulation of insulin sensitivity, energy balance, and body composition as estrogen regulates the activity and expression of the key enzymes involved in glucose transport, glycolysis, the citric acid cycle, the mitochondrial respiratory chain, and fatty acid oxidation.[52]

4.1. Estrogen and ERs in CVD

Estrogen plays an important role in the regulation of vascular tone and in the pathophysiology of CVD. Physiological effects of estrogen are mediated through ERα and ERβ, which are both expressed in VSMCs and ECs.[53] Estrogen may promote endothelium-dependent vasodilation by increasing the releases of NO, prostacyclin (PGI_2), and endothelium-derived hyperpolarizing factor (EDHF). In comparison, estrogen decreases the release of endothelin (ET-1) and Ang II, which are potent vasoconstrictors and procoagulants.[8] Furthermore, estrogen can decrease sympathetic activity, circulating levels of norepinephrine, and blood pressure.[37] The mechanisms in estrogen-induced eNOS activation and NO expression on ECs are that estrogen activates PI3K/protein kinase B (Akt) pathways, which lead to phosphorylation and activation of eNOS and increases NO production.[54] Estrogen also increases eNOS activity by causing rapid ER-dependent activation of MAPK.[55] Furthermore, NO induced by estrogen in ECs diffuses into the smooth muscle cells of the blood vessel and interacts with soluble guanylate cyclase. NO stimulates the soluble guanylate cyclase to generate the second messenger cGMP from guanosine triphosphate (GTP). The soluble cGMP activates cyclic nucleotide-dependent protein kinase G (PKG).[56] PKG, a kinase in phosphorylating a number of proteins, regulates Ca^{2+} concentrations and sensitization, hyperpolarizes cell through potassium channels, and causes actin filament and myosin dynamic alterations, and subsequently results in smooth muscle relaxation.[57] Meanwhile, estrogen may activate ATP-sensitive K^+ channels or Ca^{2+}-activated K^+ channels in VSMCs, leading to membrane hyperpolarization, and decreased Ca^{2+} entry.[8] Additionally, estrogen may promote efflux of Ca^{2+} via plasmalemmal Ca^{2+} pump in VSMCs through inhibiting PKC signal pathway and Rho-kinase activity.[8] Thus, the decrease in estrogen levels in aging women is likely to cause an increase in the expression of Ca^{2+} channels and decreased Ca^{2+} pumps, leading to increased VSMC Ca^{2+}, vasoconstriction, and thereby increase the incidence of hypertension and CVD (Fig. 9.3).

In the context of vascular function, estrogen may play a role in the extracellular matrix (ECM) remodeling and reduction of ECM degradation.[58] Matrix metalloproteinase (MMPs) are involved in the regulation of the structural integrity of the ECM. When the vessel wall is exposed to immunological stress, inflammatory cells like polymorphonuclear neutrophils (PMNs) and macrophages, produce a variety of MMPs (MMP-1, -7, -8, -13) as well as elastase. MMPs degrade the ECM by affecting the production of weaker collagen and frayed elastin fibers.[59] The activity of these enzymes is regulated by augmented gene expression, posttranslational activation of cleavage of pro-MMPs, and interaction among MMPs, plasmin, thrombin and ROS.[60] Increased MMP activity induces ECM degradation within the atherosclerotic plaque and may be involved in plaque instability and CVD events. However, postmenopausal women and animals with diminished estrogen function are more prone to vascular and cardiac stiffness and associated vascular disease and impaired cardiac relaxation.

Studies have suggested that the influence of estrogen on insulin sensitivity is largely mediated by estrogen signaling through ERα.[61] Compared to ERα wild-type animals, knockout (KO) of the ERα in mice results in obesity at the expense of white adipose tissue expansion, reduced energy expenditure, IR, and glucose intolerance.[6] It has been revealed that ERα is a positive regulator of glucose transporter type 4 (GLUT4) expressions, whereas ERβ has a suppressive role probably through regulating PPARγ activity and inhibits expression of GLUT4 in the muscle.[41,62] In the cardiovascular system, ERα is found mainly in the uterine vasculature, whereas ERβ is more abundant in ECs and VSMCs from aorta, tail, and uterine vasculature.[63] ERβ seems to be localized in the central nervous system cardiovascular control centers with inhibitory neurons in the brain that normally reduce sympathetic activity. ERβ depletion would be associated with withdrawal of inhibitory tone such as that imparted by the parasympathetic system, thereby increasing peripheral resistance and resulting in elevations in blood pressure and associated myocardial hypertrophy.[37] Meanwhile, ERβ also plays a protective role in injured arteries by inhibiting neointima formation, which is the first step in the development of atherosclerosis in mice.[64] Furthermore, both ERα and ERβ regulate inflammatory responses.

ERα mediates estrogen abrogation of cytokine-induced expression of cell adhesion molecules in ECs. Alternatively, ERβ mediated the estrogenic abrogation of tumor necrosis factor α (TNF-α)-induced inflammation in cultured smooth muscle cells.[37] Thus, ERα and ERβ play important roles in insulin and glucose metabolism with actions on the liver, adipose tissue, muscle, and pancreatic β cells.

4.2. Estrogen regulates glucose homeostasis and IR

Insulin is released from pancreatic β cells in response to high glucose levels and stimulates the uptake of glucose in skeletal muscles and adipose tissue as glycogen through glycogenesis in the liver.[65] A recent study suggested that long term estrogen exposure increases insulin release without changing β cell mass and insulin target gene expression in mice.[41] Estrogen is known to influence glucose homeostasis with dominant effects in the liver and muscle. Study has found that ERα is a positive regulator of GLUT4 expressions; whereas ERβ has a suppressive role.[66] In addition to facilitating glucose transport, estrogen also promotes aerobic glycolysis by increasing activity of the glycolytic enzymes hexokinase, phosphofructokinase, and pyruvate kinase. Estrogen increases the expression of PDH which is the primary regulatory enzyme that links glycolysis to the TCA cycle.[67] Further, estrogen prompts the expression of the complex I, complex IV, and complex V/ATP synthase in mitochondria.[38] Our previous data are consistent with above studies and clearly show that glucose intolerance induced by high-fat diet was associated with decreased expression of ERα and GLUT4 in adipose tissue, and further confirmed systemic impairment of glucose homeostasis in the older cohort of ERαKO mice.[6] Together, these findings indicate that estrogen promotes enhanced utilization of glucose and maintains glucose homeostasis.

Estrogen is known as a regulator of insulin sensitivity in both women and men. Premenopausal women are more insulin sensitive with associated improved glucose tolerance and are more resistant to develop IR compared to men.[41] In particular, estrogens increase hepatic insulin sensitivity by decreasing gluconeogenesis and glycogenolysis, and increasing insulin release in islets of Langerhans. Estrogen also prevents β-cell apoptosis, reduces proinflammatory signaling, and improves insulin action.[68] The effects of estrogen signaling through ERs on insulin production have been studied *in vivo* using aromatase KO mice, which lack the enzyme responsible for conversion of androgens to estrogens.[69] Our data further confirmed the role for ERα in modulating IR. Global ERαKO in mice resulted in elevation of IR and inflammation in skeletal muscle along with impaired glucose tolerance.[6] Conversely, ERα agonist with the selective ligand propylpyrazoletriyl in ovariectomized wild-type rats increased insulin-stimulated glucose uptake in skeletal muscle through activating AKT and AMPK signaling pathways.[70] Therefore, postmenopausal women with lower endogenous estrogen levels may contribute to the sex differences-related IR, T2DM, and CVD in CRS.

4.3. Estrogen regulates lipogenesis and lipolysis

Adiposity is a growing global epidemic that increases the risk of diabetes, CVD, and CRS. Epidemiological studies show postmenopausal women have low estrogen levels, experience a general increase in weight, and redistribute adipose tissue, which lead to increased abdominal fat deposition.[71] The increased abdominal fat in postmenopausal women tends to be visceral and not subcutaneous fat.[38] Although estrogen increases the risk for the formation of cholesterol gallstones by promoting hepatic secretion of biliary cholesterol, estrogen decreases plasma LDL, and increases plasma HDL.[72] Estrogen significantly enhances the activity of enzyme 3-hydroxy-3-methylglutaryl coenzyme A (HMG-CoA) reductase, which is the rate-limiting enzyme in hepatic cholesterol biosynthesis under high dietary cholesterol loads, suggesting that there could be an increased delivery of cholesterol to bile from de novo synthesis in the liver.[41] A recent study showed that ERα and ERβ are expressed in both subcutaneous and visceral fat tissues, and premenopausal women have a less atherogenic lipid profile than men due to higher HDL and lower TG levels, which are closely associated with central fat accumulation.[8] Further, studies found that ERα predominantly regulates adipose homeostasis *via* growth and proliferation of adipocytes.[38] With the euglycemic-hyperinsulinemic clamp, a defective suppression of hepatic glucose output has been demonstrated in ERαKO mice, along with increased expression of genes involved in lipid biosynthesis and decreased expression of genes involved in lipid transport.[6] Thus, estrogen and its receptor may play important roles in adipose tissue biology.

4.4. Estrogen regulates inflammatory responses

The incidence of CRS and T2DM are shown to rise remarkably in women following menopause. It is becoming increasingly evident that chronic activation of proinflammatory pathways may be responsible for obesity-induced

IR and T2DM. The proinflammatory cytokines, such as TNFα, Interleukin 6 (IL6), and c-reactive protein (CRP) are elevated in patients with IR and T2DM.[41] Suppression of proinflammatory responses represents a promising strategy to combat obesity and IR in CRS. Therefore, it might be expected that estrogenic treatments would modulate inflammatory responses. Currently, the most consistent reports of estrogenic modulation of inflammation in vascular tissue involve activation of cytokine pathways in inflammation. Estrogen may decrease the expression of CRP in vascular injury and increases protein S-nitrosylation in human umbilical vein ECs directly *via* ERα and indirectly by eNOS upregulation, and in turn prevents Ang II-induced upregulation of intercellular cell adhesion molecule-1.[8] Many of the intracellular signaling cascades affected by activating toll-like receptors and interleukin molecules are cross-linked with ERs.[73] Activation of nuclear factor kappa B (NF-κB) is known to mediate a variety of chronic inflammatory diseases in CRS. Estrogen has been shown to inhibit NF-κB signaling in an ER-dependent manner through both ERα and ERβ.[74] However, the complexity of hormonal effects on inflammation is depended on a variety of factors, including the amount and composition of estrogens, the type of immune stimulus, the types of cells becoming inflamed, and the presence of hormone receptors.[37,74] Therefore, the impact and the mechanisms linked to sex hormone therapy in inflammation remain to be fully elucidated.

5. CONCLUSIONS

The CRS consists of a constellation of metabolic, cardiac, and renal abnormalities. As discussed throughout this review, CRS is closely related to mitochondria dysfunction and diminished estrogen signaling in the liver, adipose tissues, skeletal muscles, and vascular tissues. Although conventional estrogen therapy might beneficially affect CRS, it is noted that some side effects such as breast cancer and thromboembolism have been found in patient treatment with estrogen. Future investigation should further identify the potential interactions and cross-talk among ERα, ERβ, and GPR 30 in the regulation of body weight, IR, and CVD in both females and males. Meanwhile, a well-controlled and -designed clinical trial is needed to elucidate whether estrogen therapy could be effective in patients with CRS. This may help to define novel pharmacological targets, and thus we may be able to select cogent and knowledge CRS therapies.

ACKNOWLEDGMENTS

The authors would like to thank Brenda Hunter for her editorial assistance. This research was supported by NIH (R01 HL73101, R01 HL107910) and the Veterans Affairs Merit System (0018) for J. R. S. The authors have no conflict of interest associated with this manuscript.

REFERENCES

1. Sowers JR, Whaley-Connell A, Hayden MR. The role of overweight and obesity in the cardiorenal syndrome. *Cardiorenal Med.* 2011;1:5–12.
2. Ford ES, Li C, Zhao G. Prevalence and correlates of metabolic syndrome based on a harmonious definition among adults in the US. *J Diabetes.* 2010;2:180–193.
3. Jindal A, Brietzke S, Sowers JR. Obesity and the cardiorenal metabolic syndrome: therapeutic modalities and their efficacy in improving cardiovascular and renal risk factors. *Cardiorenal Med.* 2012;2:314–327.
4. Gallagher EJ, Leroith D, Karnieli E. Insulin resistance in obesity as the underlying cause for the metabolic syndrome. *Mt Sinai J Med.* 2010;77:511–523.
5. Jindal A, Whaley-Connell A, Sowers JR. Obesity and heart failure as a mediator of the cerebrorenal interaction. *Contrib Nephrol.* 2013;179:15–23.
6. Manrique C, DeMarco VG, Aroor AR, et al. Obesity and insulin resistance induce early development of diastolic dysfunction in young female mice fed a Western diet. *Endocrinology.* 2013;154:3632–3642.
7. Gaspard U. Hyperinsulinaemia, a key factor of the metabolic syndrome in postmenopausal women. *Maturitas.* 2009;62:362–365.
8. Masood DE, Roach EC, Beauregard KG, Khalil RA. Impact of sex hormone metabolism on the vascular effects of menopausal hormone therapy in cardiovascular disease. *Curr Drug Metab.* 2010;11:693–714.
9. Gao L, Laude K, Cai H. Mitochondrial pathophysiology, reactive oxygen species, and cardiovascular diseases. *Vet Clin North Am Small Anim Pract.* 2008;38:137–155.
10. Nisoli E, Clementi E, Carruba MO, Moncada S. Defective mitochondrial biogenesis: a hallmark of the high cardiovascular risk in the metabolic syndrome? *Circ Res.* 2007;100:795–806.
11. Mandavia CH, Aroor AR, Demarco VG, Sowers JR. Molecular and metabolic mechanisms of cardiac dysfunction in diabetes. *Life Sci.* 2013;92:601–608.
12. Zhang Y, Sowers JR, Ren J. Pathophysiological insights into cardiovascular health in metabolic syndrome. *Exp Diabetes Res.* 2012;2012:320534.
13. Pieczenik SR, Neustadt J. Mitochondrial dysfunction and molecular pathways of disease. *Exp Mol Pathol.* 2007;83:84–92.
14. Kim JA, Wei Y, Sowers JR. Role of mitochondrial dysfunction in insulin resistance. *Circ Res.* 2008;102:401–414.
15. Nicolson GL. Metabolic syndrome and mitochondrial function: molecular replacement and antioxidant supplements to prevent membrane peroxidation and restore mitochondrial function. *J Cell Biochem.* 2007;100:1352–1369.
16. Whaley-Connell A, McCullough PA, Sowers JR. The role of oxidative stress in the metabolic syndrome. *Rev Cardiovasc Med.* 2011;12:21–29.
17. Demarco VG, Whaley-Connell AT, Sowers JR, Habibi J, Dellsperger KC. Contribution of oxidative stress to pulmonary arterial hypertension. *World J Cardiol.* 2010;2:316–324.
18. Sivitz WI, Yorek MA. Mitochondrial dysfunction in diabetes: from molecular mechanisms to functional significance and therapeutic opportunities. *Antioxid Redox Signal.* 2010;12:537–577.

19. Sowers JR, Bakris GL, Black HR, Giles TD. The cardiometabolic syndrome and calcium channel blocker combination drugs. *J Cardiometab Syndr.* 2007;2:207–212.
20. Hayden MR, Tyagi SC, Kolb L, Sowers JR, Khanna R. Vascular ossification-calcification in metabolic syndrome, type 2 diabetes mellitus, chronic kidney disease, and calciphylaxis-calcific uremic arteriolopathy: the emerging role of sodium thiosulfate. *Cardiovasc Diabetol.* 2005;4:4.
21. Gunter TE, Buntinas L, Sparagna G, Eliseev R, Gunter K. Mitochondrial calcium transport: mechanisms and functions. *Cell Calcium.* 2000;28:285–296.
22. Turner N, Heilbronn LK. Is mitochondrial dysfunction a cause of insulin resistance? *Trends Endocrinol Metab.* 2008;19:324–330.
23. Sack MN. Type 2 diabetes, mitochondrial biology and the heart. *J Mol Cell Cardiol.* 2009;46:842–849.
24. Sharef SW, Al-Senaidi K, Joshi SN. Successful treatment of cardiomyopathy due to very long-chain acyl-CoA dehydrogenase deficiency: first case report from Oman with literature review. *Oman Med J.* 2013;28:354–356.
25. Whaley-Connell A, Sowers JR. Indices of obesity and cardiometabolic risk. *Hypertension.* 2011;58:991–993.
26. Liu J, Shen W, Zhao B, et al. Targeting mitochondrial biogenesis for preventing and treating insulin resistance in diabetes and obesity: hope from natural mitochondrial nutrients. *Adv Drug Deliv Rev.* 2009;61:1343–1352.
27. de Cavanagh EM, Inserra F, Ferder L. Angiotensin II blockade: a strategy to slow ageing by protecting mitochondria? *Cardiovasc Res.* 2011;89:31–40.
28. Sowers JR. Diabetes mellitus and vascular disease. *Hypertension.* 2013;61:943–947.
29. Rains JL, Jain SK. Oxidative stress, insulin signaling, and diabetes. *Free Radic Biol Med.* 2011;50:567–575.
30. Schiff M, Loublier S, Coulibaly A, Bénit P, de Baulny HO, Rustin P. Mitochondria and diabetes mellitus: untangling a conflictive relationship? *J Inherit Metab Dis.* 2009; 32:684–698.
31. Jindal A, Whaley-Connell A, Brietzke S, Sowers JR. Therapy of obese patients with cardiovascular disease. *Curr Opin Pharmacol.* 2013;13:200–204.
32. Aroor AR, Mandavia C, Ren J, Sowers JR, Pulakat L. Mitochondria and oxidative stress in the cardiorenal metabolic syndrome. *Cardiorenal Med.* 2012;2:87–109.
33. Gleyzer N, Scarpulla RC. PGC-1-related coactivator (PRC), a sensor of metabolic stress, orchestrates a redox-sensitive program of inflammatory gene expression. *J Biol Chem.* 2011;286:39715–39725.
34. Shao D, Liu Y, Liu X, et al. PGC-1 beta-regulated mitochondrial biogenesis and function in myotubes is mediated by NRF-1 and ERR alpha. *Mitochondrion.* 2010; 10:516–527.
35. Venditti P, Bari A, Di Stefano L, et al. Involvement of PGC-1, NRF-1, and NRF-2 in metabolic response by rat liver to hormonal and environmental signals. *Mol Cell Endocrinol.* 2009;305:22–29.
36. Yu Y, Zhang XH, Ebersole B, Ribnicky D, Wang ZQ. Bitter melon extract attenuating hepatic steatosis may be mediated by FGF21 and AMPK/Sirt1 signaling in mice. *Sci Rep.* 2013;3:3142.
37. Miller VM, Duckles SP. Vascular actions of estrogens: functional implications. *Pharmacol Rev.* 2008;60:210–241.
38. Rettberg JR, Yao J, Brinton RD. Estrogen: a master regulator of bioenergetic systems in the brain and body. *Front Neuroendocrinol.* 2014;35:8–30.
39. Handgraaf S, Riant E, Fabre A, et al. Prevention of obesity and insulin resistance by estrogens requires ERα activation function-2 (ERαAF-2), whereas ERαAF-1 is dispensable. *Diabetes.* 2013;62:4098–4108.

40. Velarde MC. Pleiotropic actions of estrogen: a mitochondrial matter. *Physiol Genomics*. 2013;45:106–109.
41. Faulds MH, Zhao C, Dahlman-Wright K, Gustafsson JÅ. The diversity of sex steroid action: regulation of metabolism by estrogen signaling. *J Endocrinol*. 2012;212:3–12.
42. Guo J, Duckles SP, Weiss JH, Li X, Krause DN. 17β-Estradiol prevents cell death and mitochondrial dysfunction by an estrogen receptor-dependent mechanism in astrocytes after oxygen-glucose deprivation/reperfusion. *Free Radic Biol Med*. 2012;52:2151–2160.
43. Johannsen DL, Ravussin E. The role of mitochondria in health and disease. *Curr Opin Pharmacol*. 2009;9:780–786.
44. Vasconsuelo A, Milanesi L, Boland R. Actions of 17β-estradiol and testosterone in the mitochondria and their implications in aging. *Ageing Res Rev*. 2013;12:907–917.
45. Klinge CM. Estrogenic control of mitochondrial function and biogenesis. *J Cell Biochem*. 2008;105:1342–1351.
46. Simoes DC, Psarra AM, Mauad T, et al. Glucocorticoid and estrogen receptors are reduced in mitochondria of lung epithelial cells in asthma. *PLoS One*. 2012;7:e39183.
47. Xue B, Johnson AK, Hay M. Sex differences in angiotensin II- and aldosterone-induced hypertension: the central protective effects of estrogen. *Am J Physiol Regul Integr Comp Physiol*. 2013;305:R459–R463.
48. Bender SB, McGraw AP, Jaffe IZ, Sowers JR. Mineralocorticoid receptor-mediated vascular insulin resistance: an early contributor to diabetes-related vascular disease? *Diabetes*. 2013;62:313–319.
49. Simpkins JW, Yi KD, Yang SH, Dykens JA. Mitochondrial mechanisms of estrogen neuroprotection. *Biochim Biophys Acta*. 1800;2010:1113–1120.
50. Brinton RD. The healthy cell bias of estrogen action: mitochondrial bioenergetics and neurological implications. *Trends Neurosci*. 2008;31:529–537.
51. Bhatt MP, Lim YC, Kim YM, Ha KS. C-peptide activates AMPKα and prevents ROS-mediated mitochondrial fission and endothelial apoptosis in diabetes. *Diabetes*. 2013;62:3851–3862.
52. Martins-Maciel ER, Campos LB, Salgueiro-Pagadigorria CL, Bracht A, Ishii-Iwamoto EL. Raloxifene affects fatty acid oxidation in livers from ovariectomized rats by acting as a pro-oxidant agent. *Toxicol Lett*. 2013;217:82–89.
53. O'Lone R, Knorr K, Jaffe IZ, et al. Estrogen receptors alpha and beta mediate distinct pathways of vascular gene expression, including genes involved in mitochondrial electron transport and generation of reactive oxygen species. *Mol Endocrinol*. 2007; 21:1281–1296.
54. Lee SA, Kim EY, Jeon WK, et al. The inhibitory effect of raloxifene on lipopolysaccharide-induced nitric oxide production in RAW264.7 cells is mediated through a ROS/p38 MAPK/CREB pathway to the up-regulation of heme oxygenase-1 independent of estrogen receptor. *Biochimie*. 2011;93:168–174.
55. Jaubert AM, Mehebik-Mojaat N, Lacasa D, Sabourault D, Giudicelli Y, Ribière C. Nongenomic estrogen effects on nitric oxide synthase activity in rat adipocytes. *Endocrinology*. 2007;148:2444–2452.
56. Dey NB, Lincoln TM. Possible involvement of cyclic-GMP-dependent protein kinase on matrix metalloproteinase-2 expression in rat aortic smooth muscle cells. *Mol Cell Biochem*. 2012;368:27–35.
57. Deenadayalu V, Puttabyatappa Y, Liu AT, Stallone JN, White RE. Testosterone-induced relaxation of coronary arteries: activation of BKCa channels via the cGMP-dependent protein kinase. *Am J Physiol Heart Circ Physiol*. 2012;302:H115–H123.
58. Neugarten J, Golestaneh L. Gender and the prevalence and progression of renal disease. *Adv Chronic Kidney Dis*. 2013;20:390–395.
59. Shirwany NA, Zou MH. Arterial stiffness: a brief review. *Acta Pharmacol Sin*. 2010; 31:1267–1276.

60. Dollery CM, McEwan JR, Henney AM. Matrix metalloproteinases and cardiovascular disease. *Circ Res.* 1995;77:863–868.
61. Matic M, Bryzgalova G, Gao H, et al. Estrogen signalling and the metabolic syndrome: targeting the hepatic estrogen receptor alpha action. *PLoS One.* 2013;8:e57458.
62. Deng JY, Hsieh PS, Huang JP, Lu LS, Hung LM. Activation of estrogen receptor is crucial for resveratrol-stimulating muscular glucose uptake via both insulin-dependent and -independent pathways. *Diabetes.* 2008;57:1814–1823.
63. Khalil RA. Potential approaches to enhance the effects of estrogen on senescent blood vessels and postmenopausal cardiovascular disease. *Cardiovasc Hematol Agents Med Chem.* 2010;8:29–46.
64. Krom YD, Pires NM, Jukema JW, et al. Inhibition of neointima formation by local delivery of estrogen receptor alpha and beta specific agonists. *Cardiovasc Res.* 2007; 73:217–226.
65. Aroor AR, McKarns S, Demarco VG, Jia G, Sowers JR. Maladaptive immune and inflammatory pathways lead to cardiovascular insulin resistance. *Metabolism.* 2013; 62:1543–1552.
66. Barros RP, Machado UF, Warner M, Gustafsson JA. Muscle GLUT4 regulation by estrogen receptors ERbeta and ERalpha. *Proc Natl Acad Sci U S A.* 2006;103:1605–1608.
67. O'Mahony F, Razandi M, Pedram A, Harvey BJ, Levin ER. Estrogen modulates metabolic pathway adaptation to available glucose in breast cancer cells. *Mol Endocrinol.* 2012;26:2058–2070.
68. Meyer MR, Clegg DJ, Prossnitz ER, Barton M. Obesity, insulin resistance and diabetes: sex differences and role of oestrogen receptors. *Acta Physiol (Oxf).* 2011;203:259–269.
69. Jones ME, Thorburn AW, Britt KL, et al. Aromatase-deficient (ArKO) mice have a phenotype of increased adiposity. *Proc Natl Acad Sci U S A.* 2000;97:12735–12740.
70. Khor VK, Dhir R, Yin X, Ahima RS, Song WC. Estrogen sulfotransferase regulates body fat and glucose homeostasis in female mice. *Am J Physiol Endocrinol Metab.* 2010;299:E657–E664.
71. Pansini F, Cervellati C, Guariento A, et al. Oxidative stress, body fat composition, and endocrine status in pre- and postmenopausal women. *Menopause.* 2008;15:112–118.
72. Stubbins RE, Najjar K, Holcomb VB, Hong J, Núñez NP. Oestrogen alters adipocyte biology and protects female mice from adipocyte inflammation and insulin resistance. *Diabetes Obes Metab.* 2012;14:58–66.
73. Huang SY, Xin H, Sun J, Li R, Zhang XM, Zhao D. Estrogen receptor β agonist diarylpropionitrile inhibits lipopolysaccharide-induced regulated on activation normal T cell expressed and secreted (RANTES) production in macrophages by repressing nuclear factor κB activation. *Fertil Steril.* 2013;100:234–240.
74. Wang L, Qiu XM, Hao Q, Li DJ. Anti-inflammatory effects of a Chinese herbal medicine in atherosclerosis via estrogen receptor β mediating nitric oxide production and NF-κB suppression in endothelial cells. *Cell Death Dis.* 2013;4:e551.

CHAPTER TEN

Advances in Development of Rechargeable Mitochondrial Antioxidants

Alexander N. Lukashev*,†, Maxim V. Skulachev*,‡, Victoria Ostapenko*, Alla Yu. Savchenko§, V.V. Pavshintsev*,‡, Vladimir P. Skulachev¶

*Institute of Mitoengineering, Lomonosov Moscow State University, Moscow, Russia
†Chumakov Institute of Poliomyelitis and Viral Encephalitides, Moscow, Russia
‡Faculty of Biology, Lomonosov Moscow State University, Moscow, Russia
§I.M. Sechenov First Moscow State Medical University, Moscow, Russia
¶A.N. Belozersky Institute of Physico-Chemical Biology, Lomonosov Moscow State University, Moscow, Russia

Contents

Abstract

It has been about 15 years since the introduction of the rechargeable mitochondria-targeted antioxidants (RMA). Two major groups have been developing RMA of the MitoQ and SkQ types independently, and many additional trials have been done by other researchers. This has provided solid preclinical evidence of RMA efficacy in various models. Human trials of systemic MitoQ were not followed by further advances, but the safety of MitoQ and, most likely, other RMA in humans has been demonstrated. A prooxidant effect at higher concentrations of RMA was described. For RMA of the SkQ type, a large window between anti- and prooxidant concentrations was observed, which makes SkQs promising as potential medicines. Significant RMA-induced improvements in many diseases that do not have an accepted treatment have been described. This justifies further clinical trials of RMA.

Progress in Molecular Biology and Translational Science, Volume 127
ISSN 1877-1173
http://dx.doi.org/10.1016/B978-0-12-394625-6.00010-6

1. INTRODUCTION

Mitochondria are not just power plants of the cell, but they are among the key players in a number of regulatory pathways activated by stress, acting both as signal commuters and as early victims of adverse effects. Mitochondria are central for the metabolism of reactive oxygen species (ROS), which are crucial for many functions of a cell and of organisms [reviewed in Ref. 1]. It is therefore not surprising that mitochondria were seen as an attractive therapeutic target in numerous diseases and pathological syndromes. In fact, the first mitochondrial drug, the uncoupler dinitrophenol, was introduced over 100 years ago. Dinitrophenol was suggested to reduce body weight. Unfortunately, the drug had to be withdrawn because of some adverse effects.[2] It is now obvious that it lacked specificity to mitochondria and affected the potential on all membranes in the cell. It was not until many decades later that basic studies of mitochondria precipitated in concepts of mitochondria-targeted molecules. The two ways to aim for mitochondria are specific targeting peptides[3,4] and membranophilic cations.[5,6] The first approach is useful to direct proteins, although tetrapeptides containing dimethyltyrosine can also be used as mitochondria-targeted antioxidants[7]; the second is suitable for small molecules. An obvious choice of such small molecules was antioxidants, because they lowered the ROS level. A large fraction of ROS in cells are produced in mitochondria, and mitochondria are also very vulnerable to ROS because they contain a high concentration of cardiolipin, one of the prominent ROS targets, and because they can produce additional ROS upon damage by ROS in a positive-feedback loop.[8] To fully benefit from targeting by membranophilic cations, such antioxidants should ideally be rechargeable, i.e., they should not just trap ROS, but become reduced for the next use. These requirements are perfectly fulfilled by quinones that are reduced (and thus recharged after trapping a free radical) in center *i* of the respiratory chain complex III, localized in the inner leaflet of the inner mitochondrial membrane. The first drug using this concept was MitoQ,[9] a fusion of ubiquinone and the membranophilic cation decyltriphenylphosphonium (Fig. 10.1). Later, a similar molecule using plastoquinone instead of ubiquinone was suggested.[10] This molecule, termed SkQ1 (Fig. 10.1), was shown to have a much broader therapeutic window. All antioxidants exhibit prooxidative properties at increased concentrations, the difference between antioxidant and prooxidant concentrations being much higher for SkQ1 than for MitoQ.[10] These two molecules, SkQ1

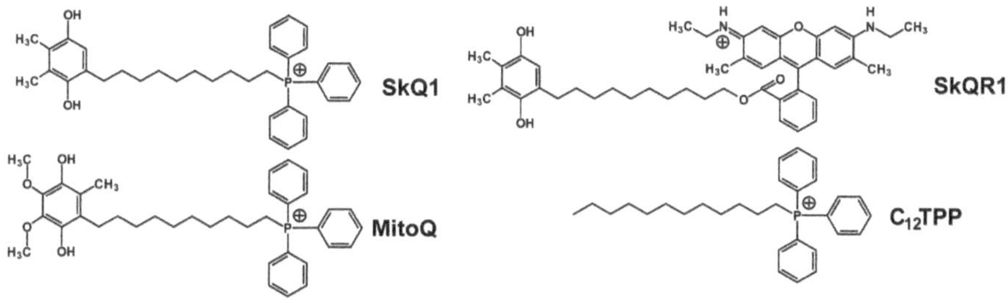

Figure 10.1 Structural formulas of SkQ1, SkQR1, MitoQ, and C_{12}TPP.

and MitoQ, further termed rechargeable mitochondrial-targeted antioxidants (RMA), were introduced over a decade ago and to date have been tested in a number of preclinical and clinical trials, which provided good reliability of the findings. This review summarizes recent advances in these trials and the prospect for the next generation of mitochondria-targeted small molecules.

2. ISCHEMIA–REPERFUSION

The most obvious therapeutic target for antioxidants is the ischemia–reperfusion syndrome, because the respiration burst after recovery of circulation may be even more damaging than ischemia *per se*. Indeed, mitochondrial antioxidants have been tested in animal models of ischemia–reperfusion. Pretreatment of rats with MitoQ (administered as 500 μM in drinking water, which roughly corresponds to 100 μmol/kg/day) significantly improved recovery of isolated rat heart after ischemia and reperfusion.[11] SkQ1 was tested in several models of ischemia and reperfusion.[12,13] Rats were pretreated with either SkQ1 or its rhodamine analog SkQR1 (Fig. 10.1). The latter compound has properties of a recyclable mitochondrial antioxidant, similar to SkQ1.[10] It is poorly suited for drug development because of limited stability, but it is convenient for experimental work because the rhodamine part allows tracking of the compound by fluorescent microscopy. Pretreatment of rats prior to ischemia–reperfusion of the heart led to significant reduction of postinfarction damage and arrhythmia. In a model of kidney ischemia (one kidney was completely removed, the other one was subjected to 1 h ischemia followed by reperfusion), SkQ1 could completely prevent death of animals (Fig. 10.2A) and significantly reduced biochemical markers of kidney insufficiency in the blood serum.

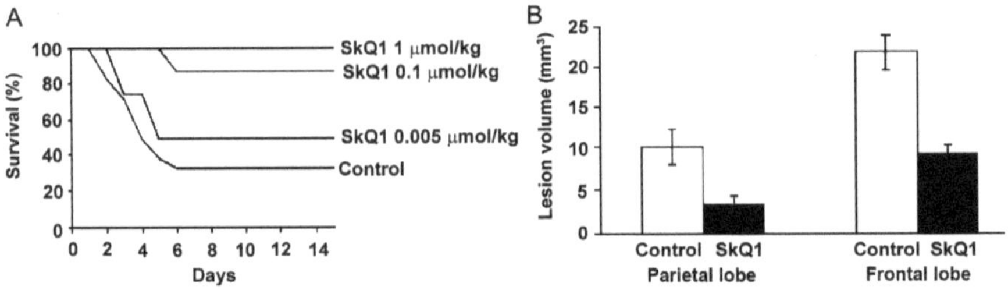

Figure 10.2 (A) Effect of intraperitoneal pretreatment with SkQ1 on survival after kidney ischemia–reperfusion in rats. (B) Effect of pretreatment with 1 μmol/kg intraperitoneal SkQR1 24 h prior to compression ischemia of brain in Wistar rats. *Panel (A) is reproduced from Ref. 14. Panel (B) is reproduced from Ref. 13.*

In a model of compression ischemia of the brain, SkQ1 reduced the volume of infarction more than twofold (Fig. 10.2B). Unfortunately, this success was hard to translate into clinical trials. First, in all *in vivo* studies pretreatment was required for the optimal effect of RMA. Such options are rarely available in humans. Second, experimental models are easy to standardize by severity of damage and timing of treatment, and pathological material is readily available for examination. In human patients, it is difficult to get precise readouts. It is likely that this line of research would get a second chance as a nonlisted indication after the drugs are approved for any other condition; however, clinical trials in a scheduled surgical operation setting are also feasible.

3. AGE-DEPENDENT DISORDERS

The grand concept behind the development of SkQ1 was to find and disable the trigger of senescence, which could be viewed as a suicide program rather than a gradual accumulation of occasional damage. It was suggested that such program might be controlled by mitochondria and mitochondrial ROS, which would thus be a valid target for "antiaging" drugs.[15] Indeed, SkQ1 prolonged the life spans of a number of species,[14,16–19] decreased manifestations of aging, especially in the OXYS rat lineage showing an accelerated senescence due to permanent oxidative stress,[20,21] and prevented or improved a plethora of age-related disorders [reviewed in Ref. 18]. Senescence *per se* is hardly suitable for clinical trials, and therefore the preclinical and clinical development of SkQ1 concentrated on particular diseases, which also include some age-related pathologies (see below).

4. LIVER PROTECTION

One of the most common clinical applications of conventional antioxidants is treatment of liver diseases, most prominently alcoholic and nonalcoholic fatty liver disease and viral hepatitis. MitoQ administered at 5 or 25 mg/kg/day significantly reduced the level of liver steatosis.[22] Our group conducted preclinical trials of SkQ1 in a number of animal models, including alcoholic fatty liver disease and carbon tetrachloride (CCl_4) and D-galactosamine (D-gal)-induced toxic hepatitis in Wistar rats (Lukashev A.N., unpublished data). There was no significant effect of SkQ1 in models of acute and chronic alcohol damage, which may be explained by lower doses (10–1250 nmol/kg/day) than those used in the MitoQ study. In other models, there was some beneficial effect of SkQ1 on the biochemical markers of liver damage (level of liver enzymes in blood serum); however, the effect was often not statistically significant and was inconsistent. A more significant result was seen upon pathological examination. SkQ1 reduced the area and severity of dystrophy in the CCl_4-induced toxic hepatitis model; the most efficient dose was 1250 nmol (670 μg) per kilogram daily administered starting 7 days before induction of hepatitis. The most pronounced effect of SkQ1 was seen in the D-gal model. D-gal sensitizes hepatocytes to trace qualities of lipopolysaccharide (LPS) that enter the blood from the intestine, and in some models D-gal is actually supplemented with LPS to increase toxicity. In our experiments, 3- to 4-month-old Wistar rats were treated with 500 mg/kg D-gal intraperitoneally. SkQ1 was given *per os* (via gavage) daily starting 7 days before induction of hepatitis. The rats were sacrificed 24 h after induction of hepatitis. SkQ1 (1250 nmol/kg/day) almost entirely prevented the development of dystrophy and significantly reduced pathological signs of inflammation (Fig. 10.3), which was accompanied by decrease in ALT, AST, and LDH serum levels. The high efficacy of SkQ1 in the D-gal model that relies on the damaging effect of LPS is consistent with the report of a protective effect of MitoQ in an LPS/peptidoglycan model of sepsis.[23]

There are no published data on preclinical trials of MitoQ in liver disease models. In 2008, there was a clinical trial of MitoQ in hepatitis C patients.[24] While the drug significantly reduced the level of AST and ALT, it had no effect on viral replication. The advent of novel direct-action antiviral drugs against hepatitis C[25] might eliminate the need for accessory liver protective

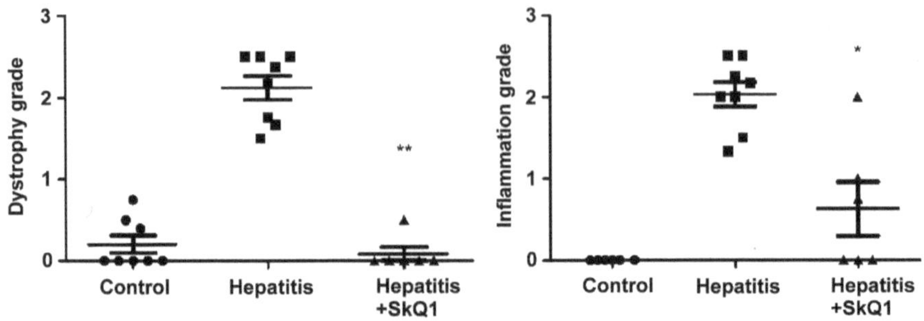

Figure 10.3 Score of pathological signs of dystrophy and inflammation in liver of Wistar rats with toxic hepatitis induced by D-galactosamine. $*p<0.05$; $**p<0.01$, Kruskal–Wallis test, Dunn's postcorrection.

therapy in hepatitis C, so the future of antioxidant treatment in viral hepatitis is unclear. Nevertheless, RMA certainly hold promise as liver protective drugs.

5. INFLAMMATION

ROS have been implicated in many stages of inflammation. While the extracellular ROS produced by NAPDH oxidases are the key mediators and effectors of inflammation, mitochondria and mitochondrial ROS play important roles in inflammation signaling.[26] Since there was a significant theoretical and practical background for use of any antioxidants to treat pathological inflammation, mechanistic and preclinical studies of RMA developed in parallel.

Even before the onset of inflammation, RMA can affect the immune status. Long-term administration of SkQ1 in mice delayed senescence-associated myeloid shift in hematopoiesis (from T cells to neutrophils).[27] Another study confirmed that SkQ1 affects hematopoiesis in adult mice. Treatment of mice with doses as low as 5 and 50 nmol (2.6 and 26 μg) SkQ1/kg/day for 21 days slightly decreased the percentage of $CD8^+$ (but not $CD4^+$) cells, and significantly decreased the percentage of granulocytes.[28] Therefore, it can be speculated that SkQ1 preconditioned the immune system to be "smart, not strong" and avoid over-reactive response. In line with this theoretical evidence, SkQ1 reduced inflammation in experimental bacterial pyelonephritis (which is often excessive for the clearance of a pathogen and may become the main damaging factor), and one proved mechanism was systemic decrease in neutrophil levels.[29] This finding might explain the mechanism underlying reduced pathological signs of inflammation (mainly the neutrophil infiltration) observed in many models. A wide range of SkQ1 doses reduced neutrophil infiltration in liver of mice with D-gal hepatitis (see above). Reduction in initial neutrophil infiltration (7 hours) also underlies the effect of topical or systemic SkQ1 administration, which strongly accelerated wound healing in elder rats.[30]

Besides a decrease in systemic neutrophil levels, RMA may affect activation of endothelium in inflammation and thus decrease infiltration. It has been shown that SkQ1 prevents TNF-α-induced overexpression of ICAM1 in endothelium *in vitro* and *in vivo*, which might decrease binding and extravasation of immune cells.[31] However, because this effect was also observed upon treatment with C_{12}TPP, it is likely that it is mediated by the uncoupling effect of SkQ1 (see below).

The earliest players in inflammation are dendritic cells (DCs). The NLRP3 inflammasome is the key regulatory complex in activation of DCs, and mitochondria play an important role in activation of the inflammasome.[32] It has been shown that MitoQ alleviated symptoms of experimental colitis in mice, at least in part by preventing activation of the NLRP3 inflammasome in DCs and release of IL-1b and IL-18.[33] The MitoQ effect consisted of both decrease in transcription of interleukin mRNA and in suppression of caspase-1 involved in their activation; therefore, the effect of RMA on DCs may be multimodal and requires further study.

The major therapeutic challenges involving pathological inflammation are autoimmune diseases. MitoQ was shown to improve clinical score in experimental autoimmune encephalomyelitis, a common experimental model of multiple sclerosis. Both pretreatment and treatment regimens were efficacious, and the beneficial effect was attributed to direct neuroprotection rather than suppression of inflammation.[34] The effect of SkQ1 in a model of collagen-induced arthritis in rats varied depending on the lineage of the rats and the administration regimen from complete prevention of disease in most animals to significant improvement of some pathological signs (Lukashev *et al.*, manuscript in preparation).

After inflammation has already developed, RMA may have the ability to alleviate the unwanted consequences on the organismal level. In a model of acute sepsis in mice induced by injection of LPS and peptidoglycan, MitoQ decreased the level of liver enzymes, IL-6, and IL-10 in the blood.[35] Mito-E (vitamin E coupled to TPP for mitochondrial targeting) and melatonin demonstrated even more pronounced beneficial effect in that study.

6. NEURODEGENERATIVE DISEASES

ROS and mitochondria are involved in the pathogenesis of a number of neurodegenerative diseases including Parkinson's disease (PD), Alzheimer's disease (AD), Huntington's disease, and amyotrophic lateral sclerosis (ALS) [reviewed in Ref. 36,37]. Therefore, neurodegenerative disorders are attractive targets for RMA.

There are many studies indicating that amyloid beta (Aβ) initiates ROS production *in vitro* and *in vivo*, and these ROS primarily originate from mitochondria. It was also found that Aβ accumulates in the mitochondrial matrix and can directly damage mitochondria.[38] Both MitoQ and SkQ1 protected neurons from Aβ in brain slices *ex vivo*.[39,40] MitoQ reduced neurological deficit in a triple-transgene model of AD,[40] while SkQ1 administered to OXYS rats improved age-related neurological symptoms, which were similar to those seen in AD patients.[41] Importantly, in both studies, RMA prevented accumulation of Aβ; therefore, at least a part of their effect was due to slowing disease progression, not just compensation of neurological deficit. The body of preclinical data, the absence of effective therapy, and the burden of AD make it one of the most promising applications for RMA.

Mitochondrial dysfunction has also been widely implicated in PD. Ubiquinone, the active part of MitoQ, has been tested in PD in a number of clinical trials [reviewed in Ref. 42]. The effects were beneficial, but limited and poorly reproducible. There are no publications on preclinical testing of RMA in PD. Clinical trials of MitoQ in PD were unsuccessful.[43]

ALS is another neurodegenerative disease with many implications of mitochondrial involvement.[36,44] The most common animal model of ALS is the superoxide dismutase (SOD1) transgenic mouse SOD1-G93A.[45] These mice develop early neurological manifestations similar to ALS and die before the age of 5–7 months. MitoQ administered in drinking water from 90 days of age (onset of symptoms) at about 100 μmol/kg/day prolonged survival of SOD1-G93A mice by 6 days (medial) and improved some pathological and clinical signs of the disease.[46]

SkQ1 was also tested in SOD1-G93A mice. Briefly, we used B6SJL-Tg (SOD1-G93A)dl1Gur/J mice (the Pushchino breeding facility) that usually develop progressive neurological deficit and die at 8–9 months of age. At day 120 (before onset of the disease), the animals were weighed and randomized into groups of 20–22 mice with even sex ratio. Males and females were analyzed separately because of the known sexual dimorphism of this model. SkQ1 was administered in drinking water at 250 nmol/kg/day (i.e. 400× less than MitoQ in a similar model). Neurophysiological parameters were assayed in the open field test weekly. In early disease stages, all behavioral and physiological parameters were significantly improved in the treated group compared to the control (Fig. 10.4). For example, at day 215, the treated female mice demonstrated twice higher mean speed and 20% higher grip strength, and 7% higher mean body weight. Unfortunately, after initial improvement, these parameters progressively deteriorated in the treated group, and before death there was no significant difference from the control group. SkQ1 did not have a significant effect on the life span of the mice, which was 270–310 days for females and 255–290 days for males in all groups. Therefore, SkQ1 did not prolong life of SOD1-G93A mice and did not halt the disease progression, but it significantly improved neurological manifestations and "quality of life" at the medium stage of the disease. The moderate effect of both RMA in the ALS model makes it an unlikely choice for the first clinical trials of RMA as neuroprotectors.

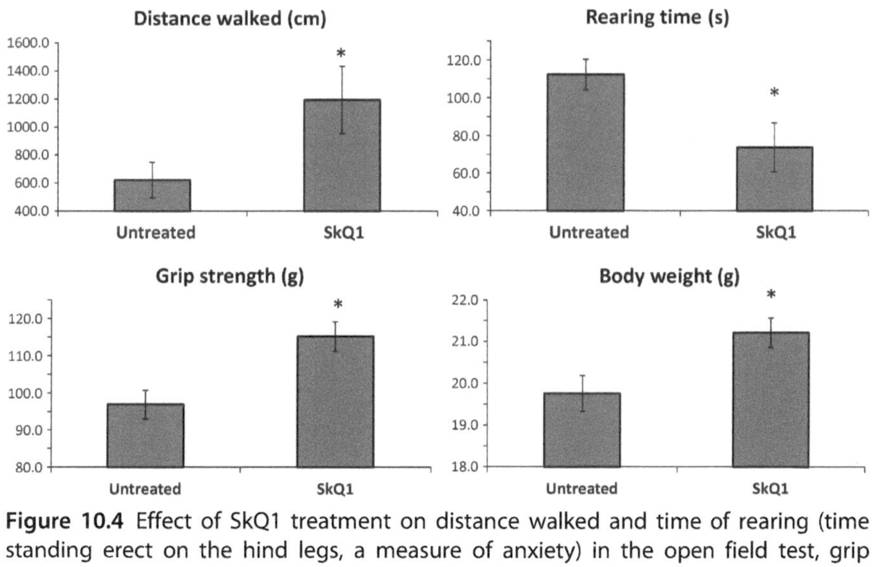

Figure 10.4 Effect of SkQ1 treatment on distance walked and time of rearing (time standing erect on the hind legs, a measure of anxiety) in the open field test, grip strength, and body weight in B6SJL-Tg(SOD1-G93A)dl1Gur/J transgenic mice (amyotrophic lateral sclerosis model). *Mann–Whitney $p < 0.05$.

7. OPHTHALMIC DISEASES

ROS have been implicated in many ophthalmic diseases, and it was technically very convenient to test RMA in ophthalmic applications because topical administration avoided many uncertainties of systemic administration of the drugs. SkQ1 administered as eye drops (250 nM solution) was very effective in age-related retinopathy in OXYS rats (Fig. 10.5). In veterinary trials, SkQ1 restored lost vision in 61 of 91 dogs, cats, and horses with recent blindness due to retinopathy, and it improved vision in about 50% animals with impaired vision.[14,18,21] This result is especially impressive knowing that the only current efficacious treatment of retinopathy is based on inhibition of vascular endothelial growth factor,[47] an expensive treatment with potential side effects. Clinical improvement was also recorded in 102 of 105 animals with uveitis and conjunctivitis. This study might be the basis for clinical trials of SkQ1 in humans. SkQ1 prevented the development of age-related cataract or reduced clinical grade of an already present cataract in OXYS rats (Fig. 10.5).[21]

A stable pharmaceutical composition of SkQ1 in the form of ophthalmic solution (eye drops) was developed. It was found that SkQ1 solutions in water are unstable at concentrations below 10 μM. However, such solutions were successfully stabilized by addition of another membranophilic cation (benzalconium chloride), hydroxypropyl methylcellulose, and a buffer solution, pH 6.5–6.8. Such eye drops containing 250 nM SkQ1 were termed Visomitin. Further studies showed that all components of Visomitin, including SkQ1, remain stable for at least 12 months when stored at 2–8 °C. It was

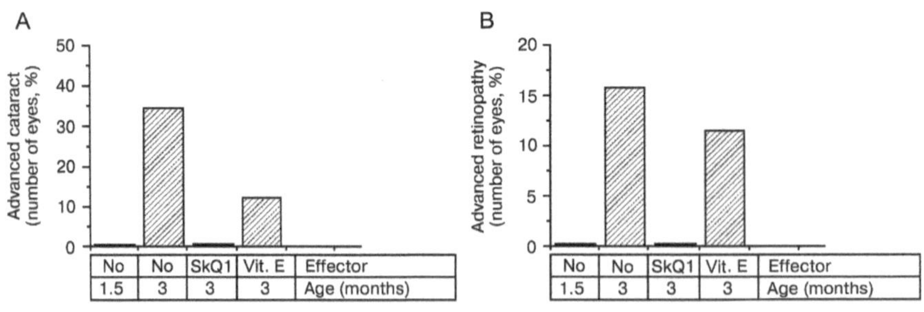

Figure 10.5 Comparison of protective effects of SkQ1 (50 nmol/kg per day) and vitamin E (500 μmol/kg per day) against cataract (A) and retinopathy (B) in OXYS rats. *Reproduced from Ref. 21.*

found that SkQ1 in this solution is very sensitive to the light, and therefore, it should be stored in darkness or in dark vials.[48]

In preclinical trials, Visomitin prevented the development of age-related cataract and reduced clinical grade of an already present cataract in OXYS rats (Fig. 10.5).[21] In another series of preclinical studies, Visomitin showed good efficacy in two models of dry eye syndrome (M. V. Skulachev *et al.*, paper in preparation).

The first human clinical trials of SkQ1 were set up for dry eye syndrome because it is a common and relatively benign condition, which nevertheless allows studying beneficial effects of the drug. Phase I–II trials compared the efficacy of Visomitin eye drops (155 ng SkQ1/ml, 2 mg hydroxypropyl methylcellulose/ml, accessory compounds) to Tears Naturale manufactured by Alcon (2 mg hydroxypropyl methylcellulose/ml, 1 mg dextran 70/ml, accessory compounds). Over 21 days of treatment, Visomitin almost completely reduced hyperemia and edema of the conjunctiva, while the reference drug had only a marginal effect (Fig. 10.6).[49] Visomitin completely healed corneal microerosions compared to a threefold reduction in the control group, and it produced a small (0.03 units) but statistically significant improvement in vision compared to none in the control group. Results of Phase III are currently being prepared for publication. Visomitin eye drops are now marketed in Russia (160,000 vials are already sold by drugstores), and a Phase II trial in dry eye syndrome has been initiated in the United States. It is likely that the list of ophthalmic indications of SkQ1 would be quickly expanded because of its efficacy in retinopathy, cataract,

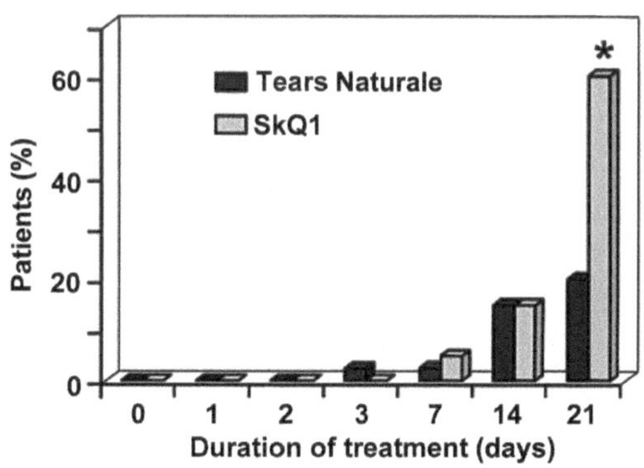

Figure 10.6 Comparison of efficacy of SkQ1-based eye drops "Visomitin" and "Tears Naturale" (Alcon) in clinical trials on patients suffering from dry eye syndrome. Percentage of successful cases is indicated. $^{*}p < 0.001$. A case was considered successful if the dry eye symptoms disappeared completely. *Reproduced from Ref. 49.*

and uveitis in experimental models and in veterinary applications[21] and anecdotal evidence of efficacy in humans with cataract (V.P. Skulachev, unpublished).

8. NOVEL MITOCHONDRIAL ANTIOXIDANTS

Over the last 15 years, mitochondrial antioxidants have evolved from targeted expendable antioxidants, which were already more efficient than general antioxidants,[50] to recyclable antioxidants that have nanomolar working concentrations and a significantly wide therapeutic window. It has been noted that the antioxidant properties of TPP-linked quinones improved upon substitution of methoxy groups in ubiquinone to methyl or hydrogen.[51] This observation led to the development of an SkQ1 analog based on thymoquinone, an herbal antioxidant that underlies the pharmacological activity of black cumin known since biblical times. This molecule, termed SkQT1, exhibited the penetrating and recyclable antioxidant properties, just as SkQ1, but had the same antioxidant activity as SkQ1 at about 10× lower concentration, while its prooxidant properties manifested at concentrations about twice higher than SkQ1.[51] Therefore, SkQT1 has a 600-fold theoretical therapeutic window, 30× better than that of SkQ1.[51]

9. MILD UNCOUPLING

Rechargeable mitochondrial antioxidants efficiently decrease the production of mitochondrial ROS level. Another strategy to achieve this goal is mild uncoupling, i.e., partial decrease in the electric potential on the mitochondrial membrane.[52] A number of membranophilic ions can perform this function, and it turned out that C_{12}-TPP and C_{12}-rhodamine, the transporter parts of RMA molecules, also decrease electric potential in mitochondria even in the absence of the quinone part.[53] In line with this theoretical evidence, C_{12}-rhodamine protected kidneys from damage in ischemia–reperfusion and rhabdomyolysis mouse models, and brain in an ischemia–reperfusion model, although it was less potent than SkQR1.[54] Therefore, mild uncoupling may add to the mitochondrial protection exhibited by RMA, and it is currently not known which of the reported effects of RMA can be attributed to uncoupling rather than direct antioxidant effect. In future, it may be recommended to include a membranophilic cation backbone (C_{12}TPP or C_{12}-rhodamine) as a control in basic experiments with RMA. As this effect is synergistic with the direct antioxidant action in most settings, this requirement may not be necessary for preclinical studies.

REFERENCES

1. Skulachev VP. Cationic antioxidants as a powerful tool against mitochondrial oxidative stress. *Biochem Biophys Res Commun*. 2013;441:275–279.
2. Grundlingh J, Dargan PI, El-Zanfaly M, Wood DM. 2,4-Dinitrophenol (DNP): a weight loss agent with significant acute toxicity and risk of death. *J Med Toxicol*. 2011;7:205–212.
3. Rosenberg LE, Fenton WA, Horwich AL, Kalousek F, Kraus JP. Targeting of nuclear-encoded proteins to the mitochondrial matrix: implications for human genetic defects. *Ann N Y Acad Sci*. 1986;488:99–108.
4. Sakhrani NM, Padh H. Organelle targeting: third level of drug targeting. *Drug Des Devel Ther*. 2013;7:585–599.
5. Liberman EA, Topaly VP, Tsofina LM, Jasaitis AA, Skulachev VP. Mechanism of coupling of oxidative phosphorylation and the membrane potential of mitochondria. *Nature*. 1969;222:1076–1078.
6. Bakeeva LE, Grinius LL, Jasaitis AA, et al. Conversion of biomembrane-produced energy into electric form. II. Intact mitochondria. *Biochim Biophys Acta*. 1970;216:13–21.
7. Zhao K, Zhao GM, Wu D, et al. Cell-permeable peptide antioxidants targeted to inner mitochondrial membrane inhibit mitochondrial swelling, oxidative cell death, and reperfusion injury. *J Biol Chem*. 2004;279:34682–34690.
8. Zorov DB, Filburn CR, Klotz LO, Zweier JL, Sollott SJ. Reactive oxygen species (ROS)-induced ROS release: a new phenomenon accompanying induction of the mitochondrial permeability transition in cardiac myocytes. *J Exp Med*. 2000;192:1001–1014.
9. Kelso GF, Porteous CM, Coulter CV, et al. Selective targeting of a redox-active ubiquinone to mitochondria within cells: antioxidant and antiapoptotic properties. *J Biol Chem*. 2001;276:4588–4596.
10. Antonenko YN, Avetisyan AV, Bakeeva LE, et al. Mitochondria-targeted plastoquinone derivatives as tools to interrupt execution of the aging program. 1. Cationic plastoquinone derivatives: synthesis and in vitro studies. *Biochemistry (Mosc)*. 2008;73:1273–1287.
11. Adlam VJ, Harrison JC, Porteous CM, et al. Targeting an antioxidant to mitochondria decreases cardiac ischemia-reperfusion injury. *FASEB J*. 2005;19:1088–1095.
12. Plotnikov EY, Silachev DN, Chupyrkina AA, et al. New-generation Skulachev ions exhibiting nephroprotective and neuroprotective properties. *Biochemistry (Mosc)*. 2010;75:145–150.
13. Bakeeva LE, Barskov IV, Egorov MV, et al. Mitochondria-targeted plastoquinone derivatives as tools to interrupt execution of the aging program. 2. Treatment of some ROS- and age-related diseases (heart arrhythmia, heart infarctions, kidney ischemia, and stroke). *Biochemistry (Mosc)*. 2008;73:1288–1299.
14. Skulachev VP, Anisimov VN, Antonenko YN, et al. An attempt to prevent senescence: a mitochondrial approach. *Biochim Biophys Acta*. 2009;1787:437–461.
15. Skulachev VP, Longo VD. Aging as a mitochondria-mediated atavistic program: can aging be switched off? *Ann N Y Acad Sci*. 2005;1057:145–164.
16. Anisimov VN, Bakeeva LE, Egormin PA, et al. Mitochondria-targeted plastoquinone derivatives as tools to interrupt execution of the aging program. 5. SkQ1 prolongs lifespan and prevents development of traits of senescence. *Biochemistry (Mosc)*. 2008;73:1329–1342.
17. Krementsova AV, Roshina NV, Tsybul'ko EA, Rybina OY, Symonenko AV, Pasyukova EG. Reproducible effects of the mitochondria-targeted plastoquinone derivative SkQ1 on *Drosophila melanogaster* lifespan under different experimental scenarios. *Biogerontology*. 2012;13:595–607.
18. Skulachev MV, Antonenko YN, Anisimov VN, et al. Mitochondrial-targeted plastoquinone derivatives. Effect on senescence and acute age-related pathologies. *Curr Drug Targets*. 2011;12:800–826.

19. Anisimov VN, Egorov MV, Krasilshchikova MS, et al. Effects of the mitochondria-targeted antioxidant SkQ1 on lifespan of rodents. *Aging (Albany NY)*. 2011;3:1110–1119.
20. Kolosova NG, Stefanova NA, Muraleva NA, Skulachev VP. The mitochondria-targeted antioxidant SkQ1 but not N-acetylcysteine reverses aging-related biomarkers in rats. *Aging (Albany NY)*. 2012;4:686–694.
21. Neroev VV, Archipova MM, Bakeeva LE, et al. Mitochondria-targeted plastoquinone derivatives as tools to interrupt execution of the aging program. 4. Age-related eye disease. SkQ1 returns vision to blind animals. *Biochemistry (Mosc)*. 2008;73:1317–1328.
22. Chacko BK, Srivastava A, Johnson MS, et al. Mitochondria-targeted ubiquinone (MitoQ) decreases ethanol-dependent micro and macro hepatosteatosis. *Hepatology*. 2011;54:153–163.
23. Lowes DA, Thottakam BM, Webster NR, Murphy MP, Galley HF. The mitochondria-targeted antioxidant MitoQ protects against organ damage in a lipopolysaccharide-peptidoglycan model of sepsis. *Free Radic Biol Med*. 2008;45:1559–1565.
24. Gane EJ, Weilert F, Orr DW, et al. The mitochondria-targeted anti-oxidant mitoquinone decreases liver damage in a phase II study of hepatitis C patients. *Liver Int*. 2010;30:1019–1026.
25. Au JS, Pockros PJ. Novel therapeutic approaches for hepatitis C. *Clin Pharmacol Ther*. 2014;95:78–88.
26. Pelletier M, Lepow TS, Billingham LK, Murphy MP, Siegel RM. New tricks from an old dog: mitochondrial redox signaling in cellular inflammation. *Semin Immunol*. 2012;24:384–392.
27. Shipounova IN, Svinareva DA, Petrova TV, et al. Reactive oxygen species produced in mitochondria are involved in age-dependent changes of hematopoietic and mesenchymal progenitor cells in mice. A study with the novel mitochondria-targeted antioxidant SkQ1. *Mech Ageing Dev*. 2010;131:415–421.
28. Yang Y, Karakhanova S, Soltek S, Werner J, Philippov PP, Bazhin AV. In vivo immunoregulatory properties of the novel mitochondria-targeted antioxidant SkQ1. *Mol Immunol*. 2012;52:19–29.
29. Plotnikov EY, Morosanova MA, Pevzner IB, et al. Protective effect of mitochondria-targeted antioxidants in an acute bacterial infection. *Proc Natl Acad Sci U S A*. 2013;110:E3100–E3108.
30. Demianenko IA, Vasilieva TV, Domnina LV, et al. Novel mitochondria-targeted antioxidants, "Skulachev-ion" derivatives, accelerate dermal wound healing in animals. *Biochemistry (Mosc)*. 2010;75:274–280.
31. Romashchenko VP, Galkin II, Zakharova VV, et al. Mitochondria-targeted compounds decrease TNF-α-dependent endothelium activation. *FEBS J*. 2013;280(Suppl 1):246.
32. Tschopp J, Schroder K. NLRP3 inflammasome activation: the convergence of multiple signalling pathways on ROS production? *Nat Rev Immunol*. 2010;10:210–215.
33. Dashdorj A, Jyothi KR, Lim S, et al. Mitochondria-targeted antioxidant MitoQ ameliorates experimental mouse colitis by suppressing NLRP3 inflammasome-mediated inflammatory cytokines. *BMC Med*. 2013;11:178.
34. Mao P, Manczak M, Shirendeb UP, Reddy PH. MitoQ, a mitochondria-targeted antioxidant, delays disease progression and alleviates pathogenesis in an experimental autoimmune encephalomyelitis mouse model of multiple sclerosis. *Biochim Biophys Acta*. 1832;2013:2322–2331.
35. Lowes DA, Webster NR, Murphy MP, Galley HF. Antioxidants that protect mitochondria reduce interleukin-6 and oxidative stress, improve mitochondrial function, and reduce biochemical markers of organ dysfunction in a rat model of acute sepsis. *Br J Anaesth*. 2013;110:472–480.

36. Chaturvedi RK, Beal MF. Mitochondrial approaches for neuroprotection. *Ann N Y Acad Sci*. 2008;1147:395–412.
37. Skulachev VP. Mitochondria-targeted antioxidants as promising drugs for treatment of age-related brain diseases. *J Alzheimers Dis*. 2012;28:283–289.
38. Pagani L, Eckert A. Amyloid-beta interaction with mitochondria. *Int J Alzheimers Dis*. 2011;2011:925050.
39. Kapay NA, Popova OV, Isaev NK, et al. Mitochondria-targeted plastoquinone antioxidant SkQ1 prevents amyloid-beta-induced impairment of long-term potentiation in rat hippocampal slices. *J Alzheimers Dis*. 2013;36:377–383.
40. McManus MJ, Murphy MP, Franklin JL. The mitochondria-targeted antioxidant MitoQ prevents loss of spatial memory retention and early neuropathology in a transgenic mouse model of Alzheimer's disease. *J Neurosci*. 2011;31:15703–15715.
41. Stefanova NA, Muraleva NA, Skulachev VP, Kolosova NG. Alzheimer's disease-like pathology in senescence-accelerated OXYS rats can be partially retarded with mitochondria-targeted antioxidant SkQ1. *J Alzheimers Dis*. 2014;38:681–694.
42. Chaturvedi RK, Beal MF. Mitochondria targeted therapeutic approaches in Parkinson's and Huntington's diseases. *Mol Cell Neurosci*. 2013;55:101–114.
43. Snow BJ, Rolfe FL, Lockhart MM, et al. A double-blind, placebo-controlled study to assess the mitochondria-targeted antioxidant MitoQ as a disease-modifying therapy in Parkinson's disease. *Mov Disord*. 2010;25:1670–1674.
44. Federico A, Cardaioli E, Da Pozzo P, Formichi P, Gallus GN, Radi E. Mitochondria, oxidative stress and neurodegeneration. *J Neurol Sci*. 2012;322:254–262.
45. Gurney ME, Pu H, Chiu AY, et al. Motor neuron degeneration in mice that express a human Cu, Zn superoxide dismutase mutation. *Science*. 1994;264:1772–1775.
46. Miquel E, Cassina A, Martinez-Palma L, et al. Neuroprotective effects of the mitochondria-targeted antioxidant MitoQ in a model of inherited amyotrophic lateral sclerosis. *Free Radic Biol Med*. 2014;70:204–213.
47. Miller JW, Le Couter J, Strauss EC, Ferrara N. Vascular endothelial growth factor a in intraocular vascular disease. *Ophthalmology*. 2013;120:106–114.
48. Tsarev DA, Zverev RS, Kazmina EM, Tashlitsky VN, Skulachev VP. Evaluating the stability of a cationic plastoquinone derivative (PDTP) in visomitin eye drops. *Pharm Chem J*. 2013;47:219–224.
49. Yani EV, Katargina LA, Chesnokova NB, et al. The first experience of using the drug Vizomitin in the treatment of "dry eyes" *Pract Med*. 2012;4:134–137.
50. Smith RA, Porteous CM, Coulter CV, Murphy MP. Selective targeting of an antioxidant to mitochondria. *Eur J Biochem*. 1999;263:709–716.
51. Severina II, Severin FF, Korshunova GA, et al. In search of novel highly active mitochondria-targeted antioxidants: thymoquinone and its cationic derivatives. *FEBS Lett*. 2013;587:2018–2024.
52. Korshunov SS, Skulachev VP, Starkov AA. High protonic potential actuates a mechanism of production of reactive oxygen species in mitochondria. *FEBS Lett*. 1997;416:15–18.
53. Severin FF, Severina II, Antonenko YN, et al. Penetrating cation/fatty acid anion pair as a mitochondria-targeted protonophore. *Proc Natl Acad Sci U S A*. 2010;107:663–668.
54. Plotnikov EY, Silachev DN, Jankauskas SS, et al. Mild uncoupling of respiration and phosphorylation as a mechanism providing nephro- and neuroprotective effects of penetrating cations of the SkQ family. *Biochemistry (Mosc)*. 2012;77:1029–1037.

INDEX

Note: Page numbers followed by "*f*" indicate figures and "*t*" indicate tables.

S

T

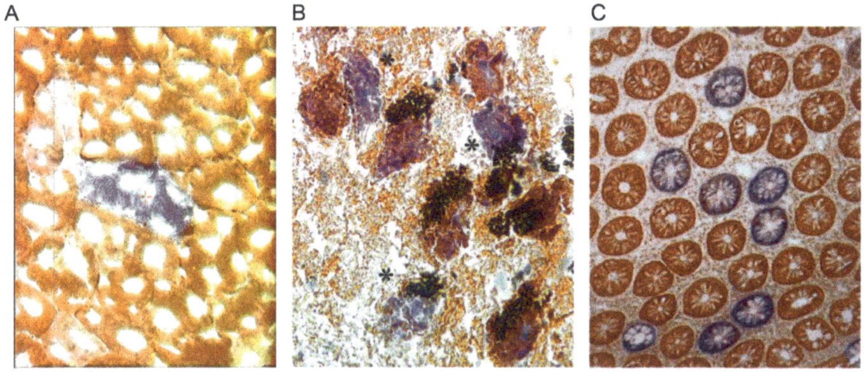

Konstantin Khrapko and Doug Turnbull, Figure 2.1

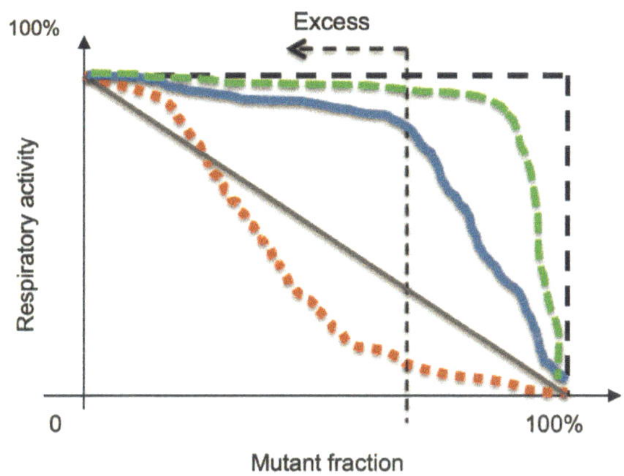

Konstantin Khrapko and Doug Turnbull, Figure 2.2

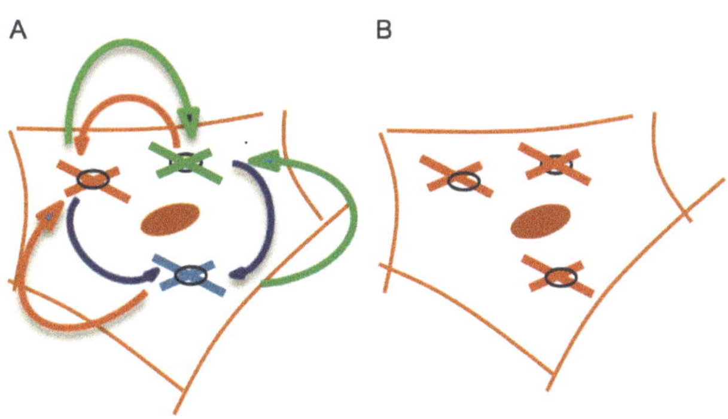

Konstantin Khrapko and Doug Turnbull, Figure 2.3

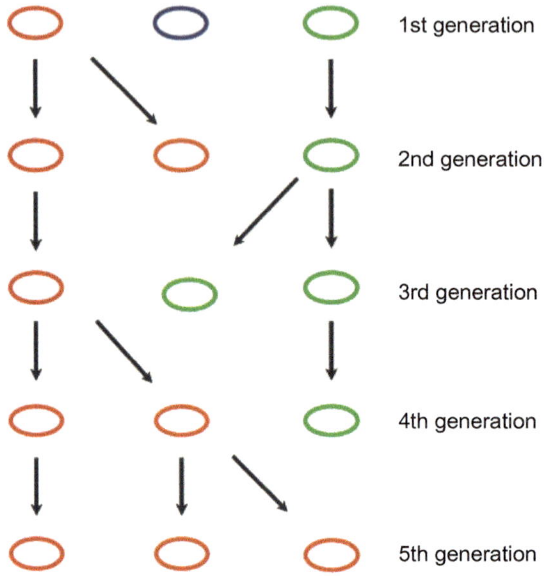

Konstantin Khrapko and Doug Turnbull, Figure 2.4

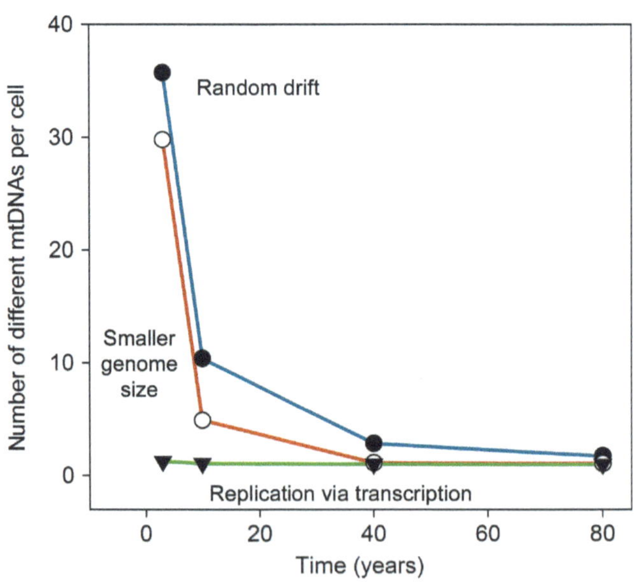

Axel Kowald and Edda Klipp, Figure 3.6

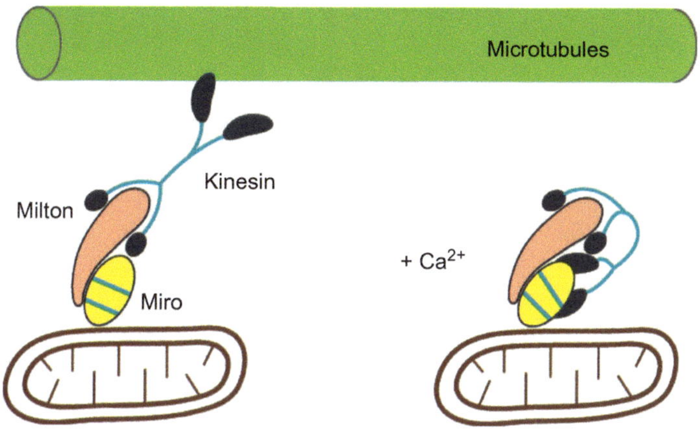

JÜRGEN BEREITER-HAHN, FIGURE 4.1

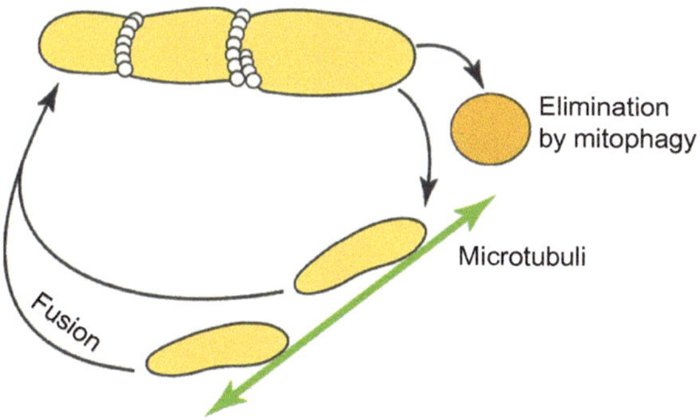

JÜRGEN BEREITER-HAHN, FIGURE 4.2

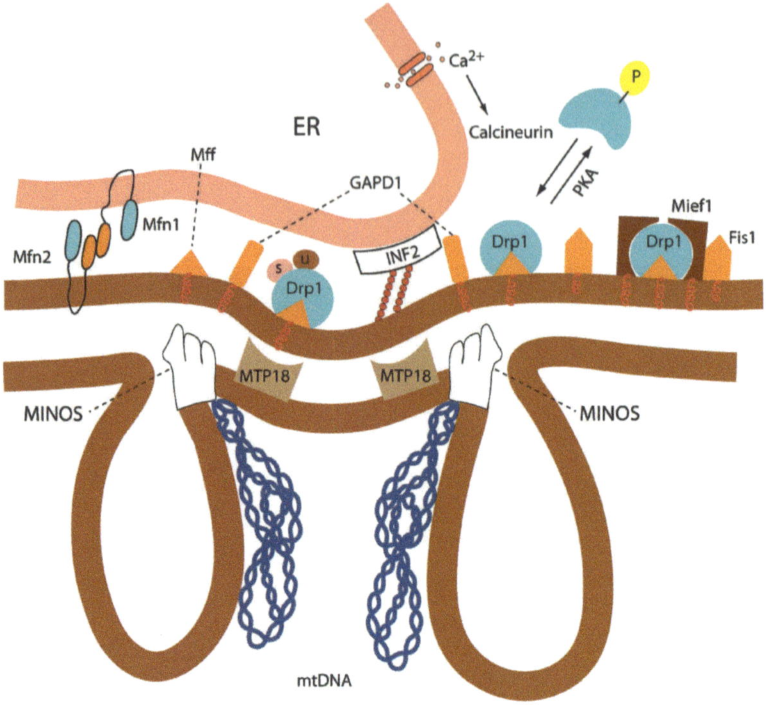

JÜRGEN BEREITER-HAHN, FIGURE 4.3

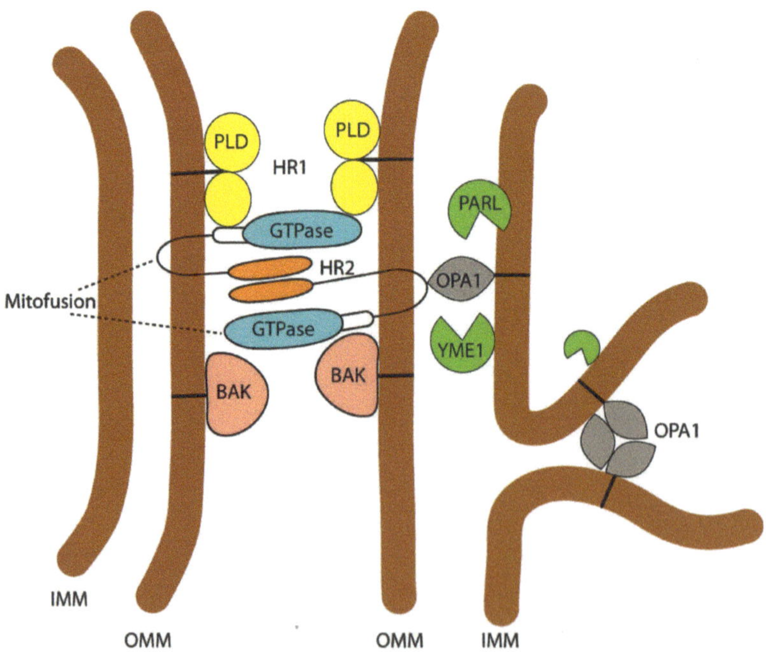

JÜRGEN BEREITER-HAHN, FIGURE 4.4

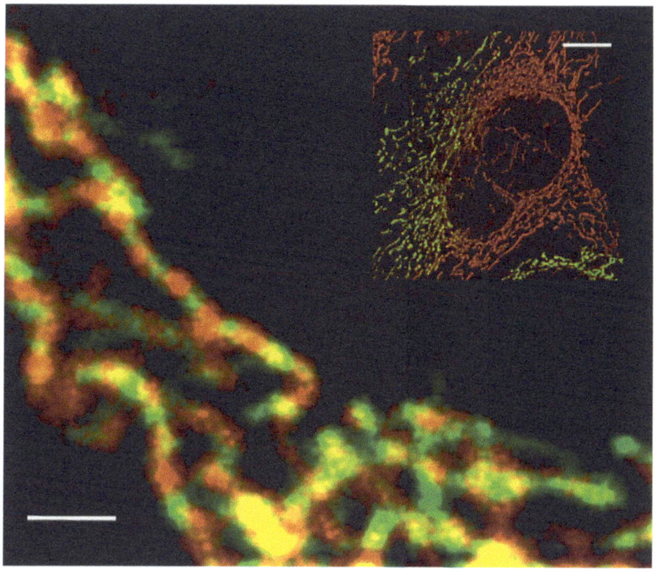

JÜRGEN BEREITER-HAHN, FIGURE 4.6

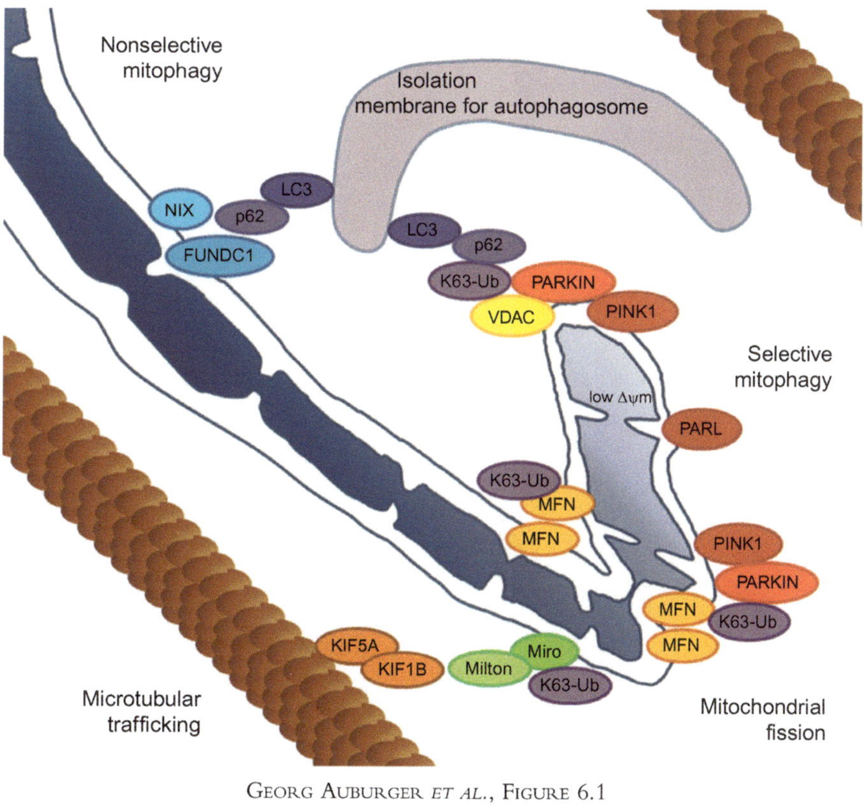

GEORG AUBURGER *ET AL.*, FIGURE 6.1

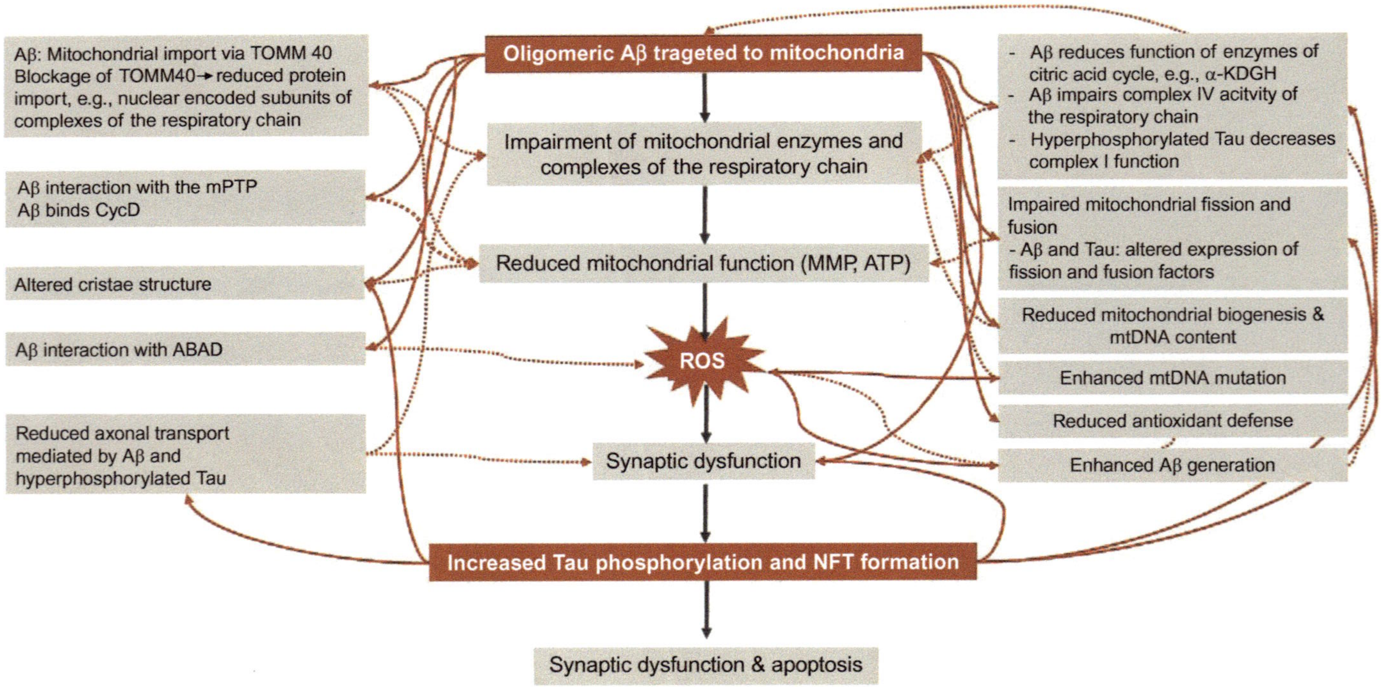

Kristina Friedland-Leuner *et al.*, Figure 7.1

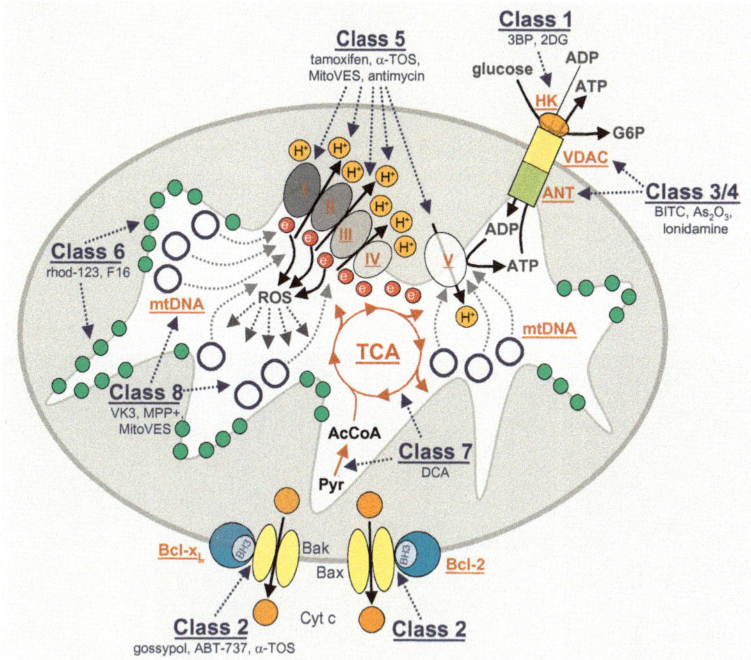

LAN-FENG DONG AND JIRI NEUZIL, FIGURE 8.1

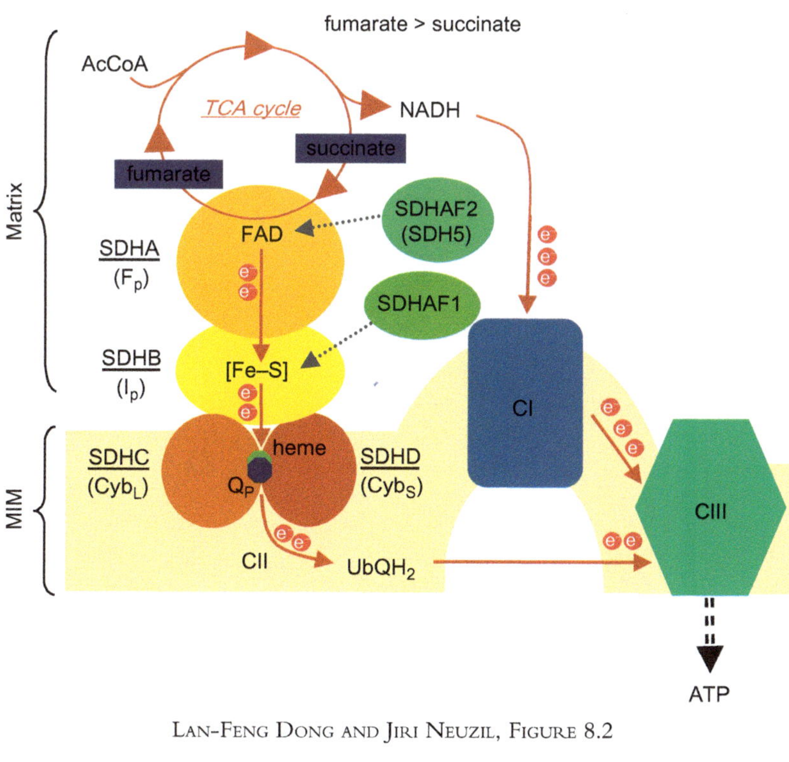

LAN-FENG DONG AND JIRI NEUZIL, FIGURE 8.2

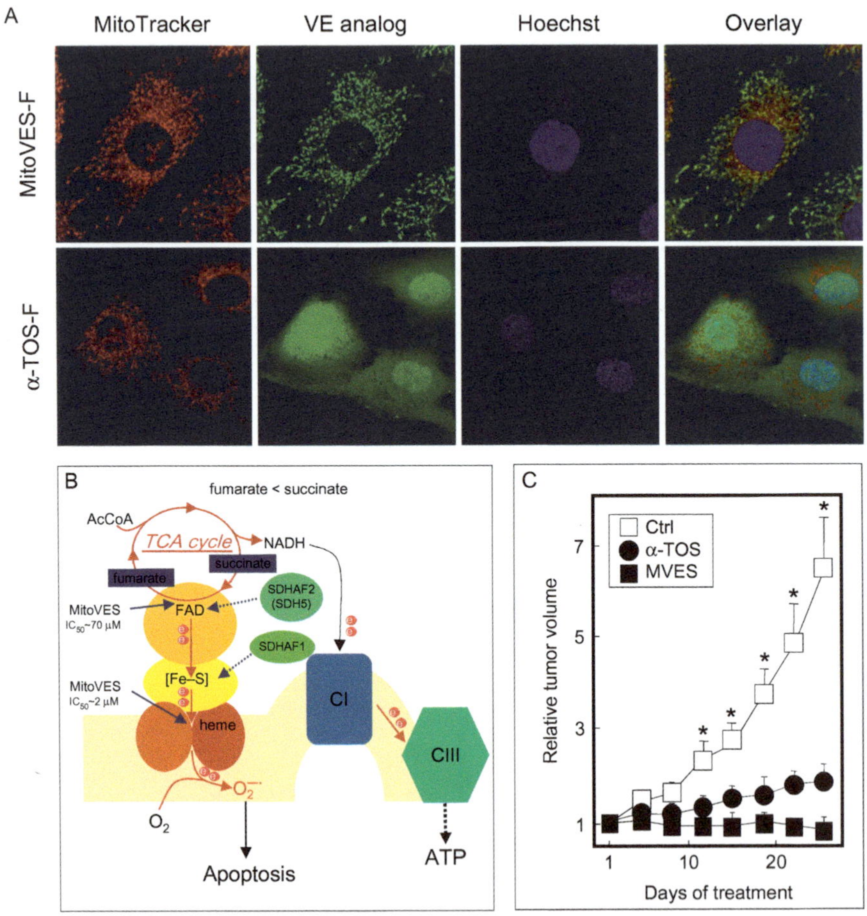

Lan-Feng Dong and Jiri Neuzil, Figure 8.3

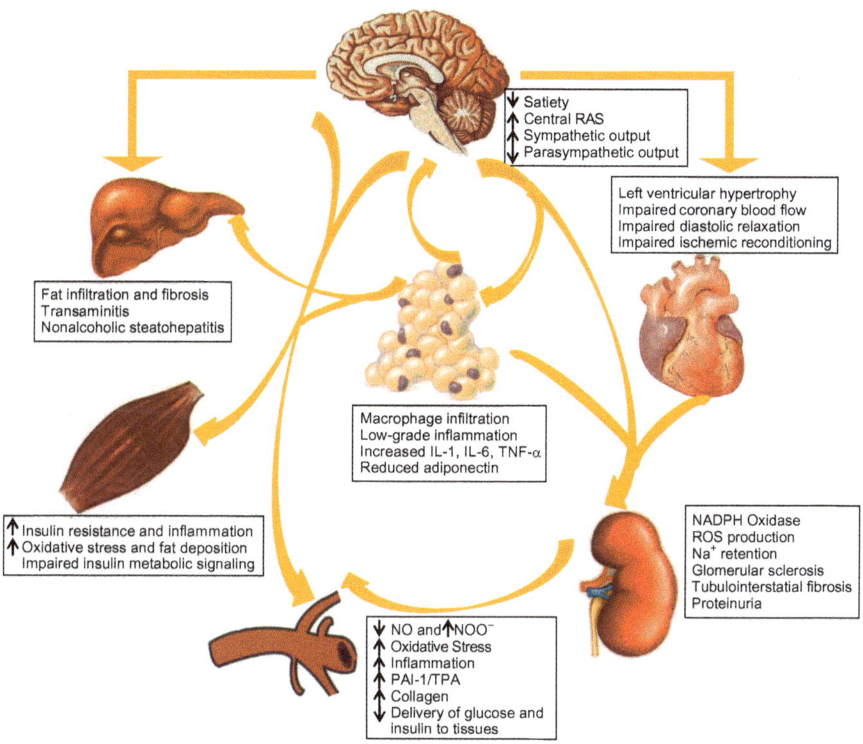

Guanghong Jia *et al.*, Figure 9.1

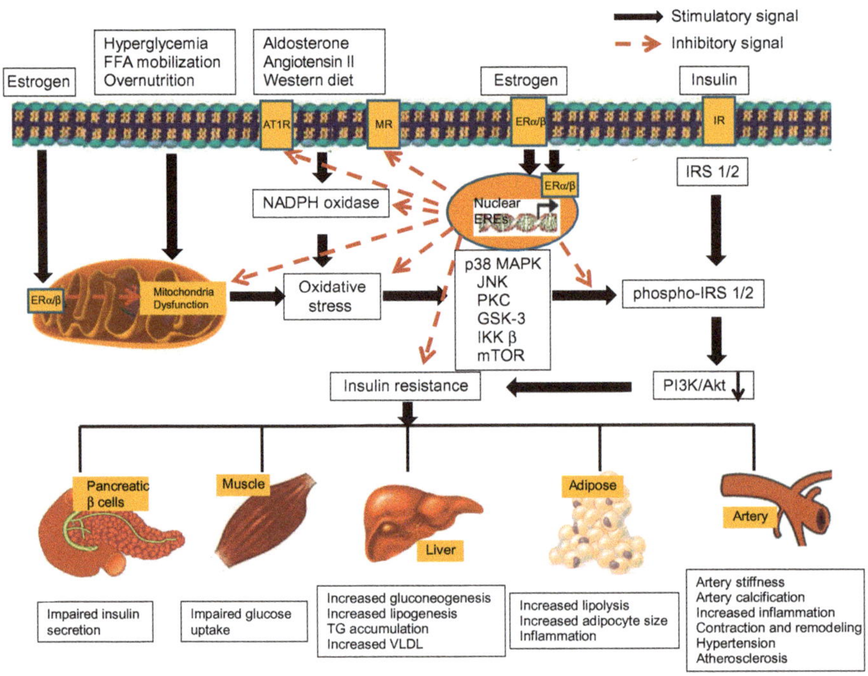

GUANGHONG JIA *ET AL.*, FIGURE 9.2

FSC
www.fsc.org
MIX
Papier | Fördert
gute Waldnutzung
FSC® C083411

Kolibri 360 GmbH
Ferdinand-Jühlke-Straße 7
99095 Erfurt, Deutschland
produktsicherheit@kolibri360.de